UNITEXT

La Matematica per il 3+2

Volume 168

Editor-in-Chief

Alfio Quarteroni, Politecnico di Milano, Milan, Italy
    École Polytechnique Fédérale de Lausanne (EPFL), Lausanne, Switzerland

Series Editors

Luigi Ambrosio, Scuola Normale Superiore, Pisa, Italy

Paolo Biscari, Politecnico di Milano, Milan, Italy

Ciro Ciliberto, Università di Roma "Tor Vergata", Rome, Italy

Camillo De Lellis, Institute for Advanced Study, Princeton, USA

Victor Panaretos, Institute of Mathematics, École Polytechnique Fédérale de Lausanne (EPFL), Lausanne, Switzerland

Lorenzo Rosasco, DIBRIS, Università degli Studi di Genova, Genova, Italy
    Center for Brains Mind and Machines, Massachusetts Institute of Technology,
    Cambridge, Massachusetts, US
    Istituto Italiano di Tecnologia, Genova, Italy

The **UNITEXT - La Matematica per il 3+2** series is designed for undergraduate and graduate academic courses, and also includes books addressed to PhD students in mathematics, presented at a sufficiently general and advanced level so that the student or scholar interested in a more specific theme would get the necessary background to explore it.

Originally released in Italian, the series now publishes textbooks in English addressed to students in mathematics worldwide.

Some of the most successful books in the series have evolved through several editions, adapting to the evolution of teaching curricula.

Submissions must include at least 3 sample chapters, a table of contents, and a preface outlining the aims and scope of the book, how the book fits in with the current literature, and which courses the book is suitable for.

For any further information, please contact the Editor at Springer: francesca.bonadei@springer.com

THE SERIES IS INDEXED IN SCOPUS

***

UNITEXT is glad to announce a new series of free webinars and interviews handled by the Board members, who rotate in order to interview top experts in their field.

Access this link to subscribe to the events:

https://cassyni.com/s/springer-unitext

Matteo Viale

The Forcing Method in Set Theory

An Introduction via Boolean Valued Logic

 Springer

Matteo Viale
Dipartimento di Matematica
Università di Torino
Torino, Italy

ISSN 2038-5714   ISSN 2532-3318   (electronic)
UNITEXT
ISSN 2038-5722   ISSN 2038-5757   (electronic)
La Matematica per il 3+2
ISBN 978-3-031-71659-1   ISBN 978-3-031-71660-7   (eBook)
https://doi.org/10.1007/978-3-031-71660-7

This Springer imprint is published by the registered company Springer Nature Switzerland AG
The registered company address is: Gewerbestrasse 11, 6330 Cham, Switzerland

If disposing of this product, please recycle the paper.

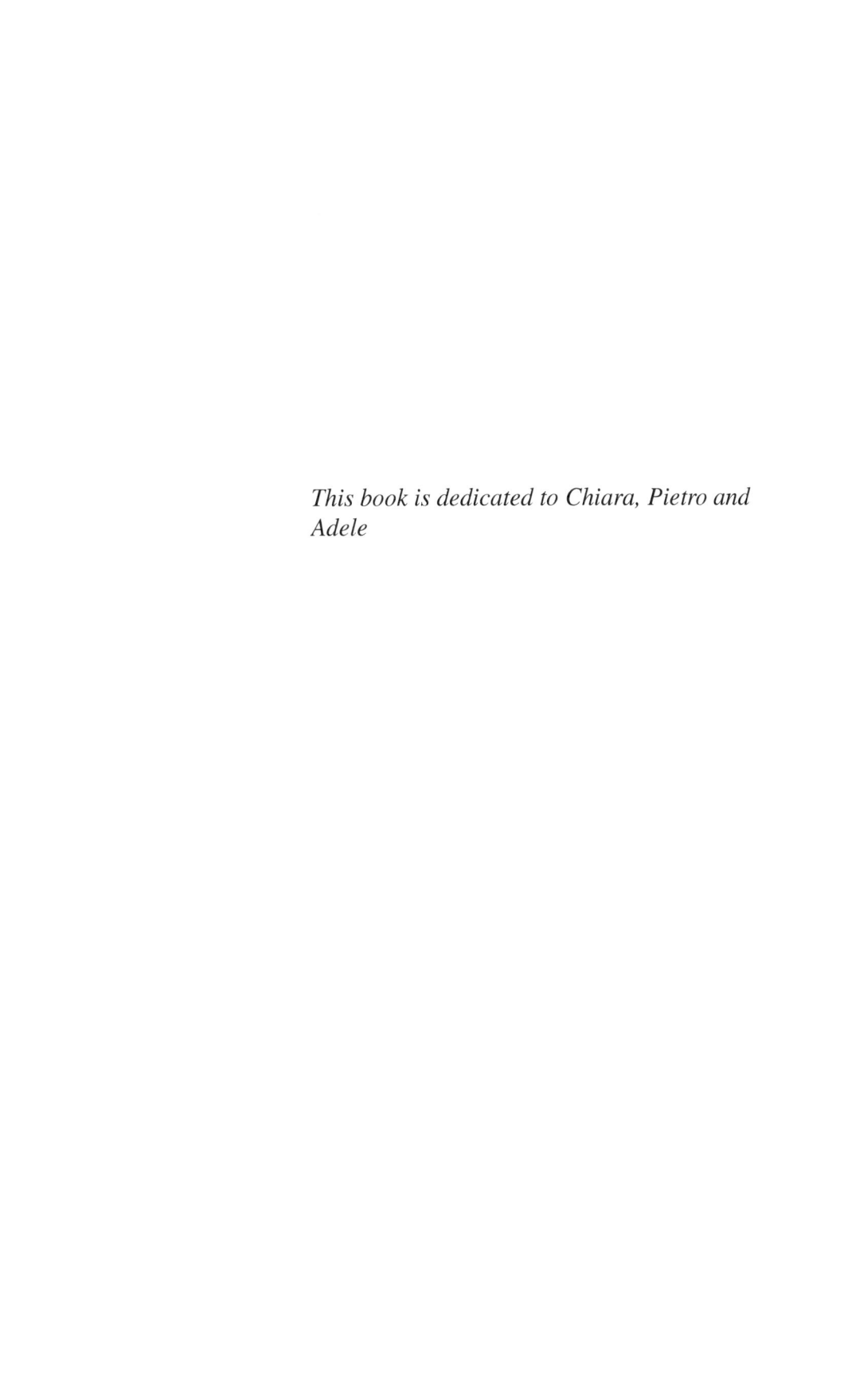

This book is dedicated to Chiara, Pietro and Adele

Preface

The main aim of this book is to give a compact self-contained presentation of the forcing technique devised by Cohen [7] to establish the independence of the continuum hypothesis from the axioms of set theory. We follow the approach to forcing via boolean valued semantics, an approach independently introduced by Vopenka and Scott/Solovay. The book develops out of notes I prepared for several master courses on this and related topics and aims to provide an alternative (and more compact) account of the forcing method with respect to the classical [5, 16, 19] or to the more recent [20, 27, 29, 38].

Our aim is to take up a reader familiar with logic and set theory at the level of an undergraduate course on both topics[1] and bring her/him to page with the use of forcing to produce independence (or undecidability results) in mathematics. Familiarity with general topology would also be quite helpful; however, the book provides a compact account of all the needed results.[2]

The presentation of the material is organized in such a way that many chapters can be read also by scholars with almost no familiarity with first order logic and/or set theory.

Torino, Italy   Matteo Viale

---

[1] E.g. familiar with most of the content of books such as [8] for what concerns first order logic, and [15] for what concerns set theory.

[2] An account of general topology which provides much more information than what will be needed here is given by the first chapter of [24].

Acknowledgments

I am warmly grateful to a number of persons who along the years and in various forms helped me in the redaction of this book. Many students have taken my courses reading preliminary versions of these notes and suffered, thus providing many inputs on how to improve the presentation; some of them contributed to the redaction of the first drafts, let me mention: Giorgio Audrito, Silvia Steila, and Filippo Calderoni. My colleague Raphaël Carroy gave me fundamental support in structuring the material and the organization of the presentation in its various parts. Luca Motto Ros and Alessandro Andretta provided helpful inputs at various stages. I express all my gratitude to the anonymous referee: her/his revision has been extremely careful and meticulous; s-he also provided very useful general comments on how to improve the book. I adopted many of them, reserving the most demanding ones to (eventual?) follow-ups of this book. I'm also grateful to many colleagues who influenced my take on the subject, they are too many to be mentioned explicitly. I am certainly indebted to my PhD advisors Alessandro Andretta and Boban Veličković who informed my views on set theory while I was learning it. I am also grateful to Jana Yagnavaragan and the Springer editing staff for their careful assistance during the revision of the proofs.

The author acknowledges support from the project: *PRIN 2017-2017NWTM8R Mathematical Logic: models, sets, computability*, PRIN 2022 "Models, sets and classifications", prot. 2022TECZJA, and GNSAGA.

Contents

Chapter 1
Introduction

There are a number of excellent books presenting the forcing method in set theory; nonetheless, I decided to add my own approach to the topic: in my opinion not enough attention has been brought in the current literature to the presentation of the forcing method by means of boolean valued semantics. This is the approach to forcing pursued in the present book.

I believe that this approach not only simplifies the exposition of the more demanding technical parts, but it also outlines and solves more elegantly the metamathematical issues inherent to the forcing machinery; furthermore—and more importantly for me—the approach via boolean valued models gives an account of this technique which naturally brings to light connections with a variety of mathematical fields other than logic and set theory, among which general topology, functional analysis, and category theory.

The book tries to present the forcing method outlining in many situations the intersections of set theory and logic with other mathematical domains. My hope is that this book can be appreciated by scholars in set theory and by readers with a mindset oriented toward areas of mathematics other than logic and a keen interest on the foundations of mathematics.

The book is divided into seven chapters (including the introduction) and five appendixes.

We give a brief account of the content of the remaining chapters with some care in elucidating the prerequisites needed to follow them:

- Chapter 2 gives preliminary material on topology, partial orders, and the standard axiomatization of set theory ZFC in the first order language $\{\in, =\}$ (which we expect most of the readers to be already familiar with).[1]

---

[1] Note however that no familiarity with first order logic is assumed till Chap. 6.

M. Viale, *The Forcing Method in Set Theory*, La Matematica per il 3+2 168, https://doi.org/10.1007/978-3-031-71660-7_1

- Chapters 3, 4, and 5 cover standard material on partial orders and boolean algebras. The reader unacquainted with first order logic and set theory can follow (almost all of) their content.
- Chapter 6 gives an overview of boolean valued semantics (a natural generalization of Tarski semantics for first order logic). From this chapter onward, we assume the reader has the required knowledge of ZFC.
- Chapter 7 is the heart of the book. It gives a detailed presentation of the forcing method via boolean valued models. It features the most celebrated application of this method, namely the undecidability of the continuum hypothesis CH.
- Appendix A analyzes in detail the logical complexity of set theoretic concepts and investigates the notion of absolute property. It is fundamental that the reader gets acquainted with this material in order to follow Chap. 7.
- Appendix B presents some classical results on the models of set theory, namely: Lévy's absoluteness, a strong form of the reflection theorem, and a weak form of Shoenfield's absoluteness. A very compact account of the basics on constructibility is also outlined. This material is not needed in other parts of the book. However, we felt that a master level course in set theory is incomplete if it does not present these results.
- Appendix C relates the presentation of forcing given here to the original method of Cohen as exposed for example in [19]. It is a very useful bridge to other more advanced textbooks in set theory, which clearly includes [5, 16, 19, 20, 27, 29, 38].
- Appendix D gives a compact presentation of classical results in general topology. The material of this Appendix complements very well the material of Chaps. 3–6 but is not needed to follow any of the other parts of the book. Furthermore, this part does not require familiarity with first order logic (it may on occasions require knowledge of some classical set theoretic results).
- A short final Appendix E puts into context the results of the present book and their relation with a variety of current research trends, suggesting possible readings and directions of study connected with the forcing method.

Overall a reader familiar with the content of some undergraduate text in set theory (such as [15, Chapters 1 to 9]) and the basics of first order logic (as developed for example in [22, Chapters 1, 2] or [8]) has enough background to follow the book in its entirety.[2] This familiarity is *not* of vital importance for the comprehension of Chaps. 2–5, while some familiarity with first order logic is assumed in Chap. 6. Chapter 7 on forcing requires the reader to be also familiar with the notion of absoluteness for set theoretic concepts. This is the content of Appendix A.

We believe that those who digest the content of this book are ready to take up any text in set theory dealing with the forcing method, and can even follow the published literature covering recent advances obtained by forcing (the unique major gap left by our account being the lack of a presentation of iterated forcing). Since

---

[2] WARNING: The material in Appendix A is not covered in [15]. An alternative satisfactory treatment of set theoretic absoluteness sufficient to follow Chap. 7 can also be found in [19, Chapters I, III, IV, V] or [16, Chapters 12, 13].

the other mentioned books on forcing already give a rather broad and well-presented exposition of the most important applications of this technique (as well as of iterated forcing), we decided here to limit ourselves to introduce the method and get the reader acquainted with it. Appendix C contains all the information needed to process the knowledge of the matter gathered here in order to follow the presentation of this same topic given elsewhere.

1.1 Detailed Content

We now give details on the content of each chapter:

- Chapter 2 recalls basic concepts on topologies and preorders and investigates in some detail:
    - On the topological side the properties of regular open sets of a topology,
    - On the partial order side some properties relating filters, density, genericity for posets to their topological counterparts.

  We also list and briefly comment the axioms of the first order theory ZFC.
- Chapters 3 and 4 introduce basic notions regarding boolean algebras. A systematic study of Stone's Duality is performed, and the basic algebraic and topological properties of boolean algebras and their Stone spaces are outlined. We also study in detail complete boolean algebras (cbas) and prove that every cba is representable as the family of regular open sets of some topological space, and that the Stone's Duality identifies cbas with compact Hausdorff extremally disconnected spaces. Furthermore, we spend some attention in explaining how to compute topologically suprema and infima of families of regular open sets. A key result for the development of forcing which we also present is the proof that every partial order admits, up to isomorphism, a unique complete boolean algebra in which it embeds as a dense suborder. We also analyze in some detail the measure algebra given by Lebesgue measurable sets modulo null sets, and we prove that it is complete, atomless, and without uncountable partitions.
- Chapter 5 introduces the minimal amount of basic combinatorial properties of partial orders which will be needed to prove the independence of the continuum hypothesis by means of the forcing method. In particular we focus on CCC partial orders, we prove the Δ-system Lemma, and we use it to prove that the notion of forcing which can be used to produce a model of ZFC where CH fails is CCC. Most (if not all) of these results can be found in [20, III.1-III.2-III.3] or [19, II]; however, these sections of both books contain a large amount of material which is not strictly necessary for our aims. We also outline in this chapter the connections between the combinatorial properties of certain partial orders and well-known topological properties of compact Hausdorff spaces associated to them (among other things we prove the Baire category theorem for compact Hausdorff spaces).

- Chapter 6 introduces the boolean valued semantics for first order theories and shows that certain function spaces which naturally occur in functional analysis produce natural examples of boolean valued models. The boolean valued semantic selects a given boolean algebra B and assigns to every statement ϕ a boolean value in (the boolean completion of) B. The boolean operations reflect the behavior of the propositional connectives, it requires more attention to give a meaning to atomic formulae and to quantifiers, and we need a certain amount of completeness for B in order to be able to interpret quantifiers in the boolean semantics. We introduce the notion of boolean valued model with the mixing property and that of full boolean valued model. We show that the former property implies the latter and that the class of boolean valued models which are of interests in developing the boolean valued semantics are the full ones, as fullness characterizes the class of boolean valued models for which a generalized form of the usual Łós theorem for ultraproducts holds.
- Chapter 7 develops the theory of forcing by means of boolean valued models giving detailed proofs of the basic properties of boolean valued models for set theory, of Cohen's forcing theorem, and the two basic applications of the method which suffice to prove the independence of CH from the standard axioms of set theory. Our presentation of forcing departs completely from the approach taken by Kunen in [19] and is more keen to that taken by Bell [5], or Jech [16]. We decide to follow this different approach for two reasons:

  1. In our eyes, the boolean valued models approach makes the metamathematical arguments needed to understand the forcing method easier to grasp and greatly simplify some proofs.
  2. The boolean valued model approach makes more transparent what is the role played by *generic filters* in the development of the forcing method and where the hypothesis that a filter is suitably generic is essential. Moreover, it enlightens the link existing between the notion of generic filter arising in forcing with the corresponding topological notion of generic point of a topological space which is at the heart of the Baire category arguments.

- Appendix A deals with the notion of absoluteness for set theoretic concepts. We first embark in a systematic analysis of the notion of Δ_1-property for a first order theory: We analyze the preservation of its semantic meaning with respect to the substructure relation, and we also prove that "Δ_1 on Δ_1 remains Δ_1." Next we analyze set theory as formalized by ZFC in the signature $\{\in, \subseteq\}$, and we outline that set theoretic properties formalized by formulae with bounded quantifiers (the so-called Δ_0-formulae) express simple (or absolute) set theoretic concepts. Leveraging on this observation, we expand the signature adding predicate symbols for the Δ_0-formulae, and we prove the standard result that the complexity of set theoretic concepts according to the Lévy hierarchy for them (as given for example in [16, Chapter 13]) is obtained by the usual stratification of the formulae in the expanded signature according to the Lévy complexity. Furthermore, we show that the transfinite recursion schema performed over an absolute property yields an absolute property.

- Appendix B presents proofs of: Lévy's absoluteness, (an apparently weaker form of) Shoenfield's absoluteness, a strong form of the reflection theorem, and a compact account of the basics on constructibility.
- Appendix C connects our presentation of the forcing method to that of [19] via posets. In this way the reader can safely orient her(him)self in the books [5, 16, 19, 20, 29, 38] or in the standard set theoretic literature, if she/he wishes to learn more sophisticated applications of the forcing technique.
- Appendix D develops some topics in general topology which complement the content of Chaps. 3–6: A systematic analysis of the notion of net is performed, and proofs of Tychonoff's compactness theorem for product spaces and of Stone–Cech's compactification theorem are given. We also provide a natural characterization of extremally disconnected compact Hausdorff spaces which is hardly traceable in the literature.
- Appendix E suggests various readings on topics in set theory.

1.2   How to Use the Book

The book develops out of notes I prepared for a master level course on forcing whose main objective is establishing the independence of the continuum hypothesis from the axioms of set theory. A one semester course (approximately 6 ECTS, 48 hours of lectures) covers (almost) all of its content with the exception of Appendix D and assumes the students are familiar with the basics of first order logic and set theory spelled out before. My usual pattern is to cover Chaps. 3–6 (eventually recalling the needed material from Chap. 2), then Appendix A, and finally Chap. 7; if time remains I cover either Appendix B or Appendix D (depending on the interests of the students).

An alternative use of the book (which I also experimented) is to cover Chaps. 2–4, and then Chap. 6 and Appendix D; this results in an half-semester course (approximately 3 ECTS, 24 hours of lectures) which relates boolean valued logic to general topology.

All over the text the book presents exercises linked to the various arguments; many of these are then used in other parts of the book; the reader is warmly invited to spend some energies on them.

1.3   Some Remarks on the Ontology of Mathematics

This section may be skipped, but we hope it can be appreciated by those readers with a mindset oriented toward philosophical matters. We assume these readers have the required familiarity with the logic and set theoretic concepts recalled below.

The first chapters (up and including Chap. 6) do not require on our side any special commitment on the ontology of mathematical entities and can be considered as a standard textbook on a mathematical theory which is developed much in the same way as one develops other fields of mathematics: We are in the situation common to most of mathematics where ontological considerations do not play a significant role. On the other hand our presentation of forcing in Chap. 7 is guided by some ontological assumptions. We bring to light these assumptions here by giving a concise account of the point of view on the philosophy of mathematics we adopt in that part of the book. We do this in the following form: We list a series of basic questions on the ontology of mathematics, and we explain in few words what are some (usual) possible stances and the one we choose to adopt:

1. **What is a mathematical reasoning?** For us a mathematical reasoning is a process expressed in a *natural language* (i.e., Italian, English, French, Chinese, whatever is most suited) which from given premises (hypotheses) produces a certain conclusion (thesis) which is *mathematically rigorous.*

2. **What does it mean *mathematically rigorous*?** For us it means that there is a first order language such that the premises and the conclusion can be formalized by first order formulae in that language ($\phi_1, \ldots \phi_n$ for the premises and ψ for the conclusion) and such that there is a sound and complete first order calculus which allows to prove ψ by premises $\phi_1, \ldots \phi_n$ on the basis of the calculus rules and axioms.

3. **What is the *meaning* of premises and conclusion?** There are various possible stances in this regard which range from:

   - *Extreme formalism:* There is no clear meaning in the premises and the conclusion of our reasoning as expressed in the natural language, since there cannot be a precise semantic interpretation of natural languages. What we know for sure is that with respect to the formalized counterpart $\phi_1, \ldots \phi_n$ of the premises and ψ of the conclusion, for what we know so far, from premises $\phi_1, \ldots \phi_n$ on the one hand, using a first order sound and complete calculus, we have not been able to derive a contradiction, and on the other hand, we have been able to derive ψ.

   - *Extreme Platonism:* There is a *hyperuranium* of mathematical entities, and the premises and the conclusion define clear mathematical properties which can be predicated of objects in this hyperuranium. Our reasoning shows that if the premises assert true properties of the hyperuranium, so does its conclusion. The fact that our reasoning (which we express in a natural language) can be formalized in a first order calculus gives a proof check of the correctness of our reasoning process establishing truths of the hyperuranium.

In this book we adopt a stance of *extreme Platonism* when dealing with mathematical reasoning.

4. **What does it mean the first order formula $\mathsf{CON}(T)$ introduced by Gödel to formalize the concept of consistency for a first order theory T in a language $\mathcal{L} = (R_i, i \in I, f_j : j \in J, c_k : k \in K)$?**

- For the extreme formalist the non-falsity[3] of $\mathsf{CON}(T)$ reflects the fact that so far nobody has been able to derive a contradiction using a sound and correct first order calculus using the axioms of T as premises.
- For a platonist the non-falsity (which for her/him is equivalent to truth) of $\mathsf{CON}(T)$ means that there is a *set* M in V and relations $R_i^M : i \in I$ on M^{n_i}, for $i \in I$, functions $f_j^M : M^{n_j} \to M$ for $j \in J$, and elements of Mc_k^M for $k \in K$ also all in V such that $(M, R_i^M : i \in I, f_j^M : j \in J, c_k^M : k \in K)$ is a Tarski model for T as well as an element of V.

5. **What is the status of the first order theory ZFC?**

- For an extreme formalist it is not different from the status of any other first order theory T to which Gödel's incompleteness theorem applies: The only sure thing we know so far is that a deduction of the false has not been found using a sound and correct first order calculus in which the premises are axioms of ZFC.
- For an extreme platonist, there is among the elements of the hyperuranium a well defined mathematical entity V consisting of all those mathematical entities which are *sets*. V is not all of the hyperuranium, for example, Russell's class $R = \{x \in V : x \notin x\}$ is a well defined mathematical entity belonging to the hyperuranium but is not an element of V (i.e., R is not a *set*!). Nonetheless, V is very large and contains as elements most (if not all) mathematical entities we commonly use to do mathematics such as the natural numbers, the complex and real numbers, most topological spaces, the spaces of functions used in functional analysis, etc. Moreover V is closed under many set theoretic operations, i.e., if $(a_i : i \in I) \in V$, then also its product $\prod_{i \in I} a_i \in V$, if $a \in V$, $\cup a$ and $\mathcal{P}(a) \in V$ as well, if $B \in V$ and $\Phi(x)$ is a property which makes sense to be asked whether it is true of mathematical entities, $\{a \in B : \Phi(a) \text{ holds}\}$ is also in V, if $F : V \to V$ is a function and $A \in V$ is a set, then $F[A]$ is also a set in V, etc. In particular the first order structure $(V, \in)$ models ZFC. So for an extreme platonist, ZFC is not only a consistent theory (since it holds in the Tarski model $(V, \in)$, though this model is not a set), but it formalizes in a first order language a true state of affairs of the large portion of the hyperuranium given by V.

6. **Are there independence results over V?**

- For an extreme platonist there are no independence results over V, given that V is a well defined coherent mathematical entity, and thus the first order

---

[3] More precisely, the unability to derive a contradiction from $\mathsf{CON}(T)$ in a sound and complete first order calculus.

theory of the Tarski structure $(V, \in)$ is complete and consistent. In particular the continuum hypothesis is either true or false in V, even though currently our imperfect knowledge of V makes it impossible to ascertain which is the case. On the other hand ZFC is just a recursive list of first order properties which reflects true properties of V, but which we know that they cannot give a complete first order axiomatization of the theory of V in the first order language $\{\in\}$ due to Gödel's incompleteness theorem. It is well possible (and it is actually the case) that there can be models (M_i, E_i) which are sets in V for $i = 0, 1$ and are first order $\in$-models of the first order $\in$-theory ZFC with the following property: There is a $\in$-formula ϕ in the language of ZFC such that $(M_0, E_0) \models \phi$ and $(M_1, E_1) \models \neg\phi$. Actually the aim of these notes is to show that this is the case for ϕ being the first order formalization of the continuum hypothesis CH in signature $\in$.

- For an extreme formalist the above question is void of content given that V is a meaningless concept.

7. **How do we proceed to prove that CH is independent from the axioms of ZFC?** We really commit ourselves to the extreme platonist stance. First of all ZFC is consistent, since $(V, \in)$ is a model of ZFC. Moreover $(V, \in)$ is a model of an $\in$-sentence formalizing the completeness theorem for first order logic. We consider just the following form of the completeness theorem in V (but with minor adjustments this can be generalized to arbitrary set sized languages): For every countable (and recursively given) first order language $\mathcal{L} = \{R_i : i \in I, f_j : j \in J, c_k : k \in K\}$ given by a set of relations, functions, and constants, there is a recursive set of natural numbers $\mathsf{Form}_{\mathcal{L}} \subseteq \omega$ in V which is a code for the formulae in $\mathcal{L}$. There is also a recursive subset $\mathsf{Sent}_{\mathcal{L}}$ of $\mathsf{Form}_{\mathcal{L}}$ consisting of the formulae without free variables (i.e., its sentences). There are also:

- A recursive predicate $\mathsf{DER}_{\mathcal{L}} \subseteq \mathsf{Form}_{\mathcal{L}}^{<\omega}$ which says that $(\phi_1, \ldots \phi_n, \psi) \in \mathsf{DER}_{\mathcal{L}}$ iff there is a derivation in first order calculus of ψ from premises $\phi_1, \ldots \phi_n$.
- A definable satisfaction predicate (i.e., a class definable in V)

$$\mathsf{Sat} : \mathsf{Form}_{\mathcal{L}} \times \mathcal{L} - \text{structures} \times V^{<\omega} \to 2 \cup \{*\}$$

such that for all $\mathcal{L}$-structures

$$\mathfrak{M} = (M, R_i^{\mathfrak{M}} : i \in I, f_j^{\mathfrak{M}} : j \in J, c_k^{\mathfrak{M}} : k \in K)$$

in V and $\vec{s} \in M^{<\omega}$,

$$(V, \in) \models \mathsf{Sat}(\phi, \mathfrak{M}, \vec{s}) = 1$$

if and only if

$$\mathfrak{M} \models \phi(\vec{s})$$

is true (in the latter case according to the rules of Tarski semantics for the $\mathcal{L}$-structure $\mathfrak{M}$, and in the former case according to the rules of Tarski semantics for the structure $(V, \in)$ to interpret the definable class function Sat).[4]

Now the correctness and completeness theorem in V says that $(V, \in)$ models the following formula in parameter $\mathsf{DER}_{\mathcal{L}}$, $\mathsf{Form}_{\mathcal{L}}$, for any set of sentences $T \subseteq \mathsf{Sent}_{\mathcal{L}}$:

> There is no $(\phi_1, \ldots \phi_n, \psi \wedge \neg\psi)$ in $\mathsf{DER}_{\mathcal{L}}$ with $\phi_1, \ldots, \phi_n \in T$ if and only if there is an $\mathcal{L}$-structure $\mathfrak{M}$ in V such that $\mathsf{Sat}(\phi, \mathfrak{M}, \emptyset) = 1$ for all $\phi \in T$.

It can be checked that the above expression can be formulated as an $\in$-formula (more on this will be said in Appendix A).

This means that in V, there is (M, E) which is a set and is model of ZFC, since we know that ZFC is consistent (given that we assume that $(V, \in)$ is a Tarski model of ZFC).

Nonetheless, in these notes we want more than this. We want that in V with the true $\in$-relation seen as a subclass of V^2 there is a transitive countable model M such that $(M, \in \cap M^2)$ is a model of ZFC. This can be achieved if for example we assume that in V there is a strongly inaccessible cardinal (more on this will be said in Appendix A). The existence of an inaccessible cardinal is an axiom which an extreme platonist considers true. So we will from now on work in the first order theory ZFC^+ extending ZFC with the statement *There is a countable transitive set $M \in V$ such that $(M, \in)$ is model of* ZFC.

We will use the forcing method to build from the transitive and countable ZFC-model M new countable transitive models $N_0, N_1 \in V$ of ZFC such that $(N_0, \in) \models \mathsf{CH}$ and $(N_1, \in) \models \neg\mathsf{CH}$.

8. **What will an extreme formalist think of this proof of the independence of CH from the axioms of set theory?** Our proof does not make any sense for a formalist! Nonetheless (even if we will not spell out the details) by means of standard logical arguments, our proof can be converted in a proof that $\mathsf{CON}(\mathsf{ZFC}^+)$ implies also $\mathsf{CON}(\mathsf{ZFC} + \mathsf{CH})$ and $\mathsf{CON}(\mathsf{ZFC} + \neg\mathsf{CH})$. This is meaningful for a formalist in the following sense: If we know that no contradiction can be derived in a sound and complete first order calculus from the axioms of ZFC^+, then such a contradiction can be derived in the same calculus neither from the axioms $\mathsf{ZFC} + \neg\mathsf{CH}$ nor from the axioms $\mathsf{ZFC} + \mathsf{CH}$. Moreover, by means of arguments which are more sophisticated (and are rooted in the reflection theorem for V), one can rework our proof of the independence of CH from the axioms of set theory and obtain a proof (also for a formalist) that $\mathsf{CON}(\mathsf{ZFC})$ implies also $\mathsf{CON}(\mathsf{ZFC} + \mathsf{CH})$ and $\mathsf{CON}(\mathsf{ZFC} + \neg\mathsf{CH})$.

---

[4] The value 0 is assigned to tuples $\phi, \mathfrak{M}, \vec{s}$ such that $\mathfrak{M} \not\models \phi(\vec{s})$ and $\vec{s} \in M^{<\omega}$. The value $*$ is assigned to tuples for which $\vec{s} \notin M^{<\omega}$.

9. **Why the existence of a ZFC-model which is a set in V is not enough for our purposes, and we want to work with countable transitive models of ZFC?**
We want to avoid to work with ill-founded models of ZFC. The pathologies we do not want to run into are well explained by the non-standard models of the first order theory of Peano's arithmetic. We know that the structure of natural numbers $\mathbb{N}$ is a set in V, for example, $(\mathbb{N}, <)$ can be presented as the model $(\omega, \in)$, and the sum, product, exponentiation of natural numbers can also be presented as suitable operations on elements of ω which can be defined in V using $\in$-formulae in the parameter ω. So let the operations $+, \cdot$ be such that $(\omega, \in, +, \cdot)$ is a representative of the isomorphism type of the structure $(\mathbb{N}, <, +, \cdot)$. Let S be the first order complete theory of the structure $(\omega, \in, +, \cdot)$ in the language $\mathcal{L} = \{+, \cdot, <\}$. It is well known that there are ill-founded models of S, i.e., structures $(M, <_M, +_M, \cdot_M) \in V$ such that the order type of $(M, <_M)$ is not isomorphic to $(\omega, \in)$ and is ill-founded: It can be shown that the order type of $(M, <_M)$ is isomorphic to an order of type $\mathbb{N} + I \times \mathbb{Z}$, where I is a dense linear order without end-points and all $a \in \mathbb{N}$ precede any $(b, c) \in I \times \mathbb{Z}$, and the order between elements in $I \times \mathbb{Z}$ is the lexicographic order. Notice that $\mathbb{N} + I \times \mathbb{Z}$ contains many non-empty sets without minimum, for Example, $I \times \mathbb{Z}$. So this is the case also for $(M, <_M)$. On the other hand $(M, <_M, +_M, \cdot_M)$ is a model of Peano's arithmetic, in particular, it models the principle of induction, which in this case amounts to say that for all formulae $\phi(x, y_1, \ldots, y_n)$ in $\mathcal{L}$ and $a_1, \ldots, a_n \in M$ if

$$(M, <_M, +_M, \cdot_M) \models \exists x \phi(x, a_1, \ldots, a_n),$$

then

$$(M, <_M, +_M, \cdot_M) \models \exists x (\phi(x, a_1, \ldots, a_n) \wedge \forall y[y < x \rightarrow \neg \phi(x, a_1, \ldots, a_n)]).$$

This means that there are ill-founded subsets of M with respect to the order $<_M$, but that the model $(M, <_M, +_M, \cdot_M)$ is not able to define any such ill-founded subset. Similar arguments can occur for models of ZFC. In particular there can be models $(N, E_N) \in V$ of ZFC which are ill-founded, i.e., there is $X = \{a_n : n \in \omega\} \in V$ subset of N such that $a_{n+1} E_N a_n$ for all $n \in \omega$. However, since (N, E_N) models the axiom of foundation, such a set $X \in V$ cannot be a definable subset of N; otherwise, this X would contradict that N models the Foundation Axiom. On the other hand, if we assume that $(N, \in \cap N^2)$ is a transitive model of ZFC, we have that $(N, \in \cap N^2)$ is really a well-founded model of ZFC in V or even in the hyperuranium. This adherence between the first order theory of $(N, \in \cap N^2)$ (where the Axiom of Foundation states that all ordered non-empty sets in N have a minimal $\in$-element) and its true properties from the point of view of V (or of the even larger hyperuranium) is important because it will enormously simplify many of our considerations and calculations on such type of ZFC-models $(N, \in \cap N^2)$.

10. **Why do we work with ZFC as a first order counterpart of the theory of** V**, rather than with Morse–Kelley MK+***choice* **or with Gödel–Bernays theory GBC?** It is just a matter of habits, since there is a well -developed study of the first order theory ZFC, in particular for what concerns the analysis of forcing. On the other hand, this is not the case for the Morse–Kelley axiomatization of set theory. Minor changes to the present approach to forcing allow to cover the development of the forcing method in the framework of MK or GBC (see, for example, [1]).

Chapter 2
Preliminaries: Preorders, Topologies, Axiomatizations of Set Theory

We recall basic facts on topologies and preorders which will be needed all over the book. On a first reading the reader may skip this chapter and come back when some of the notions introduced here are used in other parts of the book. We also warn the reader that some of the exercises of this chapter assume familiarity with standard concepts which are introduced in later chapters.

2.1 Topological Spaces

A *topology* on a given non-empty set X is a family $\tau \subseteq \mathcal{P}(X)$ with $\emptyset, X \in \tau$ which is closed under arbitrary unions and finite intersections. We call the pair (X, τ) a *topological space*.

The elements of τ are the *open sets* for the topology τ. Complements of open sets are called *closed sets*, and we denote by τ^c the family of closed sets (the family of closed sets of a topological space is closed under arbitrary intersections and finite unions). When a set A is both open and closed, we call it a *clopen set* of τ, and we denote this family by $\mathsf{CLOP}(X, \tau)$ (or just $\mathsf{CLOP}(X)$ if τ is clear from the context).

A *basis* σ for a topological space (X, τ) is a subfamily of τ with the property that every open set in τ can be written as a union of elements of σ. We say that τ is *generated* by σ. Notice that if σ is a basis for τ, any intersection of finitely many elements of σ which is non-empty contains a non-empty element of σ.

A *semibasis* σ for a topological space (X, τ) is a subfamily of τ with the property that the set of finite intersections of elements of σ is a basis. σ is a semibasis for (X, τ) if and only if τ is the weakest (i.e., smallest) topology on X containing σ. If σ is a semibasis for τ, we say that τ is generated by σ.

$U \subseteq X$ is a *neighborhood* of some $x \in X$ if $x \in U$. $F_{x,\tau}$ is the family (actually filter—see Definition 2.2.19) on the partial order (see Sect. 2.2) $(\tau \setminus \{\emptyset\}, \subseteq)$ of open neighborhoods of x.

© The Author(s), under exclusive license to Springer Nature Switzerland AG 2024
M. Viale, *The Forcing Method in Set Theory*, La Matematica per il 3+2 168,
https://doi.org/10.1007/978-3-031-71660-7_2

$x \in X$ is an *isolated* point if $\{x\}$ is open and closed.

Given a topological space (X, τ) and an arbitrary subset A of X, we denote by $\mathsf{Cl}(A)$ (the *closure* of A) the smallest closed set containing A (i.e., the intersection of all closed sets containing A). We denote by $\mathsf{Int}(A)$ (the *interior* of A) the biggest open set that is contained in A (i.e., the union of all open sets contained in A). An open set A is *regular open* if $A = \mathsf{Int}(\mathsf{Cl}(A))$. For any $A \subseteq X$ $\mathsf{Reg}(A) = \mathsf{Int}(\mathsf{Cl}(A))$ denotes the *regularization* of the set A.

Example 2.1.1 Let τ be the euclidean topology on $\mathbb{R}$; then any interval is a regular open set.

If $a < b < c$, we have that $(a; b)$, $(b; c)$ are regular open, while $(a; b) \cup (b; c)$ is not with its regularization being $(a; c)$.

Given a topological space (X, τ) and $Y \subseteq X$, the restriction $\tau \upharpoonright Y$ of τ to Y is the *subspace topology on Y* given by the family $\{A \cap Y : A \in \tau\}$ and is a topology on Y.

Given a topological space (X, τ) and $B \subseteq A \subseteq X$, B is *dense* in A if $\mathsf{Cl}(B) = \mathsf{Cl}(A)$ (with the closure computed in X with respect to τ). Remark that if B is dense in A, and $C \subseteq A$ is open in the subspace topology on A, then $B \cap C$ is dense in C.

Given a function $f : X \to Y$ and $Z \subseteq X$, we denote by $f[Z]$ the pointwise image of Z by f. A map $f : X \to Y$ between topological spaces (X, τ) and (Y, σ) is *continuous* if the preimage by f of any open set of Y is open, *open* if the (direct) image of an open set of X is open in Y, and a *homeomorphism* if it is an open and continuous bijection. Given $x \in X$ f is *continuous at x* if for every $U \in F_{f(x),\sigma}$ there is $V \in F_{x,\tau}$ such that $f[V] \subseteq U$. It can be checked that f is continuous if and only if it is continuous at every $x \in X$.

The *trivial topology* on X is given by $\{\emptyset, X\}$, and the *discrete topology* on X is given by $\mathcal{P}(X)$.

Separation Properties

A topological space (X, τ) is:

- T_0 if for any $x, y \in X$ there is some $U \in \tau$ with $x \in U$ and $y \notin U$ or there is some $V \in \tau$ with $x \notin V$ and $y \in V$.
- T_1 if for any $x, y \in X$ there is some $U \in \tau$ with $x \in U$ and $y \notin U$ and there is some $V \in \tau$ with $x \notin V$ and $y \in V$.
- T_2 or *Hausdorff* space (X, τ) if any two distinct points x and y can be *separated* by two open sets U and V in τ, that is, x is in U, y is in V, and U and V are disjoint. Recall that in a Hausdorff space X points are closed (i.e., $\{x\}$ is closed for all $x \in X$).
- T_3 if any $x \in X$ and $C \in \tau^c$ with $x \notin C$ can be separated by disjoint open sets (i.e., there are disjoint $U, V \in \tau$ with $x \in U$ and $C \subseteq V$).
- T_4 or *normal* if for any disjoint $C, D \in \tau^c$ can be separated by disjoint open sets (i.e., there are disjoint $U, V \in \tau$ with $C \subseteq U$ and $D \subseteq V$).
- (X, τ) is *0-dimensional* if τ admits a basis of clopen sets.
- (X, τ) is *extremally disconnected* if the closure of an open set is open.

Product Topologies

Let I be a set of indexes, and for all $i \in I$, let (X_i, τ_i) be a topological space and $X = \prod_{i \in I} X_i$ be the cartesian product of the sets X_i. The product topology τ on X is the weakest topology making all the projections maps

$$\pi_i : X \to X_i$$

$$f \mapsto f(i)$$

continuous. It is generated by the family of sets of the form $\prod_{i \in I} A_i$, where each A_i is open in X_i and $A_i \neq X_i$ only for finitely many i.

Lemma 2.1.2 *Let (X_i, τ_i) be topological spaces for $i \in I$ and (X, τ) be the product space with product topology. The following are equivalent:*

1. *(X, τ) is Hausdorff;*
2. *For every $i \in I$ (X_i, τ_i) is Hausdorff.*

Proof

2 implies 1   Let $f \neq g$ be in X. Then for some $i \in I f(i) \neq g(i)$. Let $U, V \in \tau_i$ separate $f(i)$ from $g(i)$. Then $\{h \in X : h(i) \in U\}$ and $\{h \in X : h(i) \in V\}$ separate f from g.

1 implies 2   Assume (X_i, τ_i) is not Hausdorff and find $x, y \in X_i$ which cannot be separated by disjoint open sets of τ_i. Let $f(j) = g(j)$ for all $j \neq i$, $f(i) = x$, $g(i) = y$. Assume U, V are basic open neighboorhoods respectively of f and g. W.l.o.g. we can suppose that $U = U_0 \cap U_1$ and $V = V_0 \cap V_1$ with $\pi_i[U_0] = \pi_i[V_0] = X_i$ and $U_1 = \pi_i^{-1}[U_2]$, $V_1 = \pi_i^{-1}[V_2]$ for some U_2, V_2 in τ_i. Then $x \in U_2$, $y \in V_2$, and $U \cap V_0 = U_3$, $V \cap V_0 = V_3$ are open neighboorhoods respectively of f, g with $U \cap V \supseteq U_3 \cap V_3 \cap U_1 \cap V_1$; the latter set is non-empty as

$$\pi_i[U_3 \cap V_3 \cap U_1 \cap V_1] = \pi_i[U_1 \cap V_1] = U_2 \cap V_2 \neq \emptyset,$$

since X_i is not Hausdorff as witnessed by x, y.

□

Compactness

A topological space (X, τ) is compact if any of the following equivalent conditions are met:

- Every family $\mathcal{F}$ of closed sets with the finite intersection property[1] has a non-empty intersection.
- Every open covering of X has a finite subcovering.

---

[1] $\mathcal{F}$ has the finite intersection property if any finite subfamily of $\mathcal{F}$ has a non-empty intersection. A family $\mathcal{A}$ of subsets of X such that $\bigcup \mathcal{A} = X$ is a covering of X.

We recall the following fundamental result (for a proof see Appendix D):

Theorem 2.1.3 (Tychonoff's Theorem) *Let (X_i, τ_i) be topological spaces for $i \in I$ and (X, τ) be the product space with product topology. The following are equivalent:*

1. *(X, τ) is compact;*
2. *For every $i \in I$ (X_i, τ_i) is compact.*

2.1.1 Key Properties of Regular Open Sets

Any clopen subset of a topological space is regular. Any open interval of $\mathbb{R}$ with the usual euclidean topology is regular, and a standard example of an open nonregular set in the euclidean topology on $\mathbb{R}$ is $(1; 2) \cup (2; 3)$: Its closure is $[1; 3]$, and the interior of its closure is $(1; 3)$. We see below that if U and V are open regular, then so is $U \cap V$. Moreover, any isolated point $x \in X$ of a topological space X is such that $\{x\}$ is clopen and thus regular.

We need several facts on regular open sets, the first of which is the following characterization:

Lemma 2.1.4 *Let (X, τ) be a topological space. For any open $A \in \tau$ we have*

$$\mathsf{Reg}(A) = \{x \in X : \exists U \in \tau \text{ open set containing } x \text{ such that } A \cap U \text{ is dense in } U\}.$$

Proof For one inclusion, take $x \in \mathsf{Reg}(A)$. The set $U = \mathsf{Reg}(A)$ is an open set containing x, and $A \cap U$ is dense in U because A is dense in $\mathsf{Cl}(A)$ and $U \subseteq \mathsf{Cl}(A)$ is open.

For the converse inclusion, take $x \in X$ and U an open set containing x such that $A \cap U$ is dense in U, then $A \cap U$ is dense also in $\mathsf{Cl}(U)$, and thus $\mathsf{Cl}(A \cap U) = \mathsf{Cl}(U)$ holds. So U is an open subset of $\mathsf{Cl}(A)$, and we obtain $x \in \mathsf{Reg}(A)$.   □

Exercise 2.1.5 The above lemma explains why (for the euclidean topology on $\mathbb{R}$) 2 belongs to $\mathsf{Reg}((1; 2) \cup (2; 3))$, while 1 and 3 do not. Work out the details of why 2 satisfies the above characterization for points of $\mathsf{Reg}((1; 2) \cup (2; 3))$, while 1 and 3 do not.

Remark 2.1.6 If U is an open neighborhood of x witnessing that $x \in \mathsf{Reg}(A)$, any $V \subseteq U$ open neighborhood of x is equally well a witness of $x \in \mathsf{Reg}(A)$, since $A \cap W = A \cap U \cap W$ is a dense open subset of W for any open $W \subseteq U$.

We also need the following crucial property:

Fact 2.1.7 *Given a topological space (X, τ), assume U, V are open sets in τ. Then $U \cap V$ is a dense open subset of V if and only if $\mathsf{Reg}(V) \subseteq \mathsf{Reg}(U)$. In particular $\mathsf{Reg}(V) = \mathsf{Reg}(U)$ iff $U \cap V \supseteq W$ for some W open dense subset of U and open dense subset of V.*

Proof Assume $U \cap V$ is a dense open subset of V. Let $x \in \mathsf{Reg}(V)$. Let W be an open neighborhood of x such that $V \cap W$ is a dense subset of W. Then $U \cap V \cap W$ is also a dense open subset of W (if $P \subseteq W$ is open non-empty, $V \cap P$ is a non-empty open subset of V, since $V \cap W$ is a dense open subset of W; thus $U \cap V \cap P$ is also a non-empty open set, given that $U \cap V$ is dense in V), and so a fortiori also $U \cap W$ is a dense open subset of W. In particular W witnesses that $x \in \mathsf{Reg}(U)$.

Conversely, assume $\mathsf{Reg}(V) \subseteq \mathsf{Reg}(U)$. Since U is a dense open subset of $\mathsf{Reg}(U)$ and V is a dense open subset of $\mathsf{Reg}(V)$, $U \cap V$ is a dense open subset of $\mathsf{Reg}(V)$, hence also of V (since the intersection of two open dense subsets of some topological space is still open dense). □

Fact 2.1.8 *Given a topological space* (X, τ), *assume* U, V *are open sets in* τ. *Then* $U \cap V$ *is non-empty if and only if so is* $U \cap \mathsf{Reg}(V)$.

Proof One implication is trivial. For the right to left direction assume $x \in U \cap \mathsf{Reg}(V)$. Find $W \in \tau$ with $x \in W$ and $W \cap V$ dense in W. Since $x \in U \cap W$, we get that $U \cap W$ is non-empty hence so is $U \cap V \cap W$ by density of $W \cap V$ in W and we are done. □

Notation 2.1.9 *Given a topological space* (X, τ) *and* $V \subseteq X$,

$$V^{\perp} = X \setminus \mathsf{Cl}(V).$$

For us priority is on the left; hence, for example, $U^{\perp\perp\perp}$ is a shorthand for $((U^{\perp})^{\perp})^{\perp}$.

Fact 2.1.10 *Given a topological space* (X, τ), *the following holds for any* $U, V \subseteq X$:

1. *For all* $U, V \subseteq X$, $U \subseteq V$ *implies* $V^{\perp} \subseteq U^{\perp}$.
2. $(U \cup V)^{\perp} = U^{\perp} \cap V^{\perp}$.
3. $\mathsf{Reg}(V) = V^{\perp\perp}$.
4. $U^{\perp\perp\perp} = U^{\perp}$.
5. *If* U, V *are open* $(U \cap V)^{\perp\perp} = U^{\perp\perp} \cap V^{\perp\perp}$.

   In particular we also have:

(A) $\mathsf{Reg} : \mathcal{P}(X) \to \mathcal{P}(X)$ *is an idempotent operator, i.e.,*

$$\mathsf{Reg}(\mathsf{Reg}(V)) = \mathsf{Reg}(V)$$

   for any $V \subseteq X$.
(B) *The intersection of any two regular open sets of* (X, τ) *is regular open.*

Proof

1: Easy exercise.
2: We need the following basic topological fact:

For any topological space (X, τ) and $A, B \subseteq X$

$$\mathsf{Cl}(A \cup B) = \mathsf{Cl}(A) \cup \mathsf{Cl}(B).$$

Proof Clearly $\mathsf{Cl}(A \cup B) \supseteq \mathsf{Cl}(A) \cup \mathsf{Cl}(B)$. For the converse inclusion observe that $A \cup B \subseteq \mathsf{Cl}(A) \cup \mathsf{Cl}(B)$ and $A \cup B$ is a dense subset of $\mathsf{Cl}(A \cup B)$. Hence

$$\mathsf{Cl}(A \cup B) = \mathsf{Cl}(\mathsf{Cl}(A) \cup \mathsf{Cl}(B)) = \mathsf{Cl}(A) \cup \mathsf{Cl}(B),$$

where:

- The first equality holds because

$$A \cup B \subseteq \mathsf{Cl}(A) \cup \mathsf{Cl}(B) \subseteq \mathsf{Cl}(A \cup B)$$

  with $A \cup B$ a dense subset of $\mathsf{Cl}(A \cup B)$.
- The second equality holds because $\mathsf{Cl}(\mathsf{Cl}(Y)) = \mathsf{Cl}(Y)$ for all $Y \subseteq X$.
  $\square$

Hence

$$(U \cup V)^{\perp} = X \setminus \mathsf{Cl}(U \cup V) = X \setminus (\mathsf{Cl}(U) \cup \mathsf{Cl}(V))$$

$$= (X \setminus \mathsf{Cl}(U)) \cap (X \setminus \mathsf{Cl}(V)) = U^{\perp} \cap V^{\perp}.$$

3:

$$x \in \mathsf{Reg}(V) \Leftrightarrow \text{there is an open neighborhood } N \text{ of } x \text{ fully contained in } \mathsf{Cl}(V)$$

$$\Leftrightarrow \text{there is an open neighborhood } N \text{ of } x \text{ disjoint from } X \setminus \mathsf{Cl}(V)$$

$$\Leftrightarrow x \notin \mathsf{Cl}(X \setminus \mathsf{Cl}(V))$$

$$\Leftrightarrow x \in V^{\perp\perp}.$$

4: Assume U is open, and then we have $U \subseteq \mathsf{Reg}(U)$. So, as $\mathsf{Reg}(U) = U^{\perp\perp}$ holds, we have

$$U \subseteq U^{\perp\perp}. \tag{2.1}$$

Now, if U is open, applying the first point to (2.1), we get $U^{\perp\perp\perp} \subseteq U^{\perp}$. Conversely, applying (2.1) to $U^{\perp}$, we get $U^{\perp} \subseteq U^{\perp\perp\perp}$, which concludes the proof.

5: We use Lemma 2.1.4.

Assume first $x \in U^{\perp\perp} \cap V^{\perp\perp} = \mathsf{Reg}(U) \cap \mathsf{Reg}(V)$. Then there are N_0, N_1 open neighborhoods of x such that $U \cap N_0$ and $V \cap N_1$ are open dense subsets respectively of N_0 and N_1. Since $x \in N_0 \cap N_1$, we get that U, V have both a dense open intersection with $N_0 \cap N_1$. Hence $N_0 \cap N_1$ witnesses that $x \in \mathsf{Reg}(U \cap V) = U^{\perp\perp}$ as $U \cap V$ has a dense open intersection with it.

For the converse inclusion let $x \in (U \cap V)^{\perp\perp} = \mathsf{Reg}(U \cap V)$. Then there is N open neighborhood of x such that $U \cap V \cap N$ is dense in N, and thus $U \cap N$ and $V \cap N$ are both dense subsets of N; this gives that $x \in U^{\perp\perp} \cap V^{\perp\perp}$, as was to be shown.

For the last two assertions:

(A). By 3 and 4

$$\mathsf{Reg}(\mathsf{Reg}(V)) = V^{\perp\perp\perp\perp} = V^{\perp\perp} = \mathsf{Reg}(V)$$

for any $V \subseteq X$.

(B). Combining (A) with 5

$$(U^{\perp\perp} \cap V^{\perp\perp})^{\perp\perp} = (U \cap V)^{\perp\perp\perp\perp} = (U \cap V)^{\perp\perp} = U^{\perp\perp} \cap V^{\perp\perp}.$$

$\square$

2.2 Preorders

A *quasi-order* or *preorder* is a set P equipped with a reflexive and transitive binary relation denoted by $\leq_P$. An antisymmetric qo is a *partial order*, or even just *po*. Every qo has an associated *strict* relation denoted by $<_P$ and defined by $x <_P y$ if and only if $x \leq_P y$ and $y \not\leq_P x$.

Driving examples of the kind of partial orders we will focus on are given by $(\tau \setminus \{\emptyset\}, \subseteq)$, where τ is a topology on some space X with no isolated points.

Exercise 2.2.1 Let (X, τ) be a topological space. Show that $(\tau \setminus \{\emptyset\}, \subseteq)$ is a partial order.

Exercise 2.2.2 Let τ be the euclidean topology on $\mathbb{R}$. Let for $A, B \in \tau$ $A \subseteq^* B$ if $A \cap B$ is a dense subset of A. Show that $(\tau \setminus \{\emptyset\}, \subseteq^*)$ is a qo but not a po. (HINT: The transitive and reflexive properties of $\subseteq^*$ are basic topological facts about density. To see that $\subseteq^*$ is not antisymmetric, consider an open interval I, and the same interval I without a point.)

Remark that if P is a partial order, then the strict relation $<_P$ is just $\leq_P \setminus \Delta_P$, where Δ_P stands for the diagonal in P^2. Remark also that this is far from being true in any qo, since for instance the total relation P^2 on P is a qo.

In a clear context we write $\leq$ instead of $\leq_P$.

When $x \leq y$ holds, we say that x is *below* y. When x is either below or above y, we say that x and y are *comparable*. An order $(P, \leq)$ is *total* or *linear* when any two elements are comparable.

Let $(P, \leq)$ be a qo.

We say that two elements x, y in P are *compatible*, and we write $x \| y$ if there is $z \in P$ such that both $z \leq x$ and $z \leq y$ hold. Otherwise x and y are *incompatible*, which is denoted by $x \perp y$.

Exercise 2.2.3 Following the notation of Exercise 2.2.2, show that $A, B \in \tau \setminus \{\emptyset\}$ are compatible for $\subseteq^*$ if and only if $A \cap B$ is non-empty.

A *chain* of a quasi-order $(P, \leq)$ is a subset of P which is linearly ordered by $\leq$.
An *antichain* of $(P, \leq)$ is a subset of P consisting of incompatible elements.
Given $\mathcal{A}$ subset of P, $p \in P$ is an *upper bound* for $\mathcal{A}$ if $p \geq a$ for all $a \in \mathcal{A}$.
p is an *exact upper bound* for $\mathcal{A}$ or a *supremum* of $\mathcal{A}$, it is an upper bound for $\mathcal{A}$ and $q \geq p$ for all upper bounds q for $\mathcal{A}$. In case P is an order, the exact upper bound of $\mathcal{A}$ is unique (when it exists); we denote it by $\bigvee \mathcal{A}$.
Exchanging $\leq$ with $\geq$, one obtains the notions of *lower bound* and *exact lower bound* or *infimum* (denoted by $\bigwedge \mathcal{A}$ when P is an order).
$p \in \mathcal{A}$ is a *maximal element* for $\mathcal{A}$ if it is an upper bound for $\mathcal{A}$.
Dually p is a *minimal element* for $\mathcal{A}$ if it is a lower bound for $\mathcal{A}$.
We recall the following equivalent of the Axiom of Choice:

Definition 2.2.4 (Zorn's Lemma) Let $(P, \leq)$ be a preorder such that all its chains have an upper bound. Then P admits a maximal element.

A subset D of P is *dense* in P if for all x in P there is some y in D below x, it is predense if its downward closure

$$\downarrow D = \{q : \exists x \in D, q \leq x\}$$

is dense, and it is a *maximal antichain* if it is a predense antichain.

Exercise 2.2.5 Let τ be the euclidean topology on $\mathbb{R}$. Following the notation of Exercise 2.2.2, show that:

- The intervals with rational end-points form a dense subset both for $(\tau \setminus \{\emptyset\}, \subseteq)$ and for $(\tau \setminus \{\emptyset\}, \subseteq^*)$.
- The set $\{(q; q + 1/n) : n \in [1; 100] \cap \mathbb{N}, q \in \mathbb{Q}\}$ is predense but not dense both for $(\tau \setminus \{\emptyset\}, \subseteq)$ and for $(\tau \setminus \{\emptyset\}, \subseteq^*)$.
- The set $\{(n; n + 1) : n \in \mathbb{Z}\}$ is a maximal antichain both in $(\tau \setminus \{\emptyset\}, \subseteq)$ and for $(\tau \setminus \{\emptyset\}, \subseteq^*)$.
- Show that any antichain of $(\tau \setminus \{\emptyset\}, \subseteq)$ or of $(\tau \setminus \{\emptyset\}, \subseteq^*)$ must be countable (HINT: An antichain $\mathcal{A}$ for both orders consists of pairwise disjoint non-empty sets; by the first item any element of $\mathcal{A}$ must contain an interval with rational end-points; if $A \neq B \in \mathcal{A}$ can they contain the same interval with rational end-points? How many such intervals are there?).

Exercise 2.2.6 Let (X, τ) be a topological space. Show that any base for τ is a dense subset of $(\tau \setminus \{\emptyset\}, \subseteq)$.

The following remark plays a crucial role in many of the arguments of these notes:

Fact 2.2.7 *Let (X, τ) be a topological space. Then:*

- *$E \subset X$ contains a dense and open set for τ if and only if $\sigma_E = \{O \in \tau : O \subseteq E\}$ is a dense and open subset of the quasi-order $(\tau \setminus \{\emptyset\}, \subseteq)$.*
- *$\sigma \subseteq \tau \setminus \{\emptyset\}$ is predense in the quasi-order $(\tau \setminus \{\emptyset\}, \subseteq)$ if and only if $\cup \sigma = D_\sigma$ is an open dense subset of X with respect to the topology τ.*

We leave the proof as an exercise for the reader.

We say that $(P, \leq)$ is *separative* if for all x and y in P, if x is not below y, then there is some z below x that is incompatible with y. Formally,

$$\forall x \in P \, \forall y \in P \, (x \nleq y \to \exists z \leq x (z \perp y)) .$$

Exercise 2.2.8 Following the notation of Exercise 2.2.2, show that neither $(\tau \setminus \{\emptyset\}, \subseteq)$ nor $(\tau \setminus \{\emptyset\}, \subseteq^*)$ are separative.

$(P, \leq)$ is *atomless* if it does not have minimal elements in the following strong sense: Given any p in P there are elements $q \perp r$ of P strictly below p.

An atom of a quasi-order $(P, \leq)$ is a minimal element $p \in P$.

Exercise 2.2.9 Following the notation of Exercise 2.2.2, show that $(\tau \setminus \{\emptyset\}, \subseteq)$ and $(\tau \setminus \{\emptyset\}, \subseteq^*)$ are atomless.

Exercise 2.2.10 Following the notation of Exercise 2.2.2, let σ be the family of open sets of τ which have non-empty intersection with $(0; 1) \cup \{2\}$. Show that the interval $(1; 3)$ is an atom of $(\sigma \setminus \{\emptyset\}, \subseteq)$ and $(\sigma \setminus \{\emptyset\}, \subseteq^*)$.

Exercise 2.2.11 Let (X, τ) be a Hausdorff topological space. Show that $a \in X$ is an isolated point if and only if $\{a\}$ is an atom of $(\tau \setminus \{\emptyset\}, \subseteq)$.

Exercise 2.2.12 Let $2^{<\omega}$ be the set of finite sequences of 0s and 1s, more precisely:

$$2^{<\omega} = \bigcup_{n \in \omega} 2^n,$$

where 2^n is the set of functions with domain n and range 2. For s, t finite strings of $0, 1$ in $2^{<\omega}$, let $s \leq t$ if $t \subseteq s$, that is, if t is an initial segment of s. Then $(2^{<\omega}, \leq)$ is a separative and atomless quasi-order. (HINT: First prove that $s \perp t$ iff $s \cup t$ is not a function and $s \| t$ iff $s \cup t = s$ or $s \cup t = t$.)

It can be seen that the quasi-orders given in Examples 2.2.12, 2.2.2 are quite similar: They give rise to isomorphic boolean completions, (see Theorem 4.2.4 and Remark 4.2.10).

Fact 2.2.13 *Assume a be a minimal element of a quasi-order $(P, \leq)$, and $D \subseteq P$ be dense. Then $a \in P$.*

Proof Since D is dense, $D \cap (\downarrow \{a\}) \neq \emptyset$. But a is a minimal element of P; hence $\{a\} = \downarrow \{a\} \subseteq D$. $\square$

Fact 2.2.14 *Assume $D \subseteq E \subseteq F$ with $(F, \leq_F)$ a quasi-order. Assume D is a dense subset of the quasi-order $(E, \leq_F)$,[2] and E is a dense subset of the quasi-order $(F, \leq_F)$. Then D is a dense subset of the quasi-order $(F, \leq_F)$. That is, the property of being dense is transitive.*

Proof Exercise for the reader. $\square$

The Order Topology

A preorder is equipped with a canonical topological structure. Let $(P, \leq)$ be a preorder. For each $p \in P$ we let:

$$\downarrow p := \downarrow \{p\} = \{q \in P : q \leq p\}.$$

The sets $\downarrow p$ form a basis for a topology τ_P on P, which we call the *preorder topology* or sometimes the *order topology*. We remark the following:

- The open sets of P in this topology are the downward closed subsets of P with respect to the order $\leq$ (dually it is easily checked that the closed sets in τ_P^c are exactly the upward closed subsets of P).
- For any $p \in P$, $\downarrow p$ is the smallest open set to which p belongs.
- A subset D of P is dense in the sense of the order iff it is dense in P with respect to the order topology.
- The family of open sets of this order topology is closed under arbitrary intersections, since the family of downward closed subsets of P has this property. In particular the order topologies are always complete and distributive sublattices of $\mathcal{P}(P)$ (see Sect. 3.10 for a definition of complete and distributive lattice).
- $(P, \leq)$ is a partial order if and only if (P, τ_P) is a T_0-space.

Remark 2.2.15 This topology is not to be confused with the one commonly associated to a linear order. For example the family of open sets for the order topology induced by the linear order $(\mathbb{R}, <)$ is given by the intervals of the form $(-\infty, a)$ or $(-\infty, a]$ as a ranges in $\mathbb{R} \cup \{+\infty, -\infty\}$, and this topology is clearly not the euclidean topology on $\mathbb{R}$, which is the one usually associated to the canonical linear order of $\mathbb{R}$. The order topology we introduced corresponds to the Alexandrov topology on a quasi-order, when reversing the order on $(P, \leq)$ (i.e., we consider as open sets what are the closed sets for the Alexandrov topology). In these notes we are interested in order topologies for orders which are *not* linear. For any quasi-order $(P, \leq)$ containing $p \neq q$ with $p \leq q$, the induced order topology is not Hausdorff: $p \in U$ for any open neighborhood of q, since $p \in N_q$.

---

[2] Here and elsewhere we take the liberty for an n-ary relation R on M^n and $Z \subseteq M$ to confuse R with $R \cap Z^n$ (for $R = \leq_F$ and $Z = E$ in this case).

2.2.1 Filters, Antichains, and Predense Sets on Quasi-Orders

The following links the notions related to the concept of density according to partial orders or the same concept according to topology:

Fact 2.2.16 *Let $(P, \leq)$ be a quasi-order and $X \subseteq P$:*

- *X is* dense *if it is dense in the order topology on P, i.e., if and only if for all $p \in P$ there exists $q \in X q \leq p$.*
- *X is* open *if it is open in the order topology on P, i.e., if it is downward closed.*
- *X is* predense *if $\downarrow X$ is a dense open set of P in the order topology.*
- *X is a* maximal antichain *if it is a predense antichain.*
- *X is* dense below $p \in P$ *if $X \cap P \upharpoonright p$ is a dense subset of the quasi-order $P \upharpoonright p$.*
- *X is* predense below $p \in P$ *if for all $q \leq p$ there is $r \in X$ compatible with q, i.e., if $\downarrow X$ is dense below p.*

Definition 2.2.17 Let $(P, \leq_P)$, $(Q, \leq_Q)$ be quasi-orders. A map $i : P \to Q$ between quasi-orders is:

- A *morphism* if it preserves the order relation.
- An *embedding* if it preserves the order and the incompatibility relations.
- A *complete embedding* if it maps predense subsets of P in predense subsets of Q.

Remark that an embedding need not be injective, examples of non-injective complete embeddings will be given later on (cfr., for example, Remark 4.2.10).

Exercise 2.2.18 Consider the space 2^ω endowed with the product topology τ. Let for $s \in 2^{<\omega} N_s = \{f \in 2^\omega : s \subseteq f\}$. Show that the map $s \mapsto N_s$ is an embedding of $(2^{<\omega}, \leq)$ into $(\tau \setminus \{\emptyset\}, \subseteq)$ with a dense image (HINT: The map $s \mapsto N_s$ is order reversing and preserve incompatibility; hence, it is an embedding of partial orders. Prove that $\{N_s : s \in 2^{<\omega}\}$ is a base for τ.).

Definition 2.2.19 Let $(P, \leq)$ be a quasi-order:

- *I is an* ideal *on P if it is a downward closed subset of P such that $a, b \in I$ entail that for some $c \in I$, $a, b \leq c$, otherwise said:*

  1. For all $p \in I$ and $q \leq p, q \in I$.
  2. For all $p, q \in I$, there is $r \in I (r \geq p, q)$.

- Dually a *filter* G on P is an upward closed subset of P such that any two elements of G are refined by some element in G, otherwise said:

  1. For all $p \in G$ and $q \geq p, q \in G$.
  2. For all $p, q \in G$, there is $r \in G (r \leq p, q)$.

- A *prefilter* on P is a subset H of P such that all its finite subsets have a lower bound in P.

Proposition 2.2.20 *Let P be a quasi-order. Let G be a filter on P and $X \subseteq P$. Then*

$$G \cap X \neq \emptyset \text{ if and only if } G \cap \downarrow X \neq \emptyset.$$

Proof If $r \in G \cap \downarrow X$, then $\exists q \geq r$ such that $q \in X$. So, since G is a filter, $q \in G \cap X$. $\square$

2.3   Axiomatizations of Set Theory

We assume throughout the platonistic stance that there is a definite mathematical entity called the universe of sets V and an $\in$-relation holding between two sets $a, b \in V$ if and only if a belongs to b. We also assume that it is meaningful to speak of the proper subclasses of V and to consider the inclusion relation $\subseteq$ holding between subclasses A, B of V if every set belonging to A belongs to B. Set theory describes the mathematical properties of this universe of sets and of its proper classes; however in this book, following standard practice, we focus on the first order axiomatization **ZFC** of set theory; **ZFC** aims to formalize true facts about the structure $(V, \in)$; in this set-up, proper classes are not object of the discourse and we can only deal with the subclasses of V which are definable by an $\in$-formula in set parameters.

We work (unless otherwise specified) in a language $\mathcal{L} = \{\in, \subseteq, =\}$ with three binary relation symbols for equality, membership, and containment. We feel free to adopt standard shorthands in the logic and set theory practice, such as:

- Ignoring the equality symbol when listing the symbols of a language (it is always assumed that $=$ is part of the language even when it is not explicitly mentioned).
- The focus is on the sublanguage $\{\in\}$ of $\{\in, \subseteq, =\}$ (as $x \subseteq y$ is definable—for any reasonable axiomatization of set theory—by the $\in$-formula $\forall z \, (z \in x \rightarrow z \in y)$).
- The use of restricted quantifiers $\forall x \in y, \exists x \in y$.
- The use of defined predicates such as $x = \{z \in y : \phi(z, x_1, \ldots, x_n)\}$ (which is a shorthand for the formula

$$\forall z \, [z \in x \leftrightarrow (z \in y \wedge \phi(z, x_1, \ldots, x_n))] \, .$$

- The use of defined constants (such as $\emptyset, \omega, \omega_1, \ldots$).
- The use of definable functions such as the rank function, etc.
- The fact that most of our reasonings about sets can be formalized in the above first order language.

Below we list the standard $\mathcal{L}$-theory **ZFC**. In Appendix A we discuss and analyze carefully a more efficient first order formalization of set theory (at least if one aims to have first order formalizations of set theory which stratify the syntactic complexity of formulae in accordance with conceptual complexity of the notion they formalize).

2.3.1 The **ZFC** Axiomatization of Set Theory

Letting $\mathcal{L} = \{\in, \subseteq, =\}$, we formalize **ZFC** as the following list of $\mathcal{L}$ formulae:

1. **Extensionality:** Two sets are equal if they have the same elements:

$$\forall x \forall y (x \subseteq y \leftrightarrow \forall z \in x (z \in y))$$

and

$$\forall x \forall y (x = y \leftrightarrow (x \subseteq y \land y \subseteq x)).$$

2. **Foundation:** Every non-empty set has an $\in$-minimal element:

$$\forall y \, [\exists x (x \in y) \rightarrow \exists x (x \in y \land \forall z \in x \neg z \in y)] \,.$$

3. **Infinity:** There is an infinite set:

$$\exists z \exists y [y \in z \land \forall x (x \in z \rightarrow \exists t [(t \in z) \land \forall w (w \in t \leftrightarrow (w \in x \lor w = x)))]].$$

4. **Pairing:** For any sets x and y there is a set containing $\{x, y\}$:

$$\forall x \forall y \exists z (x \in z \land y \in z).$$

5. **Union:** For every set a its union $\cup a$ exists:

$$\forall y \exists z \forall w (w \in z \leftrightarrow \exists x \in y \, w \in x).$$

6. **Separation (or Comprehension):**[3] For any formula $\phi(x, x_1, \ldots, x_n)$ and any set x with displayed free variables, the set

$$\{z \in x : \phi(x, x_1, \ldots, x_n)\}$$

exists; more formally,

$$\forall x_1, \ldots x_n \, \forall x \, \exists y \, [\forall z \, [z \in y \leftrightarrow (z \in x \land \phi(z, x_1, \ldots, x_n))]] \,.$$

---

[3] It would be more correct to reserve the word "Comprehension" for the corresponding axiom in class set theories such as Morse–Kelley or Gödel–Bernays stating that every property defines a class; nonetheless, we follow a standard practice and refer freely to this axiom as either the Separation axiom or the Comprehension axiom.

7. **Collection:**[4] For all formulae $\phi(x, y, x_1, \ldots, x_n)$ with displayed free variables

$$\forall x_1 \ldots \forall x_n \, \forall w \, [(\forall x \in w \, \exists y \, \phi(x, y, x_1, \ldots, x_n)) \rightarrow \exists z \, (\forall x \in w \, \exists y \in z \phi(x, y, x_1, \ldots, x_n))].$$

8. **Powerset:** For every set a its powerset $\mathcal{P}(a)$ exists:

$$\forall y \exists z \forall w (w \in z \leftrightarrow w \subseteq y).$$

9. **Choice:** There exists a well-ordering of any set.[5]

We also adopt the following notation:

- **ZF** is the axiom system consisting of all the above axioms except the Axiom of Choice.
- **ZF$^-$** is **ZF** with the exception of the Powerset Axiom.
- **ZFC$^-$** is **ZFC** with the exception of Powerset Axiom.
- **Z** is Zermelo's set theory consisting of **ZF** with the exception of Choice and Collection.
- **ZC** is Zermelo's set theory with Choice.
- **Z$^-$** (**ZC$^-$**) is Zermelo's set theory (with Choice) without the Powerset Axiom.

In various parts of the book (notably in Chap. 7 and Appendixes A and B), we will often need to consider finite tuples $\langle b_0, \ldots, b_n \rangle$ of sets. They can either be considered the set $\{\langle 0, a_0 \rangle, \ldots, \langle n, a_n \rangle\}$ in V given by a function with domain $n+1$, or an $n + 1$-tuple of sets in V^{n+1} with V being the domain of an $\in$-structure for the Tarski semantics. The latter choice is convenient when we look at satisfaction in $(V, \in)$ of formulae in parameters $b_0, \ldots, b_n$, and the former choice is convenient when we need to manipulate sets. The reader will have to decide depending on the context what is the correct viewpoint on the matter.

---

[4] We find more convenient to develop set theory using collection instead of replacement; the latter is obtained by requiring that any definable function with domain a set has range which is also a set.

[5] This is cumbersome to formalize in the signature $\{\in, \subseteq, =\}$. We leave to the reader to check that it can be done.

Chapter 3
Boolean Algebras

The core of this chapter develops the basic properties of boolean algebras: First we prove Stone's Duality theorem linking boolean algebras to the category of compact zero-dimensional Hausdorff spaces. Next we give several different presentations of these algebras in terms of their logical properties (Definition 3.1.1), of their ring structure (Theorem 3.9.3), and as partial orders (Lemma 3.10.3).

An introductory and exhaustive text on boolean algebras is [12].

3.1 Basic Definitions

We give the following *equational characterization* of a boolean algebra:

Definition 3.1.1 Let $(B, \wedge, \vee, \neg, 0, 1)$ be a sextuple consisting of a set B, two total binary operations $\wedge$ and $\vee$ on B, a total unary operation $\neg$ on B, and two elements 0 and 1 of B.

$(B, \wedge, \vee, \neg, 0, 1)$ is a boolean algebra if it satisfies the following equations for all $a, b, c \in B$:

$$a \vee (b \vee c) = (a \vee b) \vee c \qquad \text{associativity}$$
$$a \wedge (b \wedge c) = (a \wedge b) \wedge c$$

$$a \vee (b \wedge c) = (a \vee b) \wedge (a \vee c) \qquad \text{distributivity}$$
$$a \wedge (b \vee c) = (a \wedge b) \vee (a \wedge c)$$

$$a \vee b = b \vee a \qquad \text{commutativity}$$
$$a \wedge b = b \wedge a$$

© The Author(s), under exclusive license to Springer Nature Switzerland AG 2024
M. Viale, *The Forcing Method in Set Theory*, La Matematica per il 3+2 168,
https://doi.org/10.1007/978-3-031-71660-7_3

$$a \vee 0 = a \qquad\qquad\qquad\qquad\qquad \text{identity}$$

$$a \wedge 1 = a$$

$$a \vee \neg a = 1 \qquad\qquad\qquad\qquad\qquad \text{complements}$$

$$a \wedge \neg a = 0.$$

A bounded distributive lattice is a structure $(\mathsf{B}, \wedge, \vee, 0, 1)$ such that its two operations $\wedge, \vee$ satisfy the identity laws, the commutativity and associativity laws, and the distributivity laws.

Oftentimes it is convenient to attach a pedix R_B to some R among $\{\wedge, \vee, \neg, 0, 1\}$, especially when more than one boolean algebra is considered and it is not clear to which elements of which boolean algebras the operations or constants refer.

Example 3.1.2 Let 0, 1, $\vee$, $\wedge$, and $\neg$ be, respectively, $\emptyset$, X, $\cup$, $\cap$, $\neg$, and $\subseteq$; then the powerset $\mathcal{P}(X)$ of X is a (complete) boolean algebra.

Given a (non-empty) set X and a topology τ on X:

- The family τ and the family of closed sets τ^c are bounded distributive sublattices of $\mathcal{P}(X)$ (with the same operations we have on $\mathcal{P}(X)$).
- The family $\mathsf{CLOP}(X, \tau)$ of clopen set of τ (with the same operations we have on $\mathcal{P}(X)$) is a boolean subalgebra of $\mathcal{P}(X)$ (though in general it is not complete).

Notation 3.1.3 *It is often convenient to introduce further operations on a boolean algebra. For example, given a boolean algebra B and $a, b \in \mathsf{B}$ $a \setminus b = a \wedge \neg b$, and $a \triangle b = (a \setminus b) \vee (b \setminus a) = (a \vee b) \setminus (a \wedge b)$.*

Notice that if B is $\mathcal{P}(X)$, the above operations turn out to be the natural set theoretic operations on subsets of X.

The main results of this chapter are the following:

- Boolean algebras admit maximal ideals with corresponding dual ultrafilters (the Prime Ideal Theorem 3.7.4). This is crucial for many classical results on first order logic whose proof depends on the Axiom of Choice (e.g., in the proof of the compactness and/or completeness theorems for first order logic).
- Every boolean algebra is isomorphic to the algebra of clopen sets of some compact zero-dimensional space (Stone's Duality Theorem 3.8.2). This is a cornerstone result linking the logical and algebraic point of view on boolean algebras to topological compactness.
- Boolean algebras can also be described as the class of commutative rings with idempotent multiplication (see Definition 3.9.1 and Theorem 3.9.3). This characterization allows infusing the study of boolean algebra with methods coming from algebra and ring theory; for example, we will see that it greatly simplifies certain computations, among which those regarding the properties of boolean ideals and of boolean quotients.

In the remainder of this chapter we assume the reader is familiar with the basic properties of orders and topological spaces. We refer the reader to the previous Chap. 2 for the missing details.

3.2 The Order on Boolean Algebras

Note that for sets A, B in the boolean algebra $\mathcal{P}(X)$, $A \subseteq B$ if and only if $A \cap B = A$ if and only if $A \cup B = B$. We can introduce abstractly an order relation $\leq_{\mathsf{B}}$ on a boolean algebra B using these latter equations to define it.

Proposition 3.2.1 *Let* $(\mathsf{B}, \wedge, \vee, \neg, 0, 1)$ *be a boolean algebra. Define* $a \leq b$ *by* $a \wedge b = a$ *for* $a, b \in \mathsf{B}$*. Then:*

(i) $\leq$ *is an order relation on* B*.*
(ii) $a \wedge b$ *defines the infimum of* $\{a, b\}$*.*
(iii) $a \vee b$ *defines the supremum of* $\{a, b\}$*.*
(iv) $a \leq b$ *if and only if* $a \vee b = b$*.*

Proof

(i) $\leq$ **is reflexive (equivalently $\wedge$ is idempotent):**

$$a = a \wedge 1 = a \wedge (a \vee \neg a) = (a \wedge a) \vee (a \wedge \neg a) = (a \wedge a) \vee 0 = a \wedge a;$$

hence $a \leq a$.
$\leq$ **is transitive:** Assume $a \leq b$ (i.e., $a \wedge b = a$) and $b \leq c$ (i.e., $b \wedge c = b$). Then

$$a \wedge c = (a \wedge b) \wedge c = a \wedge (b \wedge c) = a \wedge b = a.$$

$\leq$ **is antisymmetric:** Assume $a \leq b \leq a$; then $a = a \wedge b = b \wedge a = b$.
(ii) First of all

$$(a \wedge b) \wedge a = (b \wedge a) \wedge a = b \wedge (a \wedge a) = b \wedge a = a \wedge b$$

and

$$(a \wedge b) \wedge b = a \wedge (b \wedge b) = a \wedge b;$$

hence $a \wedge b$ is a lower bound for a, b.
Assume $c \leq a, b$. Then $c \wedge a = c$ and $c \wedge b = c$; hence

$$c \wedge (a \wedge b) = (c \wedge a) \wedge b = c \wedge b = c,$$

therefore our thesis.

(iii) First of all we show that $a \vee b$ is an upper bound for b, a:

$$b = b \vee 0 = b \vee (a \wedge \neg a) = (b \vee a) \wedge (b \vee \neg a) \leq b \vee a, \qquad (3.1)$$

where in the latter inequality we used the fact that $c \wedge d \leq c$ (being the infimum of $\{c, d\}$ by the previous item) for all $c, d \in \mathsf{B}$; similarly we can prove $a \leq a \vee b$.

The second observation is the following:

$$\text{Assume } c, d \leq e, \text{ then } c \vee d \leq e. \qquad (3.2)$$

This holds since

$$(c \vee d) \wedge e = (c \wedge e) \vee (d \wedge e) = c \vee d.$$

By 3.1 and 3.2, we get that $a \vee b$ is the supremum of $\{a, b\}$ (3.1 grants that it is an upper bound and 3.2 that is the smallest such).

(iv) $a \leq b$ if and only if $b = \max\{a, b\} = \sup\{a, b\} = a \vee b$.

$\square$

3.3 Boolean Identities

Proposition 3.3.1 *The following hold on a boolean algebra:*

(i) $a = a \vee a = a \wedge a$ *(Idempotent laws).*
(ii) $\neg a$ *is the unique* $b \in \mathsf{B}$ *such that* $b \wedge a = 0$ *and* $b \vee a = 1$
     (Law of uniqueness for complements).
(iii) $\neg\neg a = a$ *(Double negation law).*
(iv) $\neg(a \wedge b) = \neg a \vee \neg b$ *(First De Morgan law).*
(v) $\neg(a \vee b) = \neg a \wedge \neg b$ *(Second De Morgan law).*

Proof

(i) Immediate since $a \wedge a = \min\{a, a\} = a = \max\{a, a\} = a \vee a$.
(ii) Assume $b \wedge a = 0$ and $b \vee a = 1$; we show that $b = \neg a$.

$$b = b \wedge 1 = b \wedge (a \vee \neg a) = (b \wedge a) \vee (b \wedge \neg a) = 0 \vee (b \wedge \neg a) = (b \wedge \neg a).$$

Therefore $b \leq \neg a$. On the other hand,

$$b = b \vee 0 = b \vee (a \wedge \neg a) = (b \vee a) \wedge (b \vee \neg a) = 1 \wedge (b \vee \neg a) = (b \vee \neg a).$$

Therefore $b \geq \neg a$.

(iii) By (ii) $a = \neg\neg a$, since both satisfy the equations defining the complement of
   $\neg a$.
(iv) Remark that

$$(a \wedge b) \vee (\neg a \vee \neg b) = (a \vee \neg a \vee \neg b) \wedge (b \vee \neg a \vee \neg b) \geq 1.$$

   Similarly one can prove that

$$(a \wedge b) \wedge (\neg a \vee \neg b) \leq 0.$$

   By (ii) we get that $\neg(a \wedge b) = (\neg a \vee \neg b)$.
(v) Left to the reader (along the lines of the proof of the previous item).

   □

Exercise 3.3.2 In a boolean algebra B, for any $a, b \in \mathsf{B}$,

$$a \leq b \Leftrightarrow \neg a \geq \neg b.$$

Exercise 3.3.3 Let B be a boolean algebra, and define the operation $u \to v = \neg u \vee v$ for $u, v \in \mathsf{B}$. Then

$$u \to v \geq w \Leftrightarrow u \wedge w \leq v.$$

Exercise 3.3.4 Show that for all a, b, c in a boolean algebra B

$$(a \vee b) \wedge (\neg a \vee c) \leq b \vee c.$$

Exercise 3.3.5 Show that for all a, b, c in a boolean algebra B

$$(c \wedge a \leq c \wedge b) \Leftrightarrow (c \leq a \to b).$$

3.4   Ideals and Morphisms of Boolean Algebras

Ideals on boolean algebras are the kernels of boolean algebra morphisms. They describe the concept of being "false" (when boolean operations are regarded as operations describing classical logic laws), "small" (when boolean operations are regarded as operations describing laws of sets), or the familiar notion of ring ideal (as we will see boolean algebras describe a very interesting elementary class of rings). Actually the usual algebraic properties of morphisms of rings and groups work equally well for boolean algebras (for those familiar with first order logic or universal algebra the reason being that boolean algebras are axiomatized by equational theories, as rings and groups are).

Definition 3.4.1 Let B, C be boolean algebras. A map $k : B \to C$ is a *homomorphism* of boolean algebras if it preserves the boolean operations and constants and an *isomorphism* if it is a bijective homomorphism.

A subalgebra of a boolean algebra (B, $\wedge$, $\vee$, $\neg$, 0, 1) is a subset A of B such that the inclusion map of A into B defines an injective homomorphism.

Fact 3.4.2 *A map $\phi : B \to C$ is a homomorphism of boolean algebras if it preserves $\vee$, $\neg$ or if it preserves $\wedge$, $\neg$.*

Proof Left to the reader. (**HINT:** Use De Morgan's laws.)   □

Exercise 3.4.3 Prove that a boolean morphism $\phi : B \to C$ preserves the operation of symmetric difference Δ.

Remark 3.4.4 Since boolean algebras are axiomatized by an equational theory, the class of boolean algebras is closed under homomorphic images, products, and substructures (by the easy direction of Birkhoff's theorem, see [6]). A direct proof can be a useful exercise for the reader.

Definition 3.4.5 Let B be a boolean algebra. $I \subseteq B$ is an ideal if it is closed under the $\vee$ operation and is downward closed (i.e., $b \in I$ and $a \leq b$ gives that $a \in I$ as well).

Exercise 3.4.6 I is an ideal if and only if the following two conditions are simultaneously met:

- $a \Delta b \in I$ for all $a, b \in I$ (note that $a \vee b = (a \Delta b) \Delta (a \wedge b)$).
- $a \wedge b \in I$ for all $b \in I$ and $a \in B$.

Notation 3.4.7 *Given a boolean algebra* B, $B^+ = B \setminus \{0\}$. *We will often look at* B *as the partial order* $(B^+, \leq)$.

A subset X of B^+ *is* dense *if for all $b \in B^+$ there is $a \in X$ such that $a \leq b$.*

3.5   Atomic and Finite Boolean Algebras

Definition 3.5.1 Let B be a boolean algebra. The atoms of B are the minimal elements[1] of $(B^+, \leq)$ (if they exist).

- B is atomic if its atoms form a dense subsets of B^+.
- B is atomless if it has no atoms.

---

[1] Recall that a is a minimal element of an order $(P, \leq)$ if $b < a$ for no $b \in P$.

Remark 3.5.2 The following hold:

- Let B be a boolean algebra. The following are equivalent:

  - $a \in B$ is an atom.
  - For no $b \in B$, it is the case that $0 < b < a$.
  - $a \wedge b = a$ or $a \wedge b = 0$ for all $b \in B$.

  To see this observe that if $b \leq a$, then either $b = 0$ or $b = a$, hence $a \wedge b = a$ or $a \wedge b = 0$; if $b \geq a$, $a \wedge b = a$; if $b \not\geq a$, $a \wedge b \neq a$; and since $a \wedge b \leq a$, we must have that $a \wedge b = 0$.

- Let X be a non-empty set; then $\mathcal{P}(X)$ is atomic:
  The order relation on $\mathcal{P}(X)$ given by $X \leq Y$ if $X \cap Y = X$ is the inclusion relation; all singletons $\{x\}$ for $x \in X$ are atoms of $\mathcal{P}(X)$; $\mathcal{P}(X)^+ = \mathcal{P}(X) \setminus \{\emptyset\}$, and any non-empty set $Y \subseteq X$ has some $y \in Y$ with $\{y\} \subseteq Y$.

- All finite boolean algebras B are atomic:
  Assume $b_0 \in B$ is not refined by any atom (i.e., no $a \leq b$ is an atom). Inductively define a chain

$$\{b_n : n \in \mathbb{N}\}$$

  such that $b_{n+1} < b_n$ is not refined by any atom. The procedure cannot terminate; otherwise some b_{n+1} is refined by some atom a, and hence so is b_0, contradicting our assumptions on b_0. Hence $\{b_n : n \in \mathbb{N}\}$ is an infinite subset of B, a contradiction.
  We may formalize properly this proof as follows: Fix b_0 such that no atom refines b_0. Consider the family

$$\{f : \mathrm{dom}(f) \to B \; : \quad \mathrm{dom}(f) \in \mathbb{N} \text{ or } \mathrm{dom}(f) = \mathbb{N} \text{ and } f \text{ is such that}$$

$$f(0) = b_0 \text{and } f(i) > f(i+1) \text{ for all } i + 1 \in \mathrm{dom}(f)\}$$

  ordered by inclusion. This partial order has upper bounds for all its subchains. By Zorn's Lemma the family has a maximal element h. It is easy to check that $\mathrm{dom}(h) = \mathbb{N}$ and $h(i) > h(i+1)$ for all $i \in \mathbb{N}$.

The structure of the class of finite boolean algebras is described by the following:

Proposition 3.5.3 *Assume B is a finite boolean algebra. Then $B \cong \mathcal{P}(A_B)$, where A_B is the set of atoms of B.*

The following exercise provides a concrete example of how this isomorphism can be defined for a finite boolean algebra.

Exercise 3.5.4 Consider the set $\mathrm{Div}(30) = \{1, 2, 3, 5, 6, 10, 15, 30\}$ with operations $\wedge, \vee, \neg$ given by $n \wedge m = \mathrm{MCD}(n, m)$, $n \vee m = \mathrm{mcm}(n, m)$, $\neg(n) = \frac{30}{n}$.

1. Check that:

   - $B = \langle \mathrm{Div}(30), \wedge, \vee, \neg, 1, 30 \rangle$ is a boolean algebra.

Look at the picture below to understand what is the order structure of B: The segments connecting adjacent vertexes are seen as upward directed arrows; a path is a sequence of vertexes with each of the two consecutive vertexes being end-points of an arrow; a path is coherent if it is composed of arrows that do not change direction (the paths given by vertexes $(2, 10, 30)$ or by vertexes $(30, 15, 3, 1)$ are coherent, the path $(2, 10, 5)$ is not coherent, and $(2, 3)$ is not a path); and a vertex is below another if there is a path connecting the two in which all the arrows point upward (2 is below 30 as witnessed by the path $(2, 10, 30)$, and 6 is not below 10).

- $n \leq m$ if and only if n divides m.
- The atoms of $\mathrm{Div}(30)$ are 2, 3, 5.
- The map $F : \mathrm{Div}(30) \rightarrow \mathcal{P}(\{2, 3, 5\})$ of the proposition is given by $n \mapsto \{p : p$ is a prime number and divides $n\}$ and implements an isomorphism of B with $\langle \mathcal{P}(\{2, 3, 5\}), \cap, \cup, A \mapsto \{2, 3, 5\} \setminus A, \emptyset, \{2, 3, 5\}\rangle$.

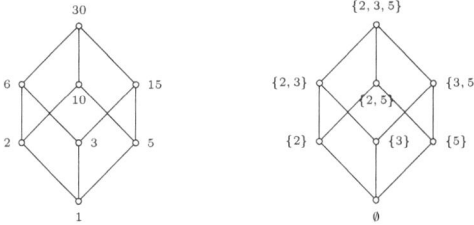

2. For which n $\mathrm{Div}(n)$ with the above operations is a boolean algebra? (**HINT:** Show that $\mathrm{Div}(18)$ is not a boolean algebra. Notice that 18 has a prime factor which divides it in power 2).

Now we prove the proposition.

Proof By the previous remark B is atomic. Define

$$F : \mathsf{B} \rightarrow \mathcal{P}(A_{\mathsf{B}})$$

$$b \mapsto \{a \in A_{\mathsf{B}} : a \leq b\}.$$

F is a morphism:

- F preserves $\wedge$:

$$F(b \wedge c) = \{a \in A_{\mathsf{B}} : a \leq b \wedge c\} = \{a \in A_{\mathsf{B}} : a \leq b \text{ and } a \leq c\}$$
$$= \{a \in A_{\mathsf{B}} : a \leq b\} \cap \{a \in A_{\mathsf{B}} : a \leq c\} = F(b) \cap F(c),$$

where the second equality follows since $b \wedge c = \inf\{b, c\}$.
- F maps 1_{B} to A_{B} and 0_{B} to $\emptyset$ (useful exercise for the reader).

- F preserves $\neg$:

$$F(\neg b) = \{a \in A_\mathsf{B} : a \leq \neg b\} = \{a \in A_\mathsf{B} : a \not\leq b\} = A_\mathsf{B} \setminus F(b),$$

where the second equality follows by the following argument: For $a \in A_\mathsf{B}$, we have that

$$a \not\leq b$$

if and only if (since a is an atom)

$$a \wedge b = 0$$

$$\text{if and only if} \tag{3.3}$$

$$a \leq \neg b.$$

For the top to bottom implication in 3.3, note that

$$a = a \wedge 1 = a \wedge (b \vee \neg b) = (a \wedge b) \vee (a \wedge \neg b) = 0 \vee (a \wedge \neg b) = a \wedge \neg b,$$

where the before last equality holds since $a \wedge b = 0$. For the bottom to top implication in 3.3, note that

$$a \wedge b = (a \wedge \neg b) \wedge b = a \wedge (\neg b \wedge b) = a \wedge 0 = 0,$$

where the first equality holds since $a \leq \neg b$.

F is an injection: If $b \neq c$, assume $b \wedge \neg c > 0_\mathsf{B}$, and let $a \in A_\mathsf{B}$ refine $b \wedge \neg c$. Then $a \in F(b) \setminus F(c)$ since $a \leq b$ while $a \not\leq c$.

F is surjective: Given $X \subseteq A_\mathsf{B}$, let $b_X = \bigvee X$. The following holds:

> For any $a \in A_\mathsf{B}$, $a \leq b_X$ if and only if $a \in X$.

If $a \in X$, clearly $a \leq b_X$. On the other hand, if $a \notin X$, then $a \wedge u = 0_\mathsf{B}$ for all $u \in X$ (since distinct atoms are pairwise incompatible), hence (by applying $|X|$ many times the distributive law)

$$a \wedge \bigvee X = \bigvee \{a \wedge u : u \in X\} = 0_\mathsf{B}.$$

Therefore $a \wedge b_X = 0_\mathsf{B} \neq a$, i.e., $a \not\leq b_X$.
We get that $F(b_X) = X$.

$\square$

3.6   Examples of Boolean Algebras

The first two examples come from propositional logic, the third and fourth from first order logic, the fifth from Lebesgue measure, and the sixth from general topology. We assume the reader is familiar with the background material needed to analyze each of these examples.

Example 3.6.1 (Lindenbaum Algebras on Finitely Many Propositional Variables) The reader should be familiar with the basic concepts of propositional calculus to follow this and the next example. Below, as a guiding example for the discussion to follow, we give the truth table of the propositional formulae

$$\phi := ((B \to A) \wedge ((B \vee C) \to A))$$

and

$$\psi := ((\neg B \wedge \neg C) \vee A)$$

in propositional variables A, B, C (also of all the propositional subformulae of ϕ):

A	B	C	$(B \to A)$	$(B \vee C)$	$((B \vee C) \to A)$	ϕ	ψ
1	1	1	1	1	1	1	1
1	1	0	1	1	1	1	1
1	0	1	1	1	1	1	1
1	0	0	1	0	1	1	1
0	1	1	0	1	0	0	0
0	1	0	0	1	0	0	0
0	0	1	1	1	0	0	0
0	0	0	1	0	1	1	1

Recall that:

- For $\phi_1, \ldots, \phi_n, \psi$ propositional formulae in propositional variables $A_1, \ldots, A_n$, $\phi_1, \ldots, \phi_n \models \psi$ if the truth tables of $\phi_1, \ldots, \phi_n$ and ψ in the variables $A_1, \ldots, A_n$ are such that every time a row assigns value 1 to each of the formulae $\phi_1, \ldots, \phi_n$, that row assigns 1 also to ψ (in our example one can check that $\phi \models (B \to A)$, while $(B \to A) \not\models \phi$).
- ϕ is logically equivalent to ψ ($\phi \equiv \psi$) if and only if $\phi \models \psi$ and $\psi \models \phi$, i.e., the two formulae have the same truth table in variables $A_1, \ldots, A_n$ (one might check by computing that this holds for ϕ and ψ in our truth table).

Let $\mathcal{A}_n = \{A_1, \ldots, A_n\}$ be a set of n-many propositional variables; the Lindenbaum algebra B_n on $\mathcal{A}_n$ is defined as follows:

- Its domain is $\{[\phi] : \phi \text{ an } \mathcal{A}_n\text{-formula}\}$, where $[\phi]$ is the equivalence class given by all formulae ψ logically equivalent to ϕ).
- $1_{B_n} = [A_1 \vee \neg A_1]$, $0_{B_n} = [A_1 \wedge \neg A_1]$.
- $[\phi] \wedge_{B_n} [\psi] = [\phi \wedge \psi]$, $[\phi] \vee_{B_n} [\psi] = [\phi \vee \psi]$, $\neg_{B_n} [\psi] = [\neg \psi]$.

Remark that

$[\phi] \leq [\psi]$ if and only if $[\phi \wedge \psi] = [\phi]$ if and only if

$$\phi \wedge \psi \models \phi \text{ and } \phi \models \phi \wedge \psi.$$

Now $\phi \models \phi \wedge \psi$ if and only if $\phi \models \psi$. Hence $[\phi] \leq [\psi]$ if and only if $\phi \models \psi$.

Since $\equiv$ is the equivalence relation on propositional $\mathcal{A}_n$-formulae induced by $\models$, we get the following very nice fact:

The order $(B_n, \leq_{B_n})$ is the quotient of the preorder given by $\models$ on the propositional $\mathcal{A}_n$-formulae.

A second nice observation is the following:

The domain of B_n is in bijective correspondence with the set of truth tables for $\mathcal{A}_n$-propositional formulae. Moreover the atoms of B_n are in bijective correspondence with the truth tables containing exactly one row in which a 1 appears. Hence B_n has 2^n-many atoms and $2^{(2^n)}$-many elements.

We can see it by the following argument: Clearly any $\mathcal{A}_n$-formula has a truth table, and two formulae are logically equivalent if and only if they have the same truth table.

Given n many propositional variables, there are 2^n-possible assignments of truth values to the propositional formulae. Hence each truth table on an $\mathcal{A}$-propositional formula has 2^n rows.

There are $2^{(2^n)}$ possible truth tables in 2^n-rows, and each such truth table identifies a unique equivalence class. Therefore,

The Lindenbaum algebra B_n has $2^{(2^n)}$-many elements.

B_n, being finite, is isomorphic to $\mathcal{P}(A_{B_n})$. We already computed the size of B_n being $2^{2^n} = |\mathcal{P}(2^n)|$; hence B_n has exactly 2^n atoms.

Let us identify which truth tables define atoms:

The atoms of B_n are identified by the truth tables with exactly one 1.

Assume $[\phi] \leq [\psi]$ for some ψ whose truth table has exactly one 1 in the relevant column. Then the truth table of ϕ can have 1 only in the places where these occur in the truth table of ψ, i.e., in at most one place, therefore either ϕ is not satisfiable (i.e., its truth table consists just of 0 in the relevant column) or ϕ is logically equivalent to ψ. This means that $[\psi]$ is an atom.

Since there are 2^n such atoms, one for each of the possible truth tables where 1 appears in exactly one row, we get that the truth tables with exactly one 1 appearing in them define all the possible atoms of $\mathcal{A}$.

Example 3.6.2 (Lindenbaum Algebras on Infinitely Many Propositional Variables) Let $\mathcal{A} = \{A_n : n \in \mathbb{N}\}$ be an infinite set of propositional variables. Recall that for $\mathcal{A}$-propositional formulae:

- $\phi_1, \ldots, \phi_m \models \psi$ if and only if, whenever n is large enough so that all the propositional variables occurring in $\phi_1 \cup \cdots \cup \phi_m \cup \psi$ are among $A_1, \ldots, A_n$, the truth tables of $\phi_1, \ldots, \phi_m$ and ψ as computed with respect to $A_1, \ldots, A_n$ are such that the following holds: every time a row assigns value 1 to each of the formulae $\phi_1, \ldots, \phi_m$, that row assigns 1 also to ψ.
- ϕ is logically equivalent to ψ ($\phi \equiv \psi$) if and only if $\phi \models \psi$ and $\psi \models \phi$, i.e., the two formulae have the same truth table in variables $A_1, \ldots, A_n$ for any large enough n.

The Lindenbaum algebra B_∞ on $\mathcal{A}$ is defined as follows:

- Its domain is $\{[\phi] : \phi \text{ an } \mathcal{A}\text{-formula}\}$, where $[\phi]$ is the equivalence class given by all formulae ψ logically equivalent to ϕ.
- $1_{\mathsf{B}_\infty} = [A_1 \vee A_1], 0_{\mathsf{B}_\infty} = [A_1 \wedge \neg A_1]$.
- $[\phi] \wedge_{\mathsf{B}_\infty} [\psi] = [\phi \wedge \psi], [\phi] \vee_{\mathsf{B}_\infty} [\psi] = [\phi \vee \psi], \neg_{\mathsf{B}_\infty} [\psi] = [\neg \psi]$.

This is an infinite atomless boolean algebra.

To prove that it is atomless, proceed as follows: Given a satisfiable formula ϕ (i.e., a formula whose column in the relevant truth tables has at least one 1), assume that $\{A_1, \ldots, A_m\}$ contains all the propositional variables occurring in ϕ. Then the truth table of ϕ in variables $A_1, \ldots, A_m$ has at least a 1 in the relevant column. We can show that $\phi \not\models \phi \wedge A_{m+1}$: Since ϕ is satisfiable, find a row l in the truth table of ϕ over $A_1, \ldots, A_m$ such that ϕ gets value 1 in row l. Consider now the truth table of $\phi \wedge A_{m+1}, \phi$ as computed in variables $A_1, \ldots, A_{m+1}$. Let k be the row which assigns 0 to A_{m+1} and to each A_j for $j = 1, \ldots, m$ exactly the same value (0 or 1) it gets in row l of the truth table of ϕ over $A_1, \ldots, A_m$. Then in this row k, $\phi \wedge A_{m+1}$ gets value 0, while ϕ gets value 1. Hence $\phi \not\models \phi \wedge A_{m+1}$. Clearly $\phi \wedge A_{m+1} \models \phi$. Therefore $[\phi \wedge A_{m+1}] < [\phi]$.

This shows that $[\phi]$ is not an atom.

The inclusion map of the Lindenbaum algebras B_n into $\mathsf{B}_\mathcal{A}$ is an injective homomorphism for all n and $\mathsf{B}_\mathcal{A}$ is the union (and direct limit) of the algebras B_n for $n \in \mathbb{N}$.

Example 3.6.3 (Lindenbaum Algebras of $\mathcal{L}$-Theories) The reader should be familiar with the basic concepts of first order logic to follow this and the next example; a possible reference is [8]. Let $\mathcal{L}$ be a first order signature and T be a satisfiable $\mathcal{L}$-theory consisting of $\mathcal{L}$-sentences. Following standard terminology, we say that

$$\phi_1, \ldots, \phi_n \models_T \psi$$

if

$$\phi_1, \ldots, \phi_n, T \models \psi$$

(i.e., there is a proof of ψ from the axioms of T and assumptions $\phi_1, \ldots, \phi_n$) and $\phi \equiv_T \psi$ if $\phi \models_T \psi$ and $\psi \models_T \phi$.

The Lindenbaum algebra B_T is defined as follows:

- Its domain is $\{[\phi]_T : \phi$ an $\mathcal{L}$-sentence$\}$, where $[\phi]_T$ is the equivalence class of ϕ with respect to the equivalence relation $\equiv_T$.
- $1_{\mathsf{B}_T} = [\phi \vee \neg\phi]_T, 0_{\mathsf{B}_T} = [\phi \wedge \neg\phi]_T$.
- $[\phi] \wedge_{\mathsf{B}_T} [\psi] = [\phi \wedge \psi]_T, [\phi] \vee_{\mathsf{B}_T} [\psi] = [\phi \vee \psi]_T, \neg_{\mathsf{B}_T} [\psi] = [\neg\psi]_T$.

Remark that $[\phi]_T = 1_{\mathsf{B}_T}$ if and only if $T \models \phi$.

Exercise 3.6.4 T is complete if and only if B_T has only two elements.

Exercise 3.6.5 Let $S \subseteq T$ be satisfiable $\mathcal{L}$-theories (consisting only of sentences) for a given language $\mathcal{L}$.

- Show that $I_T = \{[\phi]_S : T \models \neg\phi\}$ is an ideal on the Lindenbaum algebra B_S.
- Show also that $\mathsf{B}_T \cong \mathsf{B}_S/I_T$ via the map $[[\phi]_S]_{I_T} \mapsto [\phi]_T$.

Example 3.6.6 (The Algebras of Definable Subsets of an $\mathcal{L}$-Structure) Let $\mathcal{M}$ be an $\mathcal{L}$-structure with domain M.

The algebra $\mathsf{B}^n_{\mathcal{M}}$ of n-dimensional $\mathcal{L}$-definable subsets of M^n has domain

$$\left\{ T^{\mathcal{M}}_{\phi(x_1,\ldots,x_n,y_1,\ldots y_k),\langle b_1,\ldots,b_k\rangle} : \phi(x_1,\ldots,x_n,y_1,\ldots y_k) \text{ an } \mathcal{L}\text{-formula}, \right.$$

$$\{b_1,\ldots,b_k\} \subseteq M\}$$

with operations inherited as a subalgebra of $\mathcal{P}(M^n)$.

We let $T^{\mathcal{M}}_{\phi(x_1,\ldots,x_n,y_1,\ldots y_k),\langle b_1,\ldots,b_k\rangle}$ be the set

$$\left\{ \langle a_1,\ldots,a_n\rangle \in M^n : \mathcal{M} \models \phi(x_1,\ldots,x_n,y_1,\ldots,y_k)[x_i/a_i, y_j/b_j] \right\}.$$

Example 3.6.7 (The Boolean Algebras of Characteristic Functions) Let (X, τ) be a topological space, and consider the ring B_X of characteristic 2 given by the $f : X \to \mathbb{Z}_2$ which are continuous with respect to τ (where $\mathbb{Z}_2$ is endowed with the discrete topology). The boolean operations on B_X are defined as follows (as usual for $A \subseteq X$, χ_A is the characteristic function on A): $f \vee g = \max(f, g)$, $\neg f = \chi_X - f$, $f \wedge g = f \cdot g$. The top element is χ_X, and the bottom element is $\chi_\emptyset$.

When X is endowed with the discrete topology, B_X is homeomorphic to $\mathcal{P}(X)$ via the identification of subsets with their characteristic functions. When X is endowed with a connected topology, $\mathsf{B}_X = \{\chi_X, \chi_\emptyset\}$ is the algebra with two elements.

See Sect. 3.9 for more details.

Exercise 3.6.8 Assume (X, τ) is a topological space, and consider the boolean algebra B_X defined in the above Example 3.6.7. $I \subseteq \mathsf{B}_X$ is an ideal with respect to the boolean algebra structure on B_X if and only if it is an ideal with respect to the ring structure on B_X. (**HINT:** What is $f \bigtriangleup g$ in the boolean structure of B_X? Once you find out, use Exercise 3.4.6.).

Example 3.6.9 (The Algebra of Lebesgue Measurable Subsets of $[0, 1]$ **and Its Quotient Algebra Modulo the Ideal of Null Sets)** The boolean algebra $\mathcal{M}([0; 1])$ given by the Lebesgue measurable subsets of $[0; 1]$ is an example of an atomic boolean algebra properly contained in $\mathcal{P}([0; 1])$ which is not isomorphic to $\mathcal{P}([0, 1])$: $\mathcal{M}([0; 1])$ is not complete, while $\mathcal{P}([0, 1])$ is complete (see Sect. 4 for a definition of completeness). A counterexample to the completeness of $\mathcal{M}([0; 1])$ is given by any non-Lebesgue-measurable set $V \notin \mathcal{M}([0; 1])$; V is a supremum in $\mathcal{P}([0, 1])$ of the family $\{\{r\} : r \in V\}$ of atoms of $\mathcal{M}([0; 1])$.

$\mathsf{MALG} = \mathcal{M}([0; 1])/_{\mathrm{Null}}$ (where Null is the ideal of measure 0-subsets of $[0; 1]$) is an example of an atomless boolean algebra which is also complete.

See Sect. 4.3.2, Proposition 4.3.8, and Corollary 4.3.9 for more details on MALG.

Example 3.6.10 (The Clopen Sets of the Cantor Set) The clopen sets on $2^{\mathbb{N}}$ with product topology form an atomless countable boolean algebra.

The family of sets $N_s = \{f \in 2^{\mathbb{N}} : s \subseteq f\}$ as s ranges in $2^{<\mathbb{N}}$ describes a basis consisting of clopen sets of $2^{\mathbb{N}}$.

Any clopen set can be uniquely described as a finite union of sets in this basis.

It can be shown that this boolean algebra is isomorphic to the Lindenbaum algebra on infinitely many propositional variables.

3.7 The Prime Ideal Theorem

The theorem in the title is the cornerstone relating the algebraic properties of boolean algebras to topological properties of compact zero-dimensional spaces. We actually work with objects dual to prime ideal, e.g., ultrafilters. Thus a name for the theorem more adherent to our presentation would rather be the "Ultrafilter Theorem"; however we stick to the standard usage for naming it.

Definition 3.7.1 Let B be a boolean algebra.[2]

- $G \subset \mathsf{B}$ is a *prefilter* in B if and only if for every $a_1, \ldots, a_n \in G$,

$$a_1 \wedge \cdots \wedge a_n > 0_{\mathsf{B}}.$$

- $G \subset \mathsf{B}$ is *ultra* if for all $b \in \mathsf{B}$ either $b \in G$ or $\neg b \in G$.

---

[2] This definition of filter generalizes the usual definition of a filter on a set X. In that case, $\mathsf{B} = \mathcal{P}(X)$.

- $G \subset \mathsf{B}$ is a *filter* if it contains all its finite meets and is upward closed (i.e., $a \wedge b = a \in G$ entails that $b \in G$ as well).
- $G \subset \mathsf{B}$ is an *ultrafilter* if it is a filter and is ultra.
- A filter G is *principal* if $G = \{c : c \geq b\}$ for some $b \in \mathsf{B}$.
- Given $A \subseteq \mathsf{B}$, $\check{A} = \{\neg a : a \in A\}$.
- I subset of B is a prime ideal if whenever $p, q \in I$ so is $p \vee q$.

Exercise 3.7.2 Recall that the notion of ideal and filters have been defined also for partial orders and ideals have already been defined on boolean algebras.
Prove that:

- F is a filter on the partial order B^+ if and only if it is a filter on the boolean algebra B.
- F is a maximal filter if and only if it is an ultrafilter.
- $I \subseteq \mathsf{B}$ is an ideal on B if $\check{I}$ is a filter.
- I is a prime (or maximal) ideal if and only if $\check{I}$ is an ultrafilter if and only if one among $a, \neg a \in I$ (maximality) if and only if for all $a, b \in \mathsf{B}$ at least one among $a, b \in I$ whenever $a \wedge b \in I$ (primality).

Exercise 3.7.3 Assume G is a principal filter on a boolean algebra B with $a \in G$ an atom of B. Show that $G = G_a = \{b \in \mathsf{B}^+ : a \leq b\}$ and that G_a is a principal ultrafilter.

Theorem 3.7.4 (Prime Ideal Theorem) *Assume F is a prefilter on a boolean algebra B. Then F can be extended to an ultrafilter G on B.*[3]

Proof Let $\mathcal{A}$ be the family of prefilters on B containing F. We show that:

- Any chain under inclusion contained in the partial order $(\mathcal{A}, \subseteq)$ admits an upper bound in $\mathcal{A}$.
- A maximal element of $\mathcal{A}$ is an ultrafilter on B containing F.

By Zorn's Lemma, $\mathcal{A}$ has a maximal elements, hence the thesis.

A chain under inclusion of $(\mathcal{A}, \subseteq)$ admits an upper bound:
Assume $\{F_i : i \in I\} \subseteq \mathcal{A}$ is a chain (i.e., for $i, j \in I$, either $F_i \subseteq F_j$ or $F_j \subseteq F_i$). Let $H = \bigcup_{i \in I} F_i$. We show that H is a prefilter. Assume $b_1, \ldots, b_n \in H$. Then each $b_i \in F_{j_i}$ for some $j_i \in I$. Since $\{F_i : i \in I\}$ is a chain, there is some $k \leq n$ such that $F_{j_k} \supseteq F_{j_i}$ for all $i = 1, \ldots, n$. Hence each $b_i \in F_{j_k}$ for all $i = 1, \ldots, n$. Since F_{j_k} is a prefilter, $b_1 \wedge \cdots \wedge b_n > 0_\mathsf{B}$. Hence H is a prefilter.

Any maximal element of $(\mathcal{A}, \subseteq)$ is an ultrafilter:
Assume G is a maximal element of $(\mathcal{A}, \subseteq)$; we must show that G is upward closed and closed under meets and contains either b or $\neg b$ for all $b \in \mathsf{B}$. First of all we prove that if a prefilter G is ultra (i.e., such that for all $b \in \mathsf{B}$ either $b \in G$ or $\neg b \in G$), then it is an ultrafilter:

---

[3] Equivalently any ideal on B can be extended to a maximal ideal.

- Assume $b \in G$ and $a \geq b$ (i.e., $a \wedge b = b$). Then $\neg a \notin G$, since

$$\neg a \wedge b = \neg a \wedge (a \wedge b) = 0_B$$

  and G is a prefilter. Hence $a \in G$, since G is ultra.
- Assume $a, b \in G$. Then $\neg(a \wedge b) \notin G$, because $\neg(a \wedge b) \wedge a \wedge b = 0_B$. Since G is ultra, we get that $(a \wedge b) \in G$.

Now assume G is a maximal prefilter of $\mathcal{A}$. We show that G is ultra: Assume not as witnessed by b. Then $G \cup \{b\}$ and $G \cup \{\neg b\}$ are not prefilters. Hence there are $a_1, \ldots, a_n \in G$ and $b_1, \ldots, b_k \in G$ such that

$$a_1 \wedge \cdots \wedge a_n \wedge b = 0_B$$

and

$$b_1 \wedge \cdots \wedge b_k \wedge \neg b = 0_B.$$

Hence

$$0_B < a_1 \wedge \cdots \wedge a_n \wedge b_1 \wedge \cdots \wedge b_k$$
$$= a_1 \wedge \cdots \wedge a_n \wedge b_1 \wedge \cdots \wedge b_k \wedge (b \vee \neg b)$$
$$\leq (a_1 \wedge \cdots \wedge a_n \wedge b) \vee (b_1 \wedge \cdots \wedge b_k \wedge \neg b) = 0_B,$$

where the first inequality holds because G is a prefilter and $a_1, \ldots, a_n, b_1, \ldots, b_k \in G$. We reached a contradiction. Hence G is a prefilter which is ultra and thus an ultrafilter.

The theorem is proved.   □

3.8 Stone Spaces of Boolean Algebras

There is a natural functor that attaches to a boolean algebra the Stone space of its ultrafilters. We prove here that these spaces are exactly the family of compact, Hausdorff, zero-dimensional topological spaces. Finally we prove Stone's Duality theorem, which represents any boolean algebra as the family of clopen sets of its Stone space.

Let B be a boolean algebra. We define

$$\mathrm{St}(B) = \{G \subseteq B : G \text{ is an ultrafilter}\}$$

and[4]

τ_B to be the topology on $\text{St}(B)$ generated by $\{N_b = \{G \in \text{St}(B) : b \in G\} : b \in B\}$.

The topological space $(\text{St}(B), \tau_B)$ is the *Stone space* of B. We have:

1. For all $b \in B$, $N_b \cap N_{\neg b} = \emptyset$.
2. For all $b \in B$, $N_b \cup N_{\neg b} = \text{St}(B)$.
3. For all $b_1, \ldots, b_n \in B$, $N_{b_1} \cap \ldots \cap N_{b_n} = N_{b_1 \wedge \ldots \wedge b_n}$.
4. For all $b_1, \ldots, b_n \in B$, $N_{b_1} \cup \ldots \cup N_{b_n} = N_{b_1 \vee \ldots \vee b_n}$.

Exercise 3.8.1   Prove the above facts for $\text{St}(B)$.

Now we outline the key properties about the Stone space of a boolean algebra B.

Theorem 3.8.2 (Stone's Duality Theorem)   *Given a boolean algebra B, we have that:*

1. *$(\text{St}(B), \tau_B)$ is a Hausdorff zero-dimensional, compact topological space.*
2. *The map*

$$\phi : B \to \text{CLOP}(\text{St}(B))$$

$$b \mapsto N_b$$

is an isomorphism; hence the clopen sets of τ_B are the sets N_b for $b \in B$ and form a basis for τ_B.
3. *There is a natural correspondence between open (closed) subsets of $\text{St}(B)$ and ideals (filters) on B:*

   * *For $X \subseteq \text{St}(B)$,*

$$I_X = \{c \in B : N_c \subseteq X\} \text{ is an ideal on } B,$$

   and

$$F_X = \{c \in B : N_c \supseteq X\} \text{ is a filter on } B.$$

   * *$U \subseteq \text{St}(B)$ is open if and only if*

$$\bigcup \{N_c \in B : c \in I_U\} = U.$$

   * *$C \subseteq \text{St}(B)$ is closed if and only if*

$$\bigcap \{N_c \in B : c \in F_C\} = C.$$

---

[4] That is, the smallest topology that contains $\{N_b : b \in B\}$.

4. *G is an isolated point of* $\mathrm{St}(\mathsf{B})$ *if and only if* $G = G_a = \{b \in \mathsf{B} : a \leq b\}$ *is a principal ultrafilter generated by some atom* $a \in \mathsf{B}$.

Proof We prove all items as follows:

1. *Topological properties of* $\mathrm{St}(\mathsf{B})$:

   Zero-dimensional: We have already observed that $N_b \cup N_{\neg b} = \mathrm{St}(\mathsf{B})$ and $N_b \cap N_{\neg b} = \emptyset$; thus N_b, $N_{\neg b}$ are clopen; they form a semibasis by definition of τ_{B}; and since $N_{b_1} \cap \ldots \cap N_{b_n} = N_{b_1 \wedge \ldots \wedge b_n}$, this semibasis is closed under finite intersections, and hence it is a basis.

   Hausdorff: If G, H are two different points of $\mathrm{St}(\mathsf{B})$, then there is $b \in G \triangle H$; assume $b \in G \setminus H$, then $G \in N_b$ and $H \in N_{\neg b}$, since H is ultra and $b \notin H$.

   Compact: Fix $\mathcal{F}$ a family of closed sets with the finite intersection property. Let

   $$\mathcal{G} = \{N_b : \exists C \in \mathcal{F}\,(N_b \supseteq C)\}.$$

   - We first claim that $\bigcap \mathcal{F} = \bigcap \mathcal{G}$ holds.
     Since every set in $\mathcal{G}$ contains some set in $\mathcal{F}$, we get the inclusion $\subseteq$.
     Conversely, if $C \in \mathcal{F}$, since C is a closed set and $\{N_b : b \in \mathsf{B}\}$ is a basis of clopen, then we can write $C = \bigcap\{N_b : b \in A\}$ for some[5] $A \subseteq \mathsf{B}$. Thus $C \supseteq \bigcap \mathcal{G}$ (since $C = \bigcap\{N_b : b \in A\}$, every N_b with $b \in A$ contains C and hence belongs to $\mathcal{G}$ by definition of $\mathcal{G}$). Since this holds for all $C \in \mathcal{F}$, we get the other inclusion.
   - Second claim: $\mathcal{G}$ has the finite intersection property. In fact, let $N_{b_1}, \ldots, N_{b_k}$ be in $\mathcal{G}$, and let $C_1, \ldots, C_k \in \mathcal{F}$ such that $C_i \subseteq N_{b_i}$. Then

   $$\emptyset \neq \bigcap_{i=1,\ldots,k} C_i \subseteq \bigcap_{i=1,\ldots,k} N_{b_i}.$$

   Now we can conclude the proof: Let $H = \{b \in \mathsf{B} : N_b \in \mathcal{G}\}$, since $\mathcal{G}$ has the finite intersection property,

   $$N_{b_1 \wedge \ldots \wedge b_k} = N_{b_1} \cap \ldots \cap N_{b_k} \neq \emptyset \text{ for every } b_1, \ldots, b_k \in H,$$

---

[5] If C is closed, $C = \mathrm{St}(\mathsf{B}) \setminus U$ for some U open. By definition $U = \bigcup \{N_{b_j} : j \in J\}$ for some family J, since $\{N_b : b \in \mathsf{B}\}$ is a basis for τ_{B}; hence

$$C = \mathrm{St}(\mathsf{B}) \setminus \bigcup \{N_{b_j} : j \in J\} = \bigcap \{\mathrm{St}(\mathsf{B}) \setminus N_{b_j} : j \in J\} = \bigcap \{N_{\neg b_j} : j \in J\}.$$

so

$$\forall b_1, \ldots, b_k \in H (b_1 \wedge \ldots \wedge b_k > 0_\mathsf{B}).$$

Thus H is a prefilter, so—by the prime ideal theorem—there exists an ultrafilter G on B such that $G \supseteq H$. $G \in \bigcap \mathcal{G}$ because $\forall b \in H (b \in G)$, so $G \in N_b$ for all $b \in H$; hence $G \in N_b$ for all $N_b \in \mathcal{G}$. We conclude that

$$\emptyset \neq \bigcap \mathcal{G} = \bigcap \mathcal{F}.$$

□

A clopen set is of the form N_b for some $b \in \mathsf{B}$:   Let U be clopen in St(B). Then $U = \bigcup \{N_b : N_b \subseteq U\}$, since it is open. Also U is a closed subset of a compact space; hence U is compact and any of its open coverings admits a finite subcovering. Therefore there are $b_1, \ldots, b_k$ such that $U = N_{b_1} \cup \cdots \cup N_{b_k} = N_{b_1 \vee \cdots \vee b_k}$.

2. B *is isomorphic to the clopen subset of* St(B):

   ϕ **is a homomorphism:**   Observe that $N_{b \wedge c} = N_b \cap N_c$, $N_{b \vee c} = N_b \cup N_c$, $N_{\neg b} = \text{St}(\mathsf{B}) \setminus N_b$.

   ϕ **is injective:**   Assume $b \neq c$. Then either $b \wedge \neg c > 0_\mathsf{B}$ or $c \wedge \neg b > 0_\mathsf{B}$, assuming the first option, an ultrafilter G extending $\{b \wedge \neg c\}$ is in $N_b \setminus N_c$, and assuming the second option holds we can find $G \in N_c \setminus N_b$.

   ϕ **is surjective:**   Immediate since any clopen set is of the form $N_b = \phi(b)$ for some $b \in \mathsf{B}$.

3. *Correspondence between closed (open) subsets of* St(B) *with filters (ideals) on* B: This is a useful exercise for the reader, in essence it has already been proved when we established the compactness of St(B).

4. $G \in \text{St}(\mathsf{B})$ *is an isolated point (i.e., such that* $\{G\}$ *is open) of* St(B) *if and only if* G *is a principal ultrafilter:*

   G is isolated if and only if $\{G\}$ is clopen; hence $\{G\} = N_a$ for some $a \in \mathsf{B}$.

   a must be an atom: Otherwise there is $0_\mathsf{B} < b < a$. Let $c = \neg b \wedge a$; then $b \wedge c = 0_\mathsf{B}$ and $0_\mathsf{B} < c < a$. Find G_0 and G_1 in St(B) with $c \in G_0$ and $b \in G_1$. Then $a \in G_0, G_1$, but $G_0 \neq G_1$, since $b \in G_1$, $\neg b \in G_0$; we reached a contradiction with $\{G\} = N_a$.

We can also go the other way round, i.e., we take a topological space and attach to it a boolean algebra.

Proposition 3.8.3 *Let* (X, τ) *be a zero-dimensional compact topological space. Then* (X, τ) *is homeomorphic to the Stone space of* CLOP(X, τ) *via the map*

$$\pi : X \longrightarrow \text{St}(\text{CLOP}(X, \tau))$$

$$x \longmapsto G_x = \{U \in \text{CLOP}(X, \tau) : x \in U\} \in \text{St}(\text{CLOP}(X, \tau)).$$

Proof We show that π is a well defined continuous bijection which is also open (i.e., maps open sets in open sets); this suffices to prove the proposition.

Well defined and injective: The fact that G_x is an ultrafilter is an easy exercise. When x and y are distinct, since τ is Hausdorff and zero-dimensional, there is a clopen set U containing x and not y. Then $U \in G_x$ and $U \notin G_y$.

Surjective: Let $G \in \mathrm{St}(\mathsf{CLOP}(X, \tau))$, and set $C = \bigcap G$. We claim that C is a singleton. C is non-empty since X is compact and G is a family of closed sets with the finite intersection property; thus it must have a non-empty intersection. Now assume $x \neq y \in C$. Find as in previous item U clopen such that $x \in U$ and $y \notin U$. Then $U \in G$ iff $(X \setminus U) \notin G$, which gives that $x \in C$ iff $y \notin C$, a contradiction. Let x be the unique element of C. Then it is easily checked that $G = G_x$.

Continuous and open: Notice that for any clopen set U and $x \in X$, $x \in U$ if and only if $U \in G_x$; thus $\pi[U] = N_U$; from this we easily infer that π is continuous and open.

$\square$

In particular we have shown that the map $\mathsf{B} \mapsto \mathrm{St}(\mathsf{B})$ defines a natural bijection between the class of boolean algebras and the class of compact zero-dimensional, Hausdorff spaces. It can be shown that this map is a contravariant functor between these two categories which identifies homomorphisms $i : \mathsf{B} \to \mathsf{C}$ with continuous maps $f : \mathrm{St}(\mathsf{C}) \to \mathrm{St}(\mathsf{B})$. But we will not pursue this direction further here.

We give two other different presentations of the notion of boolean algebra, one axiomatizable in the first order language for rings and another in the first order language for partial orders.

3.9   Boolean Rings

Throughout this section we assume the reader is familiar with the notion of commutative ring, of an ideal on it, and of their basic properties.

We want to show that boolean algebras can also be described as commutative rings with idempotent multiplication.

Definition 3.9.1 Let $\mathsf{R} = \langle R, +, \cdot, 0, 1 \rangle$ be a commutative ring. R is boolean if it has idempotent multiplication (i.e., $a^2 = a$ for all $a \in \mathsf{R}$).

Remark 3.9.2 A commutative ring R with idempotent multiplication has automatically characteristic 2: $(a + a)^2 = a + a$ for all $a \in \mathsf{R}$, hence

$$0 = (a + a)^2 - (a + a) = a^2 + a^2 + 2a - 2a = a^2 + a^2 = a + a$$

for all $a \in \mathsf{R}$. Hence in boolean rings $a = -a$, and the sum operation and the difference operation coincide.

Theorem 3.9.3 *Let*

$$(\mathsf{B}, \wedge, \vee, \neg, 0, 1)$$

be a boolean algebra. Then

$$(\mathsf{B}, \Delta, \wedge, 0, 1)$$

is a boolean ring.
  Conversely, given a boolean ring

$$\mathsf{R} = \langle \mathsf{R}, +, \cdot, 0, 1 \rangle,$$

define for all $a, b \in \mathsf{R}$

- $a \vee b = a + b + a \cdot b$
- $a \wedge b = a \cdot b$
- $\neg a = 1 + a$

for all $a, b \in \mathsf{R}$. Then

$$\mathsf{R} = \langle \mathsf{R}, \vee, \wedge, \neg, 0, 1 \rangle$$

is a boolean algebra.

We split each of the two statement of the theorem in separate lemmata. We first show that interpreting in a boolean algebra B the boolean operation of symmetric difference Δ as a sum and that of meet $\wedge$ as a multiplication, B is naturally identified with the boolean ring of characteristic functions of clopen sets of $\mathrm{St}(\mathsf{B})$.

Exercise 3.9.4 Let (X, τ) be a topological space.

$$C(X, 2) = \{f : X \to \mathbb{Z}_2 \mid f \text{ is continuous}\},$$

where $\mathbb{Z}_2 = \{0, 1\}$ is the two elements ring endowed with discrete topology.
  Show that $C(X, 2)$ is a boolean ring when endowed with operations defined pointwise (i.e., for $p \in Xf * g(p) = f(p) * g(p)$, for $*$ among $+, \cdot$) and the constant characteristic functions $\chi_\emptyset, \chi_X$ as 0 and 1 of the ring.

Lemma 3.9.5 *Let $(\mathsf{B}, \wedge, \vee, \neg, 0, 1)$ be a boolean algebra. Then $(\mathsf{B}, \Delta, \wedge, 0, 1)$ is a boolean ring isomorphic to the ring $C(\mathrm{St}(\mathsf{B}), 2)$.*

Proof We leave to the reader to check that $C(\mathrm{St}(\mathsf{B}), 2)$ is a boolean ring. Let $\theta : \mathsf{B} \to C(\mathrm{St}(\mathsf{B}), 2)$ be defined by $b \mapsto f_b$, with $f_b(G) = 1$ iff $b \in G$.
f_b is continuous for any b since $f_b^{-1}(\{1\}) = N_b$ and $f_b^{-1}(\{0\}) = N_{\neg b}$. It is also easy to check that θ is injective. Let us now check that θ is surjective. Let $g : \mathrm{St}(\mathsf{B}) \to \{0, 1\}$ be continuous. Let $A_i = g^{-1}(\{i\})$, $i = 0, 1$. Each A_i is clopen, by continuity

of g. Clearly $A_0 \cup A_1 = \mathrm{St}(\mathsf{B})$ and $A_0 \cap A_1 = \emptyset$. Hence $A_0 = N_b$ and $N_1 = N_{\neg b}$ for some $b \in \mathsf{B}$. Then $\theta(b) = g$.
We now check that

$$f_a + f_b = f_{a \triangle b}; \quad f_a \cdot f_b = f_{a \wedge b},$$

so that the proof is completed: Given an ultrafilter $G \subseteq \mathsf{B}$, we have

$$f_a(G) + f_b(G) = 1 \iff (f_a(G) = 0 \wedge f_b(G) = 1)$$
$$\vee (f_a(G) = 1 \wedge f_b(G) = 0)$$
$$\iff (a \in G \wedge b \notin G) \vee (a \notin G \wedge b \in G)$$
$$\iff a \triangle b \in G,$$

and

$$f_a(G) \cdot f_b(G) = 1 \iff (f_a(G) = 1 \wedge f_b(G) = 1) \iff (a \in G \wedge b \in G)$$
$$\iff a \wedge b \in G.$$

$\square$

Remark 3.9.6 The above lemma greatly simplifies the proofs of certain properties of boolean operations and of boolean morphisms: For example, try to prove the associativity of the symmetric difference operation $\triangle$ on a boolean algebra B using the equational presentation of boolean algebras given in Definition 3.1.1, and compare your attempts, with the argument that $\triangle$ is associative being (modulo the above isomorphism) the sum operation of a commutative ring.

Proposition 3.9.7 *For a boolean ring* $\langle \mathsf{B}, \cdot, +, 0, 1 \rangle$*, the following holds:*

1. *I is an ideal on the boolean algebra $\langle \mathsf{B}, \wedge, \vee, \neg, 0_{\mathsf{B}}, 1_{\mathsf{B}} \rangle$ if and only if I is an ideal on the boolean ring B.*
2. *The boolean ring B does not have zero-divisors if and only if it is isomorphic to $\mathbb{Z}_2$.*
3. *The dual of a subset A of B is unambiguously defined as $\check{A} = \{ \neg a : a \in A \} = \{ 1 - a : a \in A \}$. Moreover I is an (prime) ideal on the ring B if and only if $\check{I}$ is an ultrafilter on the boolean algebra B.*

Proof

1. Assuming I is an ideal on the ring $\langle \mathsf{B}, +, \cdot, 0, 1 \rangle$, we show that I is an ideal on the boolean algebra $\langle \mathsf{B}, \wedge, \vee, \neg, 0_{\mathsf{B}}, 1_{\mathsf{B}} \rangle$: Assume $a, b \in I$, then $a + b, a \cdot b \in I$ as well, and hence $a \vee b = a + b + a \cdot b \in I$; moreover $a \in I$ and $b \leq a$ entails that $b = b \cdot a \in I$.

Conversely, assuming I is an ideal on the boolean algebra $\langle \mathsf{B}, \wedge, \vee, \neg, 0_\mathsf{B}, 1_\mathsf{B} \rangle$, we show that I is an ideal on the ring B: If $a, b \in I$,

$$a + b = a \vee b \wedge \neg(a \wedge b) \leq a \vee b \in I,$$

moreover if $a \in I$ and $b \in \mathsf{B}$, $a \cdot b = a \wedge b \leq a \in I$, and hence $a \cdot b \in I$ as well.

2. $x^2 = x$ entails that the equation $x(x - 1) = 0$ holds for all x in a boolean ring. If x belongs to a boolean ring with no zero-divisors, either $x = 0$ or $x - 1 = 0$. Hence the boolean ring is isomorphic to $\{0, 1\} = \mathbb{Z}_2$.

3. The first observation is trivial, given that sum and subtraction are the same operation on a boolean ring and $\neg a = 1 - a$ by definition.

   The second observation follows from the fact that the unique boolean ring without zero divisors is $\mathbb{Z}_2$. Now recall that I is a prime ideal on a commutative ring B if and only if $\mathsf{B}/_I$ has no zero-divisors, and I is a maximal ideal on a commutative ring B if and only if $\mathsf{B}/_I$ is a field. By the previous item, the quotient of the boolean ring B by a prime ideal is $\mathbb{Z}_2$ which is a field. This entails that all prime ideals of B are maximal, i.e. their dual is an ultrafilter.

   □

The next proposition shows that ideals on boolean algebras and kernels of boolean morphisms are the same, and its proof takes advantage of the characterization of boolean algebras as boolean rings.

Proposition 3.9.8 *Let B, C be boolean algebras, $\phi : \mathsf{B} \to \mathsf{C}$ be a boolean morphism, and $I \subseteq \mathsf{B}$ be an ideal. Then:*

- $\ker \phi = \{b \in \mathsf{B} : \phi(b) = 0_\mathsf{C}\}$ *is an ideal on B.*
- *The map $\pi_I : b \mapsto [b]_I = \{c \in \mathsf{B} : c \triangle b \in I\}$ defines a surjective morphism of B onto the quotient boolean algebra $\mathsf{B}/_I$ (with operations on $\mathsf{B}/_I$ defined by $[b]_I \wedge [c]_I = [b \wedge c]_I$, $[b]_I \vee [c]_I = [b \vee c]_I$, $\neg[b]_I = [\neg b]_I$).*
- *The map $\phi/_{\ker(\phi)} : [b]_{\ker \phi} \mapsto \phi(b)$ is a well defined injective morphism of $\mathsf{B}/_{\ker(\phi)}$ into C.*
- $\phi = (\phi/_{\ker(\phi)}) \circ \pi_{\ker(\phi)}$.

Proof By Lemma 3.9.5 and Exercises 3.4.3 and 3.4.6, it suffices to prove the proposition for the usual notions of ideal and morphism on rings, since:

- Lemma 3.9.5 shows that boolean algebras are boolean rings.
- Exercise 3.4.6 gives that I is an ideal on a boolean algebra $(\mathsf{B}, \vee_\mathsf{B}, \wedge_\mathsf{B}, \neg_\mathsf{B}, 0_\mathsf{B}, 1_\mathsf{B})$ if and only if it is an ideal on the boolean ring $(\mathsf{B}, \triangle_\mathsf{B}, \wedge_\mathsf{B}, 0, 1)$.
- Exercise 3.4.3 shows that a boolean morphism is also a ring morphism.

The proposition for ring morphisms and ring ideals is a standard result in ring theory.

□

We now prove the converse part of Theorem 3.9.3:

Lemma 3.9.9 *Assume $\mathsf{R} = \langle \mathsf{R}, +, \cdot, 0, 1 \rangle$ is a boolean ring. Let $a \vee b = a + b + a \cdot b$, $a \wedge b = a \cdot b$, $\neg a = 1 + a$ for all $a, b \in \mathsf{R}$. Then $\mathsf{R} = \langle \mathsf{R}, \vee, \wedge, \neg, 0, 1 \rangle$ is a boolean algebra.*

Proof Let us go through the equations of Definition 3.1.1 with · in the place of ∧. The associativity and commutativity laws for ·, and the identity laws for 0, 1 are ring axioms. The commutativity law for ∨ is trivially checked. We are left to check the associativity law for ∨ and the laws of complementation and distributivity:

The associativity law for ∨ holds since

$$a \vee (b \vee c) = a \vee (b + c + b \cdot c) = a + b + c + b \cdot c + a \cdot b + a \cdot c + a \cdot b \cdot c,$$

while

$$(a \vee b) \vee c = c \vee (b + a + b \cdot a) = c + b + a + b \cdot a + c \cdot b + c \cdot a + c \cdot b \cdot a.$$

The complementation laws are also immediate to check

$$a \cdot (1 - a) = a - a^2 = a - a = 0,$$

while

$$a \vee (1 - a) = a + 1 - a + a \cdot (1 - a) = 1 + 0 = 1.$$

Now

$$(a \vee b) \cdot c = (a + b + a \cdot b) \cdot c$$

$$= a \cdot c + b \cdot c + a \cdot b \cdot c = a \cdot c + b \cdot c + a \cdot b \cdot c^2$$

$$= a \cdot c + b \cdot c + (a \cdot c) \cdot (b \cdot c) = (a \cdot c) \vee (b \cdot c)$$

for all a, b, c; hence the first distributivity law $(a \vee b) \cdot c = (a \cdot c) \vee (b \cdot c)$ holds.

Also the second distributivity law $(a \cdot b) \vee c = (a \vee c) \cdot (b \vee c)$ holds:

$$(a \cdot b) \vee c = a \cdot b + c + a \cdot b \cdot c,$$

while

$$(a \vee c) \cdot (b \vee c)$$

$$= (a + c + a \cdot c) \cdot (b + c + b \cdot c)$$

$$= a \cdot b + a \cdot c + a \cdot b \cdot c + b \cdot c + c^2 + b \cdot c^2 + a \cdot b \cdot c + a \cdot c^2 + a \cdot b \cdot c^2$$

$$= a \cdot b + a \cdot c + a \cdot b \cdot c + b \cdot c + c + b \cdot c + a \cdot b \cdot c + a \cdot c + a \cdot b \cdot c$$

$$= a \cdot b + a \cdot b \cdot c + c$$

for all a, b, c.   □

3.10   Boolean Algebras as Complemented Distributive Lattices

We also give another characterization of boolean algebras in terms of their order relation. These axioms for boolean algebras can be expressed in a first order theory with a binary relation symbol for the order relation. Nonetheless we expand the language adding symbols for the operations $\wedge$, $\vee$, $\neg$ and constants 0, 1 definable in this axiom system for boolean algebras and leave to the reader to check that this is not necessary.

A *join-semilattice* $(P, \leq)$ is a partial order such that every pair of elements (x, y) of P admits a unique least upper bound denoted by $x \vee y$, the *join* of x and y.

Dually, a partial order $(P, \leq)$ is a *meet-semilattice* when any two elements x and y in P have a unique greatest lower bound denoted by $x \wedge y$, the *meet* of x and y.

A partial order $(P, \leq)$ is a *lattice* if it is both a join-semilattice and a meet-semilattice.

A lattice $(P, \leq)$ is *bounded* if it has a greatest element 1_P and a least element 0_P which satisfy $0_P \leq x \leq 1_P$ for every x in P.

A lattice $(P, \leq)$ is *distributive* if for all x, y, and z in P we have

$$x \wedge (y \vee z) = (x \wedge y) \vee (x \wedge z) \text{ and } x \vee (y \wedge z) = (x \vee y) \wedge (x \vee z).$$

Let $(P, \leq)$ be a bounded lattice. A *complement* of an element $a \in P$ is an element $b \in P$ such that $a \vee b = 1_P$ and $a \wedge b = 0_P$.

Remark 3.10.1   In a distributive lattice, if a has a complement, it is unique. In this case we denote by $\neg a$ the complement of a.

A lattice is *complemented* if it is bounded and every element has a complement.

A lattice $(P, \leq)$ is *complete* if every subset $X = \{x_i : i \in I\}$ of P has a meet (or infimum) $\bigwedge_{i \in I} x_i$ and a join (or supremum) $\bigvee_{i \in I} x_i$.

Remark 3.10.2   Notice that if $X = \emptyset$, then $\bigwedge \emptyset = 1_P$ and $\bigvee \emptyset = 0_P$, so a complete lattice is always bounded:
Formally $x = \bigwedge y$ if and only if

$$\forall z(z \in y \rightarrow x \leq z) \wedge \forall u[\forall z(z \in y \rightarrow u \leq z) \rightarrow u \leq x].$$

Hence if $y = \emptyset$, the premise $z \in y$ of the implication $(z \in y \rightarrow x \leq z)$ is always false for any z; this gives that the first conjunct is vacuously true for any x. For the same reason, the premise

$$\forall z(z \in y \rightarrow u \leq z)$$

of the principal implication in the second conjunct is always true for any $u \in P$; consequently x can make true the second conjunct if and only if $x = 1_P$.

Similarly one handles the case of $\bigvee \emptyset = 0_P$.

Lemma 3.10.3 $(B, \wedge, \vee, \neg, 0, 1)$ *is a boolean algebra if and only if* $(B, \leq)$ *is a complemented distributive lattice.*

Proof One direction is clear: Say that a boolean algebra B is a field of sets if it is a subalgebra of

$$\langle \mathcal{P}(X), \cap, \cup, A \mapsto X \setminus A, \emptyset, X \rangle$$

for some set X (i.e., the domain of B is contained in $\mathcal{P}(X)$, $\emptyset, X \in B$, and the operations on B are the restriction to B of $\cap, \cup, A \mapsto X \setminus A$). It is an easy exercise to check that boolean algebras that are fields of sets are complemented distributive lattices. By Stone's Duality theorem any boolean algebra is isomorphic to a field of sets. The converse direction is left to the reader.   □

3.11   Suprema and Infima of Subsets of a Boolean Algebra

Notation 3.11.1 *Given a boolean algebra* B, *we denote by* $\bigvee A$ *the supremum (least upper bound) under* $\leq$ *of a subset* A *of* B *(i.e., the least element* $a \in B$ *such that* $a \geq b$ *for all* $b \in A$—*if this least element exists) and by* $\bigwedge A$ *its infimum (i.e., the largest element* $a \in B$ *such that* $a \leq b$ *for all* $b \in A$—*if this largest element exists). Similarly* $\bigwedge A$ *denotes the infimum of some* $A \subseteq B$.

The following proposition gives a simple topological method to compute the supremum of a subset of a boolean algebra:

Proposition 3.11.2 *Let* B *be a boolean algebra and* $X \subseteq B$. *Then* $a = \bigvee X$ *if and only if* $\bigcup \{N_b : b \in X\}$ *is a dense open subset of* N_a *in the relative topology of* N_a *as a subset of* $(\mathrm{St}(B), \tau_B)$.

Proof Assume $A = \bigcup \{N_b : b \in X\}$ is a dense open subset of N_a, but $a \neq \bigvee X$; we will reach a contradiction.

The first assumption on A gives that $N_b \subseteq N_a$ for all $b \in X$, which occurs if and only if $a \geq b$ for all $b \in X$, i.e., a is an upper bound of X. Since $a \neq \bigvee X$, there must be some e which is still an upper bound for X with $e \not\geq a$. Now if e is an upper bound for X, then so is $c = e \wedge a$. Since $e \not\geq a$ and $e \wedge a \leq a$, we conclude that $c = e \wedge a < a$ is an upper bound for X. Hence if a is not the least upper bound for X, there must be some $0 < c < a$ which is still an upper bound for X. This gives that $a > d = a \wedge \neg c > 0$ and also that for all $b \in X N_d \cap N_b \subseteq N_d \cap N_c = \emptyset$. We get that $A \cap N_d = \emptyset$. But N_d is an open non-empty subset of N_a; hence A is not an open dense subset of N_a, the desired contradiction.

Conversely assume $A = \bigcup \{N_b : b \in X\}$ is not a dense open subset of N_a; we must argue that $a \neq \bigvee X$. If $a \not\geq b$ for some $b \in X$, certainly $a \neq \bigvee X$; therefore we can assume $N_a \supseteq N_b$ for all $b \in X$. Since A is not a dense open subset of N_a, we can find $0_B < d \leq a$ such that $N_d \cap A = \emptyset$. We conclude that $c = a \wedge \neg d$ is such that $N_a \supset N_c \supseteq N_b$ for all $b \in X$, i.e., c witnesses that a is not the least upper bound of X.   □

Corollary 3.11.3 *Let* B *be a boolean algebra; the following holds for any* X *subset of* B *and* $r \in B$:

1. $r \wedge \bigvee X = \bigvee \{r \wedge b : b \in X\}$ *if* $\bigvee X$ *is well defined.*
2. $\neg \bigvee X = \bigwedge \{\neg b : b \in X\}$ *if any among* $\bigvee X$ *or* $\bigwedge \{\neg b : b \in X\}$ *is well defined.*

Proof

1. Let $a = \bigvee X$. By the previous proposition we get that $A = \bigcup \{N_b : b \in X\}$ is a dense open subset of N_a. Therefore $N_r \cap A$ is a dense open subset of $N_r \cap N_a = N_{r \wedge a}$. Now

$$N_r \cap A = N_r \cap \bigcup \{N_b : b \in X\} = \bigcup \{N_b \cap N_r : b \in X\} = \bigcup \{N_{b \wedge r} : b \in X\}.$$

   By the previous proposition we conclude that $a \wedge r = \bigvee \{r \wedge b : b \in X\}$.
2. Remark that $a \leq b$ if and only if $\neg a \geq \neg b$. Now $a = \bigvee X$ if and only if $\bigcup \{N_b : b \in X\}$ is a dense open subset of N_a. Therefore $N_{\neg b} \supseteq N_{\neg a}$ for all $b \in X$. Now assume there is some $c > \neg a$ such that $\neg b \geq c$ for all $b \in X$. Then $\neg c < a$ and $b \leq \neg c$ for all $b \in X$, contradicting $a = \bigvee X$. We leave to the reader to handle the case in which we assume $\bigwedge \{\neg b : b \in X\}$ is well defined.   □

Exercise 3.11.4 Assume B is a complete boolean algebra. Then:

- $X \subseteq B^+$ is predense in $(B^+, \leq_B)$ if and only if $\bigvee X = 1_B$. (HINT: If not $a = \neg \bigvee X > 0_B$ and $b \wedge a = 0_B$ for all $b \in \downarrow X$, i.e. $\downarrow X$ is not dense in B^+).
- X is predense below $b \in B^+$ if and only if $\downarrow X \cap \downarrow b$ is a dense subset of $\downarrow b$ if and only if $b = \bigvee_B \downarrow X \cap \downarrow b$.
- If $D \subseteq B^+$ is dense, there exists $A \subseteq D$ maximal antichain of B^+. (HINT: Apply Zorn's Lemma to the antichains contained in D ordered by inclusion; a maximal element of this quasi-order is a maximal antichain $A \subseteq D$).

Remark 3.11.5 Recall that the Cantor space can be characterized as the unique zero-dimensional, compact separable space. We refer the reader to [18].

A possible presentation of the Cantor space C is the subset of $[0; 1]$ given by

$$C = \left\{ a_f = \sum_{i=0}^{\infty} \frac{2 \cdot f(i)}{3^{i+1}} : f \in 2^{\mathbb{N}} \right\},$$

endowed with the subspace topology inherited from the euclidean topology on $[0; 1]$.

Take the boolean algebra B of clopen subsets of C, and let $a_n = \frac{3^{n+1}-3}{3^{n+1}}$, $b_n = a_n + \frac{1}{3^{n+1}}$, $U_n = [a_n; b_n] \cap C$. We get that

$$X = \{U_n : n \in \mathbb{N}\} \subseteq B$$

is such that $C = \bigvee_B X$, but $\cup X = C \setminus \{1\}$ is a proper dense open subset of C in the euclidean topology.

Moreover if we let $X_0 = \{U_{2n} : n \in \mathbb{N}\}$, we get not only that $\bigcup X_0$ is open but also that no clopen subset of C contains X_0 as a dense subset: Let $A \supseteq X_0$ be a closed subset of C. Then it must contain 1, which is an accumulation point of X_0. If $A \subseteq C$ is also open, then $A \supseteq C \cap [\frac{3^n-1}{3^n}; 1]$ for some large enough n, since $\left\{C \cap [\frac{3^n-1}{3^n}; 1] : n \in \mathbb{N}\right\}$ is a base of clopen neighborhoods of 1 in C. This gives that $U_{2k+1} \subseteq A$ for some large enough k. But U_{2k+1} is disjoint from X_0; hence X_0 is not a dense open subset of any clopen A containing it.

In particular X_0 has no supremum in B.

Notation 3.11.6 *Given a boolean algebra B, we often consider B^+ when referring to B as an order; otherwise some definitions could indeed become trivial (also some of these definitions have already been given in Sect. 2.2 for preorders, and we specify them in the ontext of boolean algebras).*

For $X \subseteq B$, $\downarrow X = \{b : \exists c \in X\, b \leq_B c\}$, and $\uparrow X = \{b : \exists c \in X\, b \geq_B c\}$.

For $b \in B^+$, the boolean algebra $B \upharpoonright b$ is given by $\{a \in B : a \leq_B b\} = \downarrow \{b\}$, with the operations inherited from B. The top element of $B \upharpoonright b$ is b.

$X \subseteq B$ is predense if $\downarrow X$ is dense in B^+ with respect to $\leq_B$.

$X \subseteq B$ is predense below $b \in B$ if $\downarrow X$ is dense in $(B \upharpoonright b)^+$.

We will need the following property of the Stone spaces of boolean algebras:

Fact 3.11.7 *Assume B is a boolean algebra and $X \subseteq B$. The following hold:*

1. *$\bigvee X = \bigvee \downarrow X$ whenever one of the two members is well defined.*
2. *For all $r \in B$ $r \wedge \bigvee X > 0_B$ if and only if $r \wedge b > 0_B$ for some $b \in X$ whenever $\bigvee X$ is well defined.*
3. *$\bigvee X = 1_B$ iff $X \cap B^+$ is a predense subset of B^+ in the sense of the order.*
4. *For any X predense subset of B, $a = \bigvee \{q \in \downarrow X : q \leq_B a\}$.*
5. *For any set $X \subseteq B$ and $b \in B$, we have that $b = \bigvee X$ if and only if $b = \bigvee A$ for some (any) A maximal antichain contained in $\downarrow X$.*

Proof

1. $\bigvee X \leq_B \bigvee \downarrow X$ since $X \subseteq \downarrow X$. For the converse inequality, if $d \geq b$ for all $b \in X$, we also have that $d \geq c$ for all $c \in \downarrow X$; hence $\bigvee X$ is an upper bound for $\downarrow X$, and hence $\bigvee \downarrow X \leq_B \bigvee X$.
2. Left to the reader: Use Corollary 3.11.3.
3. Left to the reader: Use Corollary 3.11.3.
4. Left to the reader: Use Corollary 3.11.3.

5. Assume $b = \bigvee X$. By the first item $\bigvee X = \bigvee \downarrow X$. By Zorn's Lemma on the family of antichains contained in $\downarrow X$ ordered by inclusion, there is at least one maximal antichain A contained in $\downarrow X$. Now whenever $A \subseteq \downarrow X$ is a maximal antichain contained in $\downarrow X$, $A \cup \{\neg b\}$ is predense in B; hence—by the preceding item—$b = \bigvee \downarrow A$. Again by the first item, $b = \bigvee \downarrow A = \bigvee A$.

The proof of the converse implication is left to the reader.

$\square$

Chapter 4
Complete Boolean Algebras

We address in this chapter the theory of complete boolean algebras: We show that any complete boolean algebra (*cba* in the sequel) can be represented as the family of regular open sets of a compact topological space (Theorem 4.1.2 and Proposition 4.1.5), and we prove that every partial order can be completed to a cba, which is unique up to isomorphism (Theorem 4.2.4).

Definition 4.0.1 A boolean algebra $(\mathsf{B}, 0, 1, \vee, \wedge, \neg, \leq)$ is complete (or *cba* for short) if it admits suprema and infima for all its subsets with respect to the order relation $\leq$.

Complete boolean algebras can be split into two pieces.

Remark 4.0.2 Recall that a boolean algebra B is the disjoint sum of C and D if (as boolean rings) B is isomorphic to the product ring $\mathsf{C} \times \mathsf{D}$. Note that B is isomorphic to $\mathsf{B} \restriction \neg c \times \mathsf{B} \restriction c$ via the map $b \mapsto (b \wedge \neg c, b \wedge c)$ for any $c \in \mathsf{B}$.

Lemma 4.0.3 *A complete boolean algebra B can be split into the disjoint sum of an atomic boolean algebra and of an atomless boolean algebra. That is, there is $c \in \mathsf{B}$ such that:*

- $\mathsf{B} \restriction \neg c$ *is atomless.*
- $\mathsf{B} \restriction c$ *is atomic.*

Proof Let $A = \{a \in \mathsf{B} : a \text{ is an atom of } \mathsf{B}\}$ and $c = \bigvee A$. Then $b \wedge c = 0_{\mathsf{B}}$ entails that b is not an atom of $\mathsf{B} \restriction \neg c$ (otherwise b would also be an atom of B and thus be a refinement of c), while $b \leq c$ entails that for some atom $a \in A$, $a \wedge b > 0_{\mathsf{B}}$ which occurs only if $a \leq b$, since a is an atom. This gives that $\mathsf{B} \restriction c$ is atomic and $\mathsf{B} \restriction \neg c$ is atomless. $\qquad\square$

Definition 4.0.4 Let B, C be boolean algebras. A map $k : \mathsf{B} \to \mathsf{C}$ is a *complete homomorphism* if it maps predense subsets of B^+ to predense subsets of C^+, or equivalently if it preserves suprema and infima.

© The Author(s), under exclusive license to Springer Nature Switzerland AG 2024
M. Viale, *The Forcing Method in Set Theory*, La Matematica per il 3+2 168,
https://doi.org/10.1007/978-3-031-71660-7_4

Exercise 4.0.5 Any complete homomorphism is also a homomorphism in the usual sense.

Fact 4.0.6 *An isomorphism of boolean algebras preserves suprema and infima whenever they exist. Hence isomorphic images of complete boolean algebras are complete boolean algebras.*

Proof Left to the reader.   □

The following is less trivial and useful:

Lemma 4.0.7 *Let* B, C *be complete boolean algebras and* $i : \mathsf{B} \to \mathsf{C}$ *be an order and incompatibility preserving[1] morphism which respects suprema. Then* i *is a complete homomorphism also for the boolean algebraic structure.*

Proof Note that $i(a \vee b) = i(a) \vee i(b)$ by assumption. Now observe that $\neg a = \bigvee \{b \in \mathsf{B} : b \wedge a = 0_\mathsf{B}\}$ and $\neg i(a) = \bigvee \{c \in \mathsf{C} : c \wedge i(a) = 0_\mathsf{C}\}$; hence

$$i(\neg a) = \bigvee \{i(b) : b \in \mathsf{B}, \, b \wedge a = 0_\mathsf{B}\}$$

$$= \bigvee \{i(b) : b \in \mathsf{B}, \, i(b) \wedge i(a) = 0_\mathsf{C}\}$$

$$\leq \bigvee \{c \in \mathsf{C}, \, c \wedge i(a) = 0_\mathsf{C}\}$$

$$= \neg i(a),$$

where the second equality holds because i is incompatibility preserving.

Therefore $i(\neg a) \wedge i(a) \leq \neg i(a) \wedge i(a) = 0_\mathsf{C}$ and $i(\neg a) \vee i(a) = i(a \vee \neg a) = i(1_\mathsf{B}) = 1_\mathsf{C}$. We conclude that $\neg i(a) = i(\neg a)$.   □

4.1 Complete Boolean Algebras of Regular Open Sets

We prove that every complete boolean algebra can be represented as the family of regular open sets of some given topological space (i.e., it is isomorphic to such an algebra), and we characterize complete boolean algebras as those whose Stone spaces have the property that their regular open sets are clopen. The first step in this direction is to show that the regular open sets of a given topological space have a natural structure of complete boolean algebra.

---

[1] That is, $i(p) \leq_\mathsf{C} i(q)$ whenever $p \leq_\mathsf{B} q$ and $i(p) \wedge_\mathsf{C} i(q) = 0_\mathsf{C}$ whenever $p \wedge_\mathsf{B} q = 0_\mathsf{B}$.

Definition 4.1.1   Given a topological space (X, τ), we equip $\mathsf{RO}(X, \tau)$ (the family of regular open subsets of X according to the topology τ) with the following operations:

$$\bigvee_{i \in I} U_i = \mathsf{Reg}\left(\bigcup_{i \in I} U_i\right),$$

$$\bigwedge_{i \in I} U_i = \mathsf{Reg}\left(\bigcap_{i \in I} U_i\right),$$

$$\neg U = X \setminus \mathsf{Cl}(U).$$

We oftentimes denote $\mathsf{RO}(X, \tau)$ by $\mathsf{RO}(X)$ when τ is clear from the context. We prove the following:

Theorem 4.1.2   *Assume* (X, τ) *be a topological space. Then* $(\mathsf{RO}(X, \tau), \vee, \wedge, \neg, \emptyset, X)$ *is a complete boolean algebra.*

We prove first that $\mathsf{RO}(X)$ is a boolean algebra using Definition 3.1.1, and then we prove that it is complete.

Proposition 4.1.3   *The family* $\mathsf{RO}(X)$, *with the operations defined above, is a boolean algebra.*

Proof   We take $U, V, W \in \mathsf{RO}(X)$, and we go through the equations of Definition 3.1.1:

- Associativity of $\vee$. By Fact 2.1.7 it is enough to check that $U \cup V \cup W$ is a dense open subset of $\mathsf{Reg}(U \cup \mathsf{Reg}(V \cup W))$ and of $\mathsf{Reg}(W \cup \mathsf{Reg}(V \cup U))$. This is immediate from the definitions. Alternatively we can use the following algebraic identities:

$$
\begin{aligned}
U \vee (V \vee W) &= (U \cup (V \cup W)^{\perp\perp})^{\perp\perp} \\
&= (U^{\perp} \cap (V \cup W)^{\perp\perp\perp})^{\perp} \\
&= (U^{\perp} \cap (V \cup W)^{\perp})^{\perp} \\
&= (U^{\perp} \cap (V^{\perp} \cap W^{\perp}))^{\perp} \\
&= ((U^{\perp} \cap V^{\perp}) \cap W^{\perp})^{\perp} \\
&= ((U \cup V)^{\perp} \cap W^{\perp})^{\perp} \\
&= ((U \cup V)^{\perp\perp\perp} \cap W^{\perp})^{\perp} \\
&= ((U \cup V)^{\perp\perp} \cup W)^{\perp\perp} \\
&= (U \vee V) \vee W.
\end{aligned}
$$

- The associativity of $\wedge$ is just the associativity of $\cap$.

- Distributivity. We only show that

$$(U \wedge V) \vee (U \wedge W) = U \wedge (V \vee W)$$

holds, and the other equation is similar. To this aim observe that

$V \cup W$ is dense in $\mathsf{Reg}(V \cup W)$

$\Rightarrow U \cap (V \cup W)$ is dense in $U \cap \mathsf{Reg}(V \cup W)$

$\Rightarrow \mathsf{Reg}(U \cap (V \cup W)) = \mathsf{Reg}(U \cap \mathsf{Reg}(V \cup W)) = U \cap \mathsf{Reg}(V \cup W)$,

where the last equality holds because the intersection of open regular sets is open regular. Hence $U \cap (V \cup W) = (U \cap V) \cup (U \cap W)$ is a dense open subset both of $U \cap \mathsf{Reg}(V \cup W) = U \wedge (V \vee W)$, as well as of $\mathsf{Reg}((U \cap V) \cup (U \cap W)) = (U \wedge V) \vee (U \wedge W)$. By Fact 2.1.7 we conclude.
- Commutativity. $U \wedge V = U \cap V = V \cap U = V \wedge U$ and $U \vee V = \mathsf{Reg}((U \cup V)) = \mathsf{Reg}((V \cup U)) = V \vee U$.
- Identity. $U \vee \emptyset = \mathsf{Reg}(U \cup \emptyset) = \mathsf{Reg}(U) = U$ and $U \wedge X = U \cap X = U$.
- Complements. $U \vee \neg U = \mathsf{Reg}(U \cup (X \setminus \mathsf{Cl}(U))) = X$ (since $U \cup (X \setminus \mathsf{Cl}(U))$ is a dense subset of X: For A open, $A \cap U$ is empty iff $A \cap \mathsf{Cl}(U)$ is empty, so either $A \cap U$ is non-empty or $A \subseteq U^{\perp}$), while $U \wedge \neg U = U \cap (X \setminus \mathsf{Cl}(U)) = \emptyset$.
$\square$

It now remains to prove that $\mathsf{RO}(X)$ is complete.

Proposition 4.1.4 *The algebra* $\mathsf{RO}(X)$ *is complete.*

Proof Given a family $K = \{U_i : i \in I\}$ in $\mathsf{RO}(X)$, define $V = (\bigcup_{i \in I} U_i)^{\perp\perp}$. For any $i \in I$ we have $U_i \subseteq \bigcup_{j \in I} U_j$, so that

$$U_i = U_i^{\perp\perp} \subseteq (\bigcup_{j \in I} U_j)^{\perp\perp} = V$$

holds. This shows that V is an upper bound for the elements of K. If W is another such upper bound, then $U_i \subseteq W$, so that $\bigcup_{i \in I} U_i \subseteq W$, whence

$$V = (\bigcup_{i \in I} U_i)^{\perp\perp} \subseteq W^{\perp\perp} = W.$$

The proof for $\wedge$ is similar.
$\square$

We have shown that for a given topology τ on X there are two natural boolean algebras we can attach to it: $\mathsf{CLOP}(X, \tau)$ and $\mathsf{RO}(X, \tau)$. Observe that $\mathsf{CLOP}(X, \tau)^+$ is always contained in $\mathsf{RO}(X, \tau)^+$ and that if τ is 0-dimensional, any open set contains a clopen set, and thus $\mathsf{CLOP}(X, \tau)^+$ is a dense subset of $\mathsf{RO}(X, \tau)^+$.

The next lemma gives a necessary and sufficient condition so that $\mathsf{CLOP}(X, \tau)$ and $\mathsf{RO}(X, \tau)$ coincide.

Proposition 4.1.5 *Assume* B *is a boolean algebra.* B *is complete if and only if the regular open sets of* $\mathrm{St}(\mathsf{B})$ *overlap with the clopen subsets of* $\mathrm{St}(\mathsf{B})$.

Proof Assume B is complete. Let A be an arbitrary open set; then:

$$A = \bigcup_{i \in I} N_{b_i}$$

for a given family $\{b_i : i \in I\} \subseteq \mathsf{B}$. Since B is complete, let:

$$b = \bigvee_{i \in I} b_i.$$

Then N_b is clopen, and thus regular open. We show that $N_b = \mathsf{Cl}(A)$.

First we observe that

$$A = \bigcup_{i \in I} N_{b_i} \subseteq N_b.$$

In particular since N_b is closed $\mathsf{Cl}(A) \subseteq N_b$.

To prove the converse inclusion, we proceed as follows: First we observe that for all $c \in \mathsf{B}$

$$c \wedge b = 0 \text{ iff } c \wedge b_i = 0 \text{ for all } i \in I.$$

This gives that

$$N_c \cap N_b = \emptyset \text{ iff } N_c \cap A = \emptyset \text{ iff } N_c \cap \mathsf{Cl}(A) = \emptyset.$$

Thus

$$X \setminus \mathsf{Cl}(A) = \bigcup \{N_c : N_c \cap A = \emptyset\}$$

is disjoint from N_b. We can conclude that

$$N_b \subseteq \mathsf{Cl}(A).$$

The converse follows immediately, since $\mathsf{B} \cong \mathsf{CLOP}(\mathrm{St}(\mathsf{B})) = \mathsf{RO}(\mathrm{St}(\mathsf{B}))$, which is complete.   □

We say that a topological space (X, τ) is *extremally disconnected* if $\mathsf{CLOP}(X, \tau) = \mathsf{RO}(X, \tau)$.

4.2 Boolean Completions

In this section we prove that every preorder can be completed to a complete boolean algebra. The meaning of this assertion will be clarified by the formulation of its main results and definitions.

Notation 4.2.1 *Let $(Q, \leq_Q)$ be a preorder (i.e., $\leq_Q$ is a transitive and reflexive relation on Q).*
For $X \subseteq Q$

$$\downarrow X = \{p \in Q : \exists a \in X(p \leq_Q a)\}$$

is the downward closure of X ($\downarrow p$ stands for $\downarrow \{p\}$).

Exercise 4.2.2 Let $(Q, \leq_Q)$ be a preorder. Show that:

- The family τ_Q of downward closed subsets of Q form a family of sets closed under arbitrary unions and arbitrary intersections.
- The family $\{\downarrow q : q \in Q\}$ is a base for the topological space (Q, τ_Q) with the property that for each $q \in Q$, $\downarrow q$ is its smallest open neighborhood in τ_Q.
- (Q, τ_Q) is T_0 if and only if $\leq_Q$ is an order (i.e., $\leq_Q$ is antisymmetric).

Recall that (X, τ) is T_0 if given points $x \neq y$ in X, there is an open set which contains one but not the other.

Definition 4.2.3 Let $(Q, \leq_Q)$ be a preorder. The order topology on Q is τ_Q.

Theorem 4.2.4 *Let $(Q, \leq_Q)$ be a preorder. There exist an unique (up to isomorphism) cba B_Q and a map $j : Q \to \mathsf{B}_Q$ such that:*

1. *j preserves order and incompatibility (i.e., both $a \leq_Q b \Rightarrow j(a) \leq_{\mathsf{B}_Q} j(b)$ and $a \perp b \Leftrightarrow j(a) \wedge j(b) = 0_{\mathsf{B}_Q}$ hold).*
2. *$j[Q]$ is a dense subset of the partial order $(\mathsf{B}_Q^+, \leq)$.*

   Note that while B_Q is unique, there can be many $j : Q \to \mathsf{B}_Q$ which satisfy the above requirements.
   We split the proof into two lemmas, one for the existence part and the other for the uniqueness part.

Lemma 4.2.5 *Let $(Q, \leq_Q)$ be a preorder. The map*

$$j_Q : Q \to \mathsf{RO}(Q, \tau_Q)$$

$$q \mapsto \mathsf{Reg}(\downarrow q)$$

is such that:

1. *j_Q preserves order and incompatibility.*
2. *$j_Q[Q]$ is a dense subset of the partial order $(\mathsf{RO}(Q, \tau_Q)^+, \subseteq)$.*

Proof By Lemma 2.1.4, we have that for all open sets $A \in \tau_Q$

$$\mathsf{Reg}(A) = \{p \in Q : \downarrow p \cap A \text{ is a dense subset of } \downarrow p\} \tag{4.1}$$

since $\downarrow p$ is the smallest open neighborhood of p (if the property given in Lemma 2.1.4 holds for some open neighborhood of p it holds as well for $\downarrow p$).

We will repeatedly use the above characterization of regular open sets.

j_Q is order preserving: If $p \leq_Q q$, then $\downarrow p = \downarrow q \cap \downarrow p$ and clearly $\downarrow p$ is dense in $\mathsf{Reg}(\downarrow p)$, so $j_Q(p) \leq j_Q(q)$ by Fact 2.1.7.

j_Q is incompatibility preserving: Note that p, q are compatible if and only if $\downarrow p, \downarrow q$ (which are open sets of τ_P) have non-empty intersection, if and only if

$$\mathsf{Reg}(\downarrow p) \cap \mathsf{Reg}(\downarrow q)$$

is non-empty (by Fact 2.1.8). Hence the thesis.

j_Q has a dense image: Let $X \subseteq Q$ be non-empty. Let $p \in X$; clearly $\downarrow X \supseteq \downarrow p$; hence $j_Q(p) \leq \mathsf{Reg}(\downarrow X)$.

$\square$

We are left to show the uniqueness of this boolean completion. It suffices to prove the following:

Lemma 4.2.6 *Assume* B *is a cba and* $k : Q \to \mathsf{B}$ *preserves order and incompatibility and is such that* $k[Q]$ *is dense in* B^+. *Then the map:*

$$\pi : \mathsf{RO}(Q) \longrightarrow \mathsf{B}$$

$$A \longmapsto \bigvee \{k(p) : p \in A\}$$

is an isomorphism.

Assume the lemma holds and $j_i : Q \to \mathsf{B}_i$ for $i < 2$ preserve order and incompatibility and are such that $j_i[Q]$ is dense in B_i^+ (with both B_i cbas), and we can compose the isomorphisms given by the lemma to get an isomorphism of B_1 onto B_2.

We prove the lemma.

Proof We prove the lemma in several steps as follows:

π is order preserving: By definition.

π is incompatibility preserving: Assume A and B are incompatible in $\mathsf{RO}(Q)$. Then $A \cap B = \emptyset$; therefore

$$\pi(A) \wedge \pi(B) = \bigvee \{k(p) : p \in A\} \wedge \bigvee \{k(p) : p \in B\}$$

$$= \bigvee \{k(p) \wedge k(q) : p \in A, q \in B\}$$

$$= \bigvee \{k(r) : r \in Q, \ p \in A, \ q \in B, \ k(r) \le k(p) \wedge k(q)\}$$

$$= \bigvee \emptyset$$

$$= 0_\mathsf{B}.$$

The above holds since k preserves order and incompatibility, and hence has a dense image; this gives that for all $p \in A$ and $q \in B$

$$k(p) \wedge k(q) = \bigvee \{k(r) : r \in Q, \ k(r) \le k(p) \wedge k(q)\};$$

note also that no $p \in A$ and $q \in B$ are compatible in Q, giving that

$$k(p) \wedge k(q) = 0_\mathsf{B}.$$

$\pi \circ j_Q = k$:   Let $q \in Q$. Then

$$\pi \circ j_Q(q) = \bigvee \{k(r) : r \in \mathsf{Reg}(\downarrow q)\} \ge k(q)$$

since $q \in \mathsf{Reg}(\downarrow q)$. Assume the inequality is strict. Then

$$\pi \circ j_Q(q) \wedge \neg k(q) > 0_\mathsf{B}.$$

Now $k[Q]$ is dense in B^+, and thus we can find $r \in Q$ such that $k(r) \wedge k(q) = 0_\mathsf{B}$ and $k(r) \le \pi \circ j_Q(q)$.
Since $k(r) \wedge k(q) = 0_\mathsf{B}$, r and q are orthogonal in Q.
On the other hand we also have that $k(r) \wedge k(s) > 0_\mathsf{B}$ for some $s \in \mathsf{Reg}(\downarrow q)$, since

$$k(r) \le \pi \circ j_Q(q) = \bigvee \{k(s) : s \in \mathsf{Reg}(\downarrow q)\}.$$

This occurs only if $\downarrow r \cap \downarrow s \ne \emptyset$. Since $s \in \mathsf{Reg}(\downarrow q)$, we have that $\downarrow s \cap \downarrow q$ is a dense subset of $\downarrow s$. In particular $\downarrow s \cap \downarrow q \cap \downarrow r$ is non-empty. Thus there is $t \le r, q$. This contradicts the orthogonality of q, r in Q.
π is surjective: Let $b \in \mathsf{B}_2$, and by the density of $k[Q]$ we have that (see Proposition 3.11.2 and note that $k[Q]$ is dense in B^+)

$$b = \bigvee \{k(p) \in Q : k(p) \le b\}.$$

It is enough to show that $A = \{p \in Q : k(p) \le b\}$ is regular open to get that $\pi(A) = b$. Clearly A is downward closed and thus open. Now assume $r \in \mathsf{Reg}(A) \setminus A$. Then $k(r) \nleq b$. This gives that $k(r) \wedge \neg b > 0$ and thus that some s is such that $k(s) \le k(r) \wedge \neg b$. Since $k(s)$ and $k(r)$ are compatible in B^+, we have that some $t \in Q$ refines r and s. In particular $t \le r$ and $0_\mathsf{B} < k(t) \le k(r) \wedge \neg b$.

Since $r \in \text{Reg}(A) \setminus A$, $A \cap \downarrow r$ is dense in $\downarrow r$; since $t \leq r$, we can find $t^* \in A$ such that $t^* \leq t$. In conclusion $t^* \in A$ is incompatible with all elements of A since $k(t^*) \leq k(t)$ is incompatible with b, a contradiction.

π is injective: If $A \neq B$ are regular open, we may assume w.l.o.g. that there is $q \in Q$ such that $j_Q(q) \subseteq A$ and $j_Q(q)$ is orthogonal to B, which occurs if and only if $\downarrow q \cap B = \emptyset$. The latter gives that q is orthogonal to all elements in B. Then $\pi(A) \geq \pi(j_Q(q))$ and $\pi(j_Q(q)) = k(q)$ is orthogonal to $\bigvee k[B] = \pi(B)$, since k is order and incompatibility preserving. We get that $\pi(A) \neq \pi(B)$.

By Lemma 4.0.7 the proof of is completed.   □

The proof of Theorem 4.2.4 is completed.

Corollary 4.2.7 $(Q, \leq)$ *is a separative partial order if and only if the map* $j :$ $Q \to \text{RO}(Q)$ *of Theorem 4.2.4 is an isomorphism of* $(Q, \leq)$ *with* $(J[Q], \subseteq)$ *seen as a suborder of* $(\text{RO}(Q)^+, \subseteq)$ *if and only if* $\downarrow q = \text{Reg}(\downarrow q)$ *for all* $q \in Q$.

Proof The nontrivial equivalence is to show that Q is separative if and only if $\downarrow p = \text{Reg}(\downarrow p)$ for all $p \in Q$.

($\Rightarrow$):   Assume Q is separative, and toward a contradiction, let $r \in \text{Reg}(\downarrow p) \setminus \downarrow p$. Then we can refine r to an $s \perp p$ still in $\text{Reg}(\downarrow p)$ since Q is separative and $r \not\leq p$. This is the desired contradiction.

($\Leftarrow$):   Assume $\downarrow p = \text{Reg}(\downarrow p)$. And let $p \not\leq q$. Assume toward a contradiction that for all $r \leq p$, r and q are compatible in Q, i.e., $\downarrow q \cap \downarrow r$ is non-empty. Then $\downarrow q \cap \downarrow p$ is dense in $\downarrow p$, which (by Fact 2.1.7) gives that

$$\downarrow p = \text{Reg}(\downarrow p) \subseteq \text{Reg}(\downarrow q) = \downarrow q,$$

i.e., $p \in \downarrow q$, contradicting our assumption that $p \not\leq q$.

                                                                           □

Corollary 4.2.8 *Using the terminology of 4.2.4,* $(Q, \leq)$ *is an atomless preorder iff* B_Q *is atomless.*

Proof It follows since the map j of the theorem preserves the order and incompatibility relation and has a dense image.   □

Exercise 4.2.9 Recall that for $s \in 2^{<\omega} N_s = \{f \in 2^\omega : s \supseteq f\}$:

- Show that if G is an ultrafilter on the boolean algebra $\text{RO}(2^\omega)$, then $\{s : N_s \in G\}$ is a filter on $(2^{<\omega}, \supseteq)$.
- Conversely, for any $f \in 2^\omega$, show that $G_f = \{N_s : s \subseteq f\}$ is a prefilter on the boolean algebra $\text{CLOP}(2^\omega)$, whose upward closure in $\text{CLOP}(2^\omega)$ is a ultrafilter in $\text{St}(\text{CLOP}(2^\omega))$.
- Show also that for any $f \in 2^\omega$ the upward closure in $\text{RO}(2^\omega)$ of G_f is just a filter on the boolean algebra $\text{RO}(2^\omega)$. (HINT: To show that G_f does not generate a ultrafilter on $\text{RO}(2^\omega)$, look at Fact 5.2.3 to argue that *Even* and *Odd* are regular open sets not in $\uparrow G_{c_0}$, where c_0 is the constant sequence of 0.)

We conclude this section with the following observation:

Remark 4.2.10 There is a nice theorem asserting that up to isomorphism there is a unique atomless complete boolean algebra B such that B^+ contains a countable dense subset. Here is a list of partial orders $(P, \leq)$ and topological spaces (X, τ) such that B is isomorphic to the regular open sets in the relevant topology:

1. The atomless partial order $(\tau \setminus \{\emptyset\}, \subseteq)$, where τ is the standard euclidean topology on $\mathbb{R}$.
2. The partial order $(D, \subseteq)$ given by open intervals with rational endpoints: D is countable and is a dense subset of $\tau \setminus \{\emptyset\}$ under inclusion. This gives that $\mathsf{RO}(D)$ and $\mathsf{RO}(\tau \setminus \{\emptyset\})$ are isomorphic atomless complete boolean algebras admitting a countable dense subset.
3. The boolean completion $\mathsf{RO}(2^{<\omega})$ of the partial order $(2^{<\omega}, \supseteq)$ (the latter is a separative countable atomless partial order and is contained in its boolean completion as a dense subset).
4. The regular open subsets of $\mathbb{R}$ in the euclidean topology τ (the map $A \mapsto \mathsf{Reg}(A)$ surjects the partial order $(\tau \setminus \{\emptyset\}, \subseteq)$ onto $\mathsf{RO}(\mathbb{R}, \tau)^+$ preserving order and incompatibility).
5. The regular open sets of the product topology τ^* on 2^ω (the map $s \mapsto N_s = \{f \in 2^\omega : s \subseteq f\}$ is order and incompatibility preserving and maps $2^{<\omega}$ in a dense subset of $\mathsf{RO}(2^\omega, \tau^*)^+$).

Notice that in each case the regular open sets considered refer to different topological spaces: The first three algebras of regular open sets are induced by the order topology respectively on $(D, \subseteq)$, on $(\tau \setminus \{\emptyset\}, \subseteq)$, on $(2^{<\omega}, \supseteq)$, while in the fourth and fifth cases these algebras are given by the regular open sets of the topological space $(\mathbb{R}, \tau)$ or of the space $(2^\omega, \tau^*)$. Remark also that the map $j : \tau \setminus \{\emptyset\} \to \mathsf{RO}(\tau \setminus \{\emptyset\})$ given by Theorem 4.2.4 identifies two open sets iff they have dense intersections. In particular in this case the relevant j is not injective, but it is still order and incompatibility preserving.

In conclusion we get that the same atomless complete boolean algebra can be obtained as the algebra of regular open sets of five distinct topologies on five distinct topological spaces whose topologies (when seen as partial orders under inclusion) are not always isomorphic, but whose algebras of regular open sets—on the other hand—are all isomorphic.

This reflects a common state of affairs for all complete atomless boolean algebras.

4.2.1 Some Remarks on Partial Orders and Their Boolean Completions

Summing up, in these first sections, we have proved among other things, the following results:

Let $(Q, \leq_Q)$ a preorder. There exists an unique (up to isomorphism) cba B_Q such that there exists a map $j : Q \to \mathsf{B}_Q$ such that:

1. j preserves order and incompatibility.
2. $j[Q]$ is dense in B_Q^+.
3. Q is a separative partial order iff j is an injection.
4. Q is atomless iff B_Q is atomless.
5. B_Q is the cba given by the regular open sets of many topological spaces.
6. $\mathrm{St}(\mathsf{B}_Q) = \{G \subseteq \mathsf{B}_Q : G \text{ is an ultrafilter}\}$ with the topology τ_{B_Q} generated by the sets $\{N_b : b \in \mathsf{B}_Q\}$ is such that

$$\mathsf{RO}(\mathrm{St}(\mathsf{B}_Q), \tau_{\mathsf{B}_q}) = \mathsf{CLOP}(\mathrm{St}(\mathsf{B}_Q), \tau_{\mathsf{B}_q}) \cong \mathsf{RO}(Q, \tau_Q)$$

and $(\mathrm{St}(\mathsf{B}_Q))$ is an extremally disconnected compact Hausdorff topological space.
7. $\mathsf{B}_Q \cong C(\mathrm{St}(\mathsf{B}_Q), 2)$ where the latter is

$$\{f : \mathrm{St}(\mathsf{B}_Q) \to \mathbb{Z}_2, \ f \text{ is continous}\}$$

(with $\mathbb{Z}_2 = \{0, 1\}$ endowed of the discrete topology).

4.3 Miscellanea: Completeness, Chain Conditions, and the Measure Algebra

4.3.1 The κ-Chain Condition

Definition 4.3.1 Given a preorder $(P, \leq)$ and a cardinal κ:

- P is κ-CC if all its antichains have size at most κ (CCC stands for $\aleph_0$-CC).
- P is $< \kappa$-CC if it is λ-CC for all cardinals $\lambda < \kappa$.

A topological space has the κ-CC if the quasi-order $(\tau \setminus \{\emptyset\}, \subseteq)$ has the κ-CC (accordingly we define being $< \kappa$-CC).

Note that every countable quasi-order has the CCC.
As a consequence:

- $2^{<\omega}$ has the CCC.
- Separable topological spaces (such as $\mathbb{R}^n$ with euclidean topology) have the CCC.

On the other hand the partial order $(\omega_1)^{<\omega}$ does not have the CCC: Indeed, the set

$$\{\{(0, \alpha)\} : \alpha < \omega_1\}$$

is an uncountable antichain.

The following holds:

Lemma 4.3.2 *Assume P is a quasi-order with the CCC. Then $\mathsf{RO}(P)^+$ has the CCC as well.*

Proof Assume $A \subseteq \mathsf{RO}(P)^+$ is an antichain. For each $a \in A$ find $p_a \in P$ such that $i(p_a) \leq a$ where $i : P \to \mathsf{RO}(P)$ is the canonical immersion of P in its boolean completion. Since i is order and incompatibility preserving, $\{p_a : a \in A\}$ is an antichain in P and thus is countable. Moreover, the map $a \mapsto p_a$ is injective since $a \neq b$ entails $a \wedge b = 0$, which gives that p_a and p_b are incompatible in P. We conclude that A is countable as well.   □

Definition 4.3.3 A boolean algebra B is:

- κ-complete if all its subsets of size at most κ have an exact upper bound
- $< \kappa$-complete if it is λ-complete for all $\lambda < \kappa$

Lemma 4.3.4 *Let B a boolean algebra. If $(\mathsf{B}^+, \leq_\mathsf{B})$ is $< \kappa$-CC and B is $< \kappa$-complete, then B is complete.*

Proof Given $X \subseteq \mathsf{B}$, find (by Zorn's lemma) Y a maximal antichain contained in $\downarrow X$. Since Y has size less than κ, $b = \bigvee Y$ exists. By Fact 3.11.7, $\bigvee X = b = \bigvee Y$, and hence $\bigvee X$ exists.   □

4.3.2 The Algebra of Lebesgue Measurable Sets Modulo Null Sets

Recall that a subset of $[0, 1]$ is *Borel* if it can be obtained in countably many steps starting from the basic open intervals applying the operations of countable unions and taking the complement. We say that a $A \subseteq [0, 1]$ is *null* (or *measure zero*) if for every $\epsilon > 0$ there exists a family $\{I_i : i < \omega\}$ of open intervals such that $A \subseteq \bigcup \{I_i : i < \omega\}$ and $\sum_{i<\omega} I_i < \epsilon$. For every $A \subseteq [0, 1]$, we say A is *Lebesgue measurable* if and only if $A \triangle X$ is null, for some Borel set $X \subseteq [0, 1]$. For every Lebesgue measurable $A \subseteq [0, 1]$, we denote the Lebesgue measure of A with $\mu(A)$, and we define it as the infimum of $\sum_{n \in \mathbb{N}} I_n$, where $\{I_i : i < \omega\}$ is a covering of A consisting of basic open intervals.

Exercise 4.3.5 Let $\mathcal{M}([0, 1])$ be the boolean algebra of Lebesgue measurable subsets of $[0, 1]$ with usual boolean operations of union, intersection, and taking the complement. The set Null of all null subsets of $[0, 1]$ is an ideal of $\mathcal{M}([0, 1])$.

We can consider $\mathcal{M}([0, 1])/\text{Null}$, the boolean algebra of the Lebesgue measurable subsets of $[0, 1]$ modulo the ideal of null sets. The elements of $\mathcal{M}([0, 1])/\text{Null}$ are equivalence classes of Lebesgue measurable subsets of the unit interval

$$[X]_{\text{Null}} = \{Y \subseteq [0, 1] : X \triangle Y \text{ is null}\}.$$

This is also known as the *measure algebra*, and sometimes it is denoted with MALG.

Proposition 4.3.6 *The measure algebra* MALG *is CCC, i.e.,* MALG *has no uncountable antichains.*

Proof Let $\mathcal{A}$ be an antichain of MALG. This means that $[A]_{\text{Null}} \cap [B]_{\text{Null}} \in \text{Null}$, i.e., that $\mu(A \cap B) = 0$ for all $[A]_{\text{Null}}, [B]_{\text{Null}} \in \mathcal{A}$. For every $n \in \omega$, let

$$\mathcal{A}_n = \{[X]_{\text{Null}} \in A : \mu(X) \geq 1/n\}.$$

We claim that $|\mathcal{A}_n| \leq n$. For, if $|\mathcal{A}_n| > n$ and $[X_1]_{\text{Null}}, \ldots, [X_{n+1}]_{\text{Null}} \in \mathcal{A}_n$ are pairwise distinct classes, then

$$\mu(\bigcup_{j=1}^{n+1} X_j) > 1,$$

though $\bigcup_{j=1,\ldots,n+1} X_j \subseteq [0, 1]$, which has measure 1, a contradiction. So, $\mathcal{A} = \bigcup_{n<\omega} \mathcal{A}_n$ is a countable union of finite sets which implies that $\mathcal{A}$ is countable. $\square$

Proposition 4.3.7 *The measure algebra* MALG *is countably complete, i.e., if* $\{A_n : n \in \omega\} \subseteq$ MALG, $\bigvee_{n\in\omega} A_n$ *exists in* MALG.

Proof Let for each n, $A_n = [B_n]_{\text{Null}}$ for some measurable set $B_n \subseteq [0, 1]$. Check that $[\bigcup_{n\in\omega} B_n]_{\text{Null}}$ is in MALG and is an exact upper bound of $\{A_n : n \in \omega\}$. $\square$

Proposition 4.3.8 *The measure algebra* MALG *is atomless.*

Proof It suffices to show that if $\mu(A) > 0$, then A can be split into two pieces of positive measure. Assume not and build by induction sets A_n and intervals $I_n = [i_n/2^n, (i_n + 1)/2^n]$ such that $\mu(A_n) = \mu(A)$ and $A_n = A \cap [i_n/2^n, (i_n + 1)/2^n]$ as follows:

- $A_0 = A$, $i_0 = 0$, hence $I_0 = [0, 1]$.
- Given A_n and $I_n = [i_n/2^n, (i_n + 1)/2^n]$, let $j = 2i_n + 1$. Then $A_n = (A \cap [i_n/2^n, j/2^{n+1})]) \cup (A \cap [j/2^{n+1}, (i_n + 1)/2^n])$ with

$$\mu(A \cap [i_n/2^n, j/2^{n+1})] \cap (A \cap [j/2^{n+1}, (i_n + 1)/2^n]) = \mu(\{j/2^{n+2}\}) = 0.$$

Hence by assumption on A either $\mu(A) = \mu(A \cap [i_n/2^n, j/2^{n+1})])$ or $\mu(A) = \mu(A \cap [j/2^{n+1}, (i_n + 1)/2^n])$. We let $i_{n+1} = i_n$ and $I_{n+1} = [i_n/2^n, j/2^{n+1})]$

if the first case occurs, and $i_{n+1} = j$ and $I_{n+1} = [j/2^{n+1}, (i_n + 1)/2^n)]$ if the second case occurs. We let $A_{n+1} = A_n \cap I_{n+1}$.

We obtain that $\mu(A_n) = \mu(A)$ for all n and that $\bigcap_{n \in \mathbb{N}} A_n \subseteq \{x\}$ for a unique point x given by the intersection of all intervals I_n. By the countable completeness of μ, we get that

$$0 = \mu(\{x\}) \geq \mu(\bigcap_{n \in \mathbb{N}} A_n) = \inf_{n \in \mathbb{N}} (\mu(A_n)) = \mu(A) > 0,$$

a contradiction. Hence A can be split into two pieces of positive measure, concluding the proof.   □

Corollary 4.3.9 *The measure algebra is complete, atomless, and CCC.*

Chapter 5
More on Preorders

In this chapter we analyze certain combinatorial properties of preorders; in particular we focus on the one hand on the relations existing between a partial order and its boolean completion and on the other hand on the quasi-order introduced by Cohen to obtain the consistency of the failure of CH by means of forcing, and we outline the key combinatorial features used to prove this result. The material of this chapter overlaps with some parts of [20, Chapter III] or [19, Chapter II].

5.1 Generic Filters

Definition 5.1.1 Let $(P, \leq)$ be a quasi-order. Let $\mathcal{F} = \{D_i : i \in I\}$ be a family of subsets of P. Let G be a filter. G is $\mathcal{F}$-*generic* if $G \cap D_i \neq \emptyset$, for all $i \in I$.

The following is a useful equivalent of Baire's category theorem:

Lemma 5.1.2 (Generic Filter Lemma) *Let $(P, \leq)$ be a quasi-order and $\mathcal{F} = \{D_i : i \in \omega\}$ be a family of predense subsets of P. Then, for every $p \in P$, there exists a filter G on P $\mathcal{F}$-generic with $p \in G$.*

Proof Using AC and recursion on ω, choose $p_n \in P$ for $n \in \omega$ so that $p_0 = p$, $p_{n+1} \leq p_n$, and $p_{n+1} \in \downarrow D_n$. Let

$$G = \uparrow \{p_n : n \in \omega\}.$$

G is upward closed by definition. We check now if it is a filter. Let $r_0, r_1 \in G$, and let m_i be such that $r_i \geq p_{m_i}$, for $i = 0, 1$. Then $r_i \geq p_n$ for all $n \geq m_0, m_1$ and $i = 0, 1$. □

© The Author(s), under exclusive license to Springer Nature Switzerland AG 2024
M. Viale, *The Forcing Method in Set Theory*, La Matematica per il 3+2 168,
https://doi.org/10.1007/978-3-031-71660-7_5

Corollary 5.1.3 (Baire's Category Theorem) *Assume (X, τ) is a compact Hausdorff space. Then the intersection of any countable family of dense open subsets is dense.*

Proof Let $\{D_n : n \in \mathbb{N}\}$ be a countable family of dense open subsets of X. Let A be an open non-empty subset of X; we must find a point $x \in A \cap \bigcap_{n \in \mathbb{N}} D_n$.

We use the following property of compact Hausdorff spaces (normality): Any non-empty open set O admits an open subset B such that $\mathsf{Cl}(B) \subseteq O$.

So fix B non-empty and open such that $\mathsf{Cl}(B) \subseteq A$. Let σ be the restriction of τ to $\mathsf{Cl}(B)$ so that $(\mathsf{Cl}(B), \sigma)$ is also a compact Hausdorff space. Notice that $E_n = D_n \cap B$ is a dense open subset of $\mathsf{Cl}(B)$ for all $n \in \mathbb{N}$. Consider now the quasi-order $(\sigma \setminus \{\emptyset\}, \subseteq)$ and the sets

$$F_n = \left\{ O \in \sigma \setminus \{\emptyset\} : \mathsf{Cl}(O) \subseteq E_n \right\}.$$

Claim 5.1.3.1 F_n *is open dense in* $(\sigma \setminus \{\emptyset\}, \subseteq)$.

$\square$

Proof Clearly F_n is open. Let $C \in \sigma$ be open non-empty. Hence $E_n \cap C$ is an open non-empty subset of C. Since $(\mathsf{Cl}(B), \sigma)$ is compact Hausdorff, there is $U \in \sigma \setminus \{\emptyset\}$ such that $\mathsf{Cl}(U) \subseteq E_n \cap C$. Then $U \in F_n$ refines C. Hence F_n is dense since C was chosen arbitrarily in $\sigma \setminus \{\emptyset\}$. $\square$

Now let G be a filter on $(\sigma \setminus \{\emptyset\}, \subseteq)$ such that $G \cap F_n \neq \emptyset$ for all $n \in \mathbb{N}$, which exists by Lemma 5.1.2. Notice that each $B_n \in G \cap F_n$ is such that $\mathsf{Cl}(B_n) \subseteq \mathsf{Cl}(B) \cap E_n \subseteq A \cap D_n$. Notice also that the family $\left\{ \mathsf{Cl}(B_n) : n \in \mathbb{N} \right\}$ has the finite intersection property, since any finite subset of this family $\mathsf{Cl}(B_{i_1}) \ldots \mathsf{Cl}(B_{i_k})$ is such that

$$\mathsf{Cl}(B_{i_1}) \cap \cdots \cap \mathsf{Cl}(B_{i_k}) \supseteq B_{i_1} \cap \cdots \cap B_{i_k} \supseteq U \neq \emptyset$$

for some $U \in G$, since G is a filter and $B_{i_1}, \ldots B_{i_k} \in G$. Since $\mathsf{Cl}(B)$ is compact, $\bigcap \left\{ \mathsf{Cl}(B_n) : n \in \mathbb{N} \right\}$ is non-empty. Any point in this intersection belongs to $A \cap \bigcap_{n \in \mathbb{N}} D_n$.

The two exercises below show that the generic filter lemma is nontrivial only if we are considering atomless quasi-orders. We will see in Chap. 7 that the forcing method invented by Cohen stems from a careful analysis of the notion of generic filter.

The following exercises show that atoms of preorders give rise to trivial generic filters.

Exercise 5.1.4 Let P be a preorder and a an atom of P. Then $G_a = \uparrow \{a\}$ is a $\mathcal{D}$-generic filter, where $\mathcal{D}$ is the collection of dense subsets of P. (HINT: An atom of P belongs to all dense subsets of P).

The following exercise outlines in more detail the relations existing between atoms of a boolean algebra and the notion of genericity.

Exercise 5.1.5 Assume C is a boolean algebra. Then $St(C) \setminus \{G\}$ is open dense for any $G \in St(C)$ which is a nonprincipal ultrafilter (i.e., such that $a \notin G$ for any a atom of $St(C)$). (HINT: Recall (or prove) that G is a nonprincipal ultrafilter if and only if G is not an isolated point of $St(C)$; moreover any non-isolated point of a Hausdorff topological space has a complement which is open dense in $St(C)$).

Show the following:

1. Assume C is atomless; then the intersection of *all* dense open subsets of $St(C)$ is empty (HINT: Already the intersection of

$$\{St(C) \setminus \{G\} : G \in St(C)\}$$

   is empty).
2. If $a \in C$ is an atom, then $G_a = \{b \in B : a \leq b\}$ is an ultrafilter in $St(C)$ meeting all the dense open subsets of $St(C)$.

We now come to a basic application of the generic filter lemma which is at the heart of Cohen's forcing method.

Exercise 5.1.6 Show that the following sets are dense open in $2^{<\omega}$:

- For $f \in 2^{\omega}$, $D_f = \{s \in 2^{<\omega} : s \perp f\}$.
- $E_n = \{s \in 2^{<\omega} : n \in \mathrm{dom}(s)\}$.

Prove that there is no filter G on $2^{<\omega}$ which is $\{D_f : f \in 2^{\omega}\} \cup \{E_n : n \in \omega\}$-generic.

(HINT: Assume toward a contradiction that there exists a filter G such that $G \cap D_f \neq \emptyset$ for every $f \in 2^{\omega}$ and $G \cap E_n \neq \emptyset$ for every $n \in \omega$. Let $\bigcup\{s : s \in G\} = g \in 2^{\omega}$. Then $G \cap D_g \neq \emptyset$ and so there should be $t \in G$ such that $t \perp g$, i.e., $\exists n (t(n) \neq g(n))$. But

$$G \ni t \subseteq g = \bigcup G,$$

a contradiction.)

We can even show that certain quasi-orders have a family of $\aleph_1$-many dense sets which cannot be met in a filter:

Fact 5.1.7 *Consider the partial order $((\omega_1)^{<\omega}; \supseteq)$ ordered by reverse inclusion. There exists a family $\{D_\alpha : \alpha < \omega_1\}$ of dense sets such that for every filter $G \subseteq \omega_1^{<\omega}$ there exists α such that $G \cap D_\alpha = \emptyset$.*

Proof Set

$$B_\alpha = \{s \in (\omega_1)^{<\omega} : \exists n\ s(n) = \alpha\}$$

$$E_n = \{s \in (\omega_1)^{<\omega} : |s| \geq n\}.$$

For all α and n, B_α and E_n are open by definition, and let us see that they are dense. Take $s \in (\omega_1)^{<\omega}$. If there exists $n < |s|$ such that $s(n) = \alpha$, then $s \in B_\alpha$; otherwise $s^\frown \alpha = s \cup \{(|s|, \alpha)\} \in B_\alpha$. Hence B_α is dense. We leave to the reader the proof that E_n is dense for every n.

Assume now that there exists a filter G such that $G \cap B_\alpha \neq \emptyset$ for every $\alpha \in \omega_1$ and $G \cap E_n \neq \emptyset$ for every $n \in \omega$; then $\bigcup G : \omega \to \omega_1$ is a surjection, a contradiction. So the family $\{B_\alpha : \alpha < \omega_1\} \cup \{E_n : n \in \omega\}$ is the one we were looking for. $\square$

At this point we can already bring forward something that we will formalize in the last chapter of these notes. Let M be a transitive countable model of ZFC, and assume that $P \in M$ is atomless and separative. It can be seen that the family of dense sets of P is uncountable. On the other hand there are only countably many dense sets of P which can belong to M. The generic filter lemma guarantees that there exists a filter G that intersects all the dense sets of P which are in M.

Now observe the following:

Fact 5.1.8 *Assume M is a countable transitive model of ZFC, $P \in M$ is atomless and separative, and G is an M-generic filter, i.e., G meets all the dense subsets of P which belong to M. Then $G \notin M$.*

Proof It is always the case that $P \setminus G$ is an open dense subset of P whenever G is a filter on P and P is atomless and separative (given any $p \in P$ find $r, q \leq p$ and incompatible, then at least one between r and q is not in G). Thus $G \in M$ implies $P \setminus G \in M$. However $G \cap (P \setminus G) = \emptyset$; thus G cannot be M-generic. $\square$

Hence, whenever M is a countable transitive model of ZFC, $P \in M$ is atomless and separative, and G is an M-generic filter, we can define

$$M[G] = \bigcap \{N \supseteq M : N \text{ is transitive} \wedge N \vDash \text{ZFC} \wedge G \in N\}.$$

Our arguments show already that $M[G]$ strictly contains M (since $G \in M[G] \setminus M$), provided that there is some transitive set $N \supseteq M \cup \{G\}$ which is a model of ZFC. We will further show that $M[G]$ is itself a model of ZFC and that (depending on the choice of the $P \in M$ for which G is M-generic) we can define $M[G]$ so that it satisfies CH or its negation by carefully choosing P.

5.2 The Quasi-Orders Fn(X, Y)

Definition 5.2.1 Given sets X, Y and a cardinal κ, let Fn(X, Y, κ) be the quasi-order of functions with domain a subset of X of size less than κ and ranging in Y. The order on Fn(X, Y, κ) is given by the reverse inclusion.

We write simply Fn(X, Y) instead of Fn(X, Y, ω), and for any $p \in$ Fn(X, Y), we put

$$\downarrow p = \{f \in Y^X : p \subset f\}.$$

So $2^{<\omega}$ is the set of functions in Fn($\omega, 2$) whose domain is a natural number.

Remark 5.2.2

1. The order $(2^{<\omega}, \subseteq)$ is a dense suborder of (Fn($\omega, 2$), $\subseteq$), in particular they have the same boolean completion, which can be represented as $\mathsf{RO}(2^\omega)$ the family of regular open sets in 2^ω with the product topology. The map $s \mapsto N_s = \{f \in 2^\omega : s \subseteq f\}$ implements an order and incompatibility preserving embedding of Fn($\omega, 2$) into $\mathsf{RO}(2^\omega)$ with a dense image, since the family

$$\left\{N_s : s \in 2^{<\omega}\right\}$$

   forms a basis of clopen sets (and thus regular open) for the product topology on 2^ω. We leave to the reader to check that this map is order and incompatibility preserving.

2. If Y is finite, then the space Y^X endowed with the product topology is a compact zero-dimensional Hausdorff space with no isolated points; in particular any clopen set in 2^X is a finite union of sets of the form $\downarrow p$ for some $p \in$ Fn($X, 2$). For any $p \in$ Fn($X, 2$), we can write $\downarrow p$ as a closed set:

$$N_p = \bigcup\{2^X \setminus N_t \mid t \neq p, \operatorname{dom}(t) = \operatorname{dom}(p)\},$$

   since there are only finitely many t ranging in 2 with the same domain as p.
   The compact Hausdorff space 2^ω endowed with the product topology is also known in the literature as the *Cantor space*.

3. The family of clopen sets in the product topology on 2^X is a boolean algebra with the standard set theoretic operations and the sets $\downarrow p$ as p ranges in Fn($X, 2$) form a dense subset of the positive elements of this boolean algebra. Its boolean completion is the space of regular open sets of 2^X with the product topology.

Fact 5.2.3 *Some regular open sets of 2^ω are not closed.*

Proof A counterexample is given by Odd and Even, where Odd (respectively, Even) is the set of sequences in 2^ω that differ from 0^ω and start with an odd (respectively, even) number of zeros.

These two sets are open and disjoint, and their closures intersect only in 0^ω.

In particular, 0^ω is the unique point in the closure of Odd and Even such that no open set containing it has a dense intersection with Odd or a dense intersection with Even, while any element of Odd (respectively, Even) has a clopen neighborhood fully contained in Odd (respectively, Even). This means that Odd and Even are regular and open, but they are not closed.   □

Exercise 5.2.4

- The map

$$i : 2^{<\omega} \longrightarrow 2^{\omega}$$

$$s \longmapsto s^\frown 1^\frown 0^\omega$$

  is continuous and injective and has a dense image in the Cantor space.
- The map

$$i^* : \mathsf{RO}(2^{<\omega}) \longrightarrow \mathsf{RO}(2^{\omega})$$

$$A \longmapsto \bigcup \{N_s \mid s \in A\}.$$

  is an isomorphism of complete boolean algebras.

In particular, $\mathsf{RO}(2^\omega)$ is another possible representation of the boolean completion of the quasi-orders $2^{<\omega}$, $\mathrm{Fn}(\omega, 2)$, and as a boolean algebra $\mathsf{RO}(2^\omega)$ is a proper superalgebra of the boolean algebra given by the clopen subsets of 2^ω. The positive elements of the latter however form a dense suborder of $\mathsf{RO}(2^\omega)^+$.

The latter observation outlines a distinction between 2^ω and the Stone space of the Boolean completion of the quasi-order $2^{<\omega}$, a distinction that is common to the Stone spaces of a boolean algebra, and the Stone space of its boolean completion. We spell out the details in the following observation:

Remark 5.2.5 Let $\mathsf{B} = \mathsf{RO}(2^\omega)$ be the boolean completion of $2^{<\omega}$ and $\mathrm{St}(\mathsf{B})$ its associated Stone space. Then $\mathrm{St}(\mathsf{B})$ is a zero-dimensional compact Hausdorff space, and there is a natural projection

$$\pi : \mathrm{St}(\mathsf{B}) \to 2^\omega$$

$$G \mapsto f_G = \bigcup \{s \in 2^{<\omega} : N_s \in G\}.$$

This projection is:

- Continuous closed and open, since $N_s \in G$ iff $s \subset f_G$ for all $s \in 2^{<\omega}$.
- Surjective: Given $f \in 2^\omega$, consider an ultrafilter G that contains N_s for every $s \subset f$; then $\pi(G) = f$.
- However π is not injective: For example, there are G and H ultrafilters in $\mathrm{St}(\mathsf{B})$ such that $\pi(G) = \pi(H) = 0^\omega$, but $Odd \in G$, $Even \in H$.

This occurs since B can be identified with the family of regular open sets of 2^ω and $\mathrm{St}(\mathsf{B})$ is a Stone space whose clopen sets overlap with its regular open sets, while we already remarked that the clopen subsets of 2^ω form a strictly proper subalgebra of the regular open subsets of 2^ω.

5.2.1 The Quasi-Order Fn($\omega_2 \times \omega, 2$)

In this section we introduce the order introduced by Cohen to produce by means of forcing a counterexample to CH.

Definition 5.2.6 We define the following quasi-order:

$$\text{Fn}(\omega_2 \times \omega, 2) = \{s : s : \omega_2 \times \omega \to 2 \wedge \text{dom}(s) \text{ is finite}\}.$$

We can naturally identify

$$(2^\omega)^{\omega_2} = \{f : \text{dom}(f) = \omega_2 \wedge \forall i \in \text{dom}(f)(f(i) \in 2^\omega)\}$$

with the space $2^{\omega_2 \times \omega}$. With this identification its product topology is generated by the family $\{N_s : s \in \text{Fn}(\omega_2 \times \omega, 2)\}$, where in this case we use this natural identification to let

$$N_s = \{f \in (2^\omega)^{\omega_2} : \forall(\alpha, n) \in \text{dom}(s) \ f(\alpha)(n) = s(\alpha, n)\}.$$

Moreover the following holds:

Lemma 5.2.7 *The map $s \mapsto N_s$ defines a dense embedding of the quasi-order* Fn($\omega_2 \times \omega, 2$) *into* RO($2^{\omega_2 \times \omega}$). *In particular* RO(Fn($\omega_2 \times \omega, 2$)) *and* RO($2^{\omega_2 \times \omega}$) *are isomorphic complete boolean algebras.*

Proof Notice that the family $\{N_s : s \in \text{Fn}(\omega_2 \times \omega, 2)\}$ is a base for the product topology on $2^{\omega_2 \times \omega}$ consisting of clopen (and thus also regular open) sets.

In particular this gives that the target of the map is dense. It is an easy exercise to check that the map is also order and incompatibility preserving.   □

We define the following subsets of RO($2^{\omega_2 \times \omega}$):

- $D_{n,\alpha} = \{N_s : s \in \text{Fn}(\omega_2 \times \omega, 2) \ (\alpha, n) \in \text{dom}(s)\}$.
- $E_{\alpha,\beta} = \{N_s : s \in \text{Fn}(\omega_2 \times \omega, 2) \ \exists n \ s(\alpha, n) \neq s(\beta, n)\}$.

Let $\mathcal{D}$ be the family

$$\{D_{n,\alpha} : n \in \omega, \alpha \in \omega_2\} \cup \{E_{\alpha,\beta} : \alpha \neq \beta \in \omega_2\}.$$

Assume that we could find a filter G which is $\mathcal{D}$-generic, then, letting $g_\alpha = \bigcup\{\langle n, s(\alpha, n)\rangle : s \in G, n \in \omega\}$, we would have that $\{g_\alpha : \alpha < \omega_2\}$ are different elements of 2^ω, and this would entail the failure of CH.

Exercise 5.2.8 Show that $\{N_s : s \in \text{Fn}(\omega_2 \times \omega, 2)\}$ is a dense subset of RO($2^{\omega_2 \times \omega}$) and that the map $s \mapsto N_s$ is injective and order and incompatibility preserving.

Show also that $E_{\alpha,\beta}$ and $D_{n,\alpha}$ are dense in RO($2^{\omega_2 \times \omega}$) for all $\alpha \neq \beta < \omega_2$ and $n < \omega$.

5.3   Quasi-Orders with the Countable Chain Condition and the Δ-System Lemma

Recall that P has the countable chain condition (CCC) if every antichain of P is countable.

Definition 5.3.1 If S is a set of finite sets, then it is a Δ-**system** if there is some (possibly empty) r such that for any $a, b \in S$, if $a \neq b$, then $a \cap b = r$. r is the *root* of the system.

Lemma 5.3.2 (Δ-**system lemma**) *Let κ be an uncountable regular cardinal, and let $\mathcal{A}$ be a family of finite sets with $|\mathcal{A}| = \kappa$. Then there is a $\mathcal{D} \in [\mathcal{A}]^\kappa$ such that $\mathcal{D}$ forms a Δ-system.*

Proof Since $cf(\kappa) = \kappa > \omega$ and there are only $\aleph_0$ possible $|X|$ for $X \in \mathcal{A}$, we may fix $n \in \omega$ and $\mathcal{D} \in [\mathcal{A}]^\kappa$ such that $|s| = n$ for all $s \in \mathcal{D}$. Now, we prove it by induction on n.

1. $n = 1$: Then $\mathcal{D}$ is already a Δ-system with empty root.
2. $n > 1$: For each $p \in X$, let $\mathcal{D}_p = \{X \in \mathcal{D} : p \in X\}$. There are two cases.

   - Case I: $|\mathcal{D}_p| = \kappa$ for some p. Fix p, and let $E = \{X \setminus \{p\} : X \in \mathcal{D}_p\}$, which is a family of κ sets of size $n - 1$. Applying the lemma inductively, fix $C \in [E]^\kappa$ that forms a Δ-system with some root r. Then $\{Z \cup \{p\} : Z \in C\} \in [\mathcal{D}]^\kappa$ forms a Δ-system with root $r \cup \{p\}$.
   - Case II: $|\mathcal{D}_p| < \kappa$ for all p. Then, for any set S with $|S| < \kappa$, $\{X \in \mathcal{D} : X \cap S \neq \emptyset\} = \bigcup_{p \in S} \mathcal{D}_p$ has size less than κ, since κ is regular; thus, there is an $X \in \mathcal{D}$ such that $X \cap S = \emptyset$. Then, by recursion on β, we may choose $X_\beta \in \mathcal{D}$ for $\beta < \kappa$ so that for each β, $X_\beta \cap \bigcup_{\alpha < \beta} X_\alpha = \emptyset$. But then $\{X_\beta : \beta < \kappa\}$ is a Δ-system with empty root.

The proof is completed.   □

Corollary 5.3.3 *Assume $X \subseteq [\omega_1]^{<\omega}$ has cardinality $\aleph_1$. Then there are $Y \subseteq X$ (with $|Y| = \omega_1$) and $r \in [\omega_1]^{<\omega}$ such that $\forall a, b \in Y (a \cap b = r)$.*

We can now prove the following:

Proposition 5.3.4 *For every set X, $\mathrm{Fn}(X, 2)$ has the CCC.*

First of all remark the following:

Fact 5.3.5 *Assume $f : X \to Y$ is a bijection. Then $\hat{f} : \mathrm{Fn}(X, 2) \to \mathrm{Fn}(Y, 2)$ is an isomorphism of quasi-orders, where for any $s \in \mathrm{Fn}(X, 2)$ $\hat{f}(s)$ is the sequence with domain $f[\mathrm{dom}(s)]$ such that $\hat{f}(s)(y) = s \circ f^{-1}(y)$ for all y in its domain.*

Proof A useful exercise for the reader.   □

Proof In view of the above fact, it is enough to show that $\mathrm{Fn}(\kappa, 2)$ has the CCC for all cardinals κ. If $\kappa \leq \aleph_0$, we are done, since $\mathrm{Fn}(\kappa, 2)$ is countable in this case, so we suppose $\kappa > \omega$. Take $\{s_\alpha : \alpha < \omega_1\} \subseteq \mathrm{Fn}(\kappa, 2)$ with $s_\alpha \neq s_\beta$ if $\alpha \neq \beta$. We claim that there are at least two compatible elements in $\{s_\alpha : \alpha < \omega_1\}$. First, we find a set $X \subseteq \kappa$ such that $|X| \leq \aleph_1$ and $\text{dom}(s_\alpha) \subseteq X$ for any $\alpha < \omega_1$. Let, for any $\alpha < \omega_1$,

$$\text{dom}(s_\alpha) = \{\beta_0^\alpha, \ldots, \beta_{k_\alpha}^\alpha\},$$

with k_α less than ω and $\beta_i^\alpha < \beta_j^\alpha$ for $i < j$. Let

$$X = \{\beta_j^\alpha : \alpha < \omega_1 \ \wedge \ j \leq \kappa_\alpha\}.$$

Notice that $|X| \leq \aleph_1$ and $\text{dom}(s_\alpha) \subseteq X$ for any $\alpha < \omega_1$. We have to distinguish two cases:

1. $|X| \leq \aleph_0$. We will prove that this case leads to a contradiction. For all $r \in [X]^{<\omega}$, let $Z_r = \{s_\alpha : \text{dom}(s_\alpha) = r\}$. Obviously $Z_r \subseteq 2^r$ and $|2^r| = 2^{|r|} < \omega$. Thus $\forall r \in [X]^{<\omega}(|Z_r| < \omega)$. We have that

$$\{s_\alpha : \alpha < \omega_1\} = \bigcup_{r \in [X]^{<\omega}} Z_r.$$

But $|\bigcup_{r \in [X]^{<\omega}} Z_r| \leq \aleph_0$, since the Z_r's are finite and $[X]^{<\omega}$ is countable. However an uncountable set cannot be equal to a countable one, so we reached a contradiction.

2. $|X| = \aleph_1$. For all α, let

$$s_\alpha = \{(\beta_0^\alpha, i_0^\alpha), \ldots, (\beta_{n-1}^\alpha, i_{n_\alpha - 1}^\alpha)\}$$

all listed so that $\beta_i^\alpha < \beta_j^\alpha$ for $i < j < n_\alpha$, and define

$$t_\alpha = \langle i_j^\alpha : j < n_\alpha \rangle.$$

Then each $t_\alpha \in 2^{<\omega}$ belongs to a countable set, and the map $\alpha \mapsto t_\alpha$ has countable range. By regularity of ω_1, we can select an uncountable $W \subseteq \omega_1$ and a $t \in 2^n \subseteq 2^{<\omega}$ such that $t_\alpha = t$ for all $\alpha \in W$.

Remark that for all $\alpha \in W$ it also holds that $\text{dom}(s_\alpha) \in [X]^n$, i.e. $|s_\alpha| = n$. Now define

$$\mathcal{D} = \{\text{dom}(s_\alpha) : \alpha \in W\}.$$

We claim that $|\mathcal{D}| = \aleph_1$. To this aim, consider the function $\varphi : W \to \mathcal{D}, \alpha \mapsto \text{dom}(s_\alpha)$. φ is injective, since if $\text{dom}(s_\alpha) = \text{dom}(s_\gamma) = \{\beta_0, \ldots, \beta_{n-1}\}$, then

$$s_\alpha = \{(\beta_0, t(0)), \ldots, (\beta_{n-1}, t(n-1))\} = s_\gamma,$$

thus $\alpha = \gamma$. Thus we can apply the Δ-system lemma to $\mathcal{D}$, and we obtain a set $\mathcal{B} \subseteq \mathcal{D}$ of size $\aleph_1$ and a root $r = \{\beta_{i_0}, \ldots, \beta_{i_k}\} \in [X]^k$ for some $k < n - 1$ listed in increasing order such that (defining $Z = \varphi^{-1}[\mathcal{B}]$) for all $\alpha \neq \gamma \in Z$ $\mathrm{dom}(s_\alpha) \cap \mathrm{dom}(s_\gamma) = r$.

Then for all $\alpha \neq \gamma \in Z$, $s_\alpha \cup s_\gamma$ is a function: For any β_{i_j} in the common domain $r = \mathrm{dom}(s_\alpha) \cap \mathrm{dom}(s_\gamma)$, $s_\alpha(\beta_{i_j}) = t(i_j) = s_\gamma(\beta_{i_j})$.

$\square$

Corollary 5.3.6 *The boolean algebra* $\mathsf{RO}(2^{\omega_2 \times \omega})$ *has the CCC.*

Proof By Lemma 4.3.2, $P = \mathrm{Fn}(\omega_2 \times \omega, 2)$ embeds as a dense suborder of $\mathsf{RO}(2^{\omega_2 \times \omega})$ via the map $s \mapsto N_s$. In particular $\mathsf{RO}(P)$ and $\mathsf{RO}(2^{\omega_2 \times \omega})$ are isomorphic boolean algebras, by Theorem 4.2.4. We conclude that $\mathsf{RO}(2^{\omega_2 \times \omega})$ is CCC using Lemma 4.3.2 for $\mathsf{RO}(P)$.

$\square$

Chapter 6
Boolean Valued Models

This chapter consists of three sections:

1. In the first section we give the formal definition of boolean semantic for any first order language, and we present the soundness theorem for the semantic for the language of set theory. The boolean valued semantic selects a given boolean algebra B and assigns to every statement ϕ a boolean value in B. The boolean operations will reflect the behavior of the propositional connectives; it will require more attention to give a meaning to atomic formulae and to quantifiers, and we need that B has a high degree of completeness in order to be able to interpret quantifiers in boolean semantics. The standard Tarski semantics will be recovered when we choose the boolean algebra $\{0, 1\}$ as B.

2. The second section carves a bit more into the theory of B-valued models M and their Tarski quotient $M/_G$ induced by an ultrafilter $G \in \mathrm{St}(\mathsf{B})$. We supply some guiding examples of such models, among which we analyze the space of analytic functions over the real numbers $C^\omega(\mathbb{R})$. We show that this is a boolean valued model which is not properly behaving; this will lead us to the key property of *fullness*.

3. In the third section we state a necessary and sufficient condition (that of being a *full* model) on a B-valued model M which gives a complete control on how truth in $M/_G$ is determined by the topological properties of G as a point of $\mathrm{St}(\mathsf{B})$ via a Łoś theorem for full boolean valued models. We also prove a version of the forcing theorem relating the boolean value of a formula ϕ in a B-valued model M to the topological density of the family of G such that $M/_G \models \phi$. We then provide three interesting distinct examples of full boolean valued models and obtain that Łoś theorem for ultraproducts $\prod_{x \in X} M_x/_G$ of Tarski models M_x by an ultrafilter G on $\mathcal{P}(X)$ is a special application of the Łoś theorem for full $\mathcal{P}(X)$-valued models. We also introduce Cohen's forcing relation on a B-valued model M and compare it to the B-valued semantics for M.

© The Author(s), under exclusive license to Springer Nature Switzerland AG 2024   81
M. Viale, *The Forcing Method in Set Theory*, La Matematica per il 3+2 168,
https://doi.org/10.1007/978-3-031-71660-7_6

All over this chapter we assume the reader familiar with the basics of first order logic, namely the soundness and completeness theorem for Tarski semantics and—in the last part of the chapter—familiarity with the ultrapower construction and with Łoś theorem for Tarski semantics.

6.1 Boolean Valued Models and Boolean Valued Semantics

In this section we give the formal definition of a boolean valued model for any first order *relational* language (i.e., a language containing no function symbols), and we introduce a sound and complete boolean semantics for these languages with respect to first order calculus.

We limit ourselves to analyze relational languages to avoid some technicalities arising in the semantical interpretation of function symbols in boolean valued models.

Assume we have fixed an $\mathcal{L}$-structure $\mathfrak{M}$; its Tarski semantic can be seen as a function that takes an $\mathcal{L}$-statement φ and assigns 1 or 0 to φ according to the fact that $\mathfrak{M} \vDash \varphi$ or $\mathfrak{M} \nvDash \varphi$. We generalize this framework letting the evaluation function be defined on arbitrary B-valued models and assign its values inside (the boolean completion of) B.

This brings to the following:

Definition 6.1.1 Let $\mathcal{L} = \{R_i : i \in I, c_j : j \in J\}$ be a language with no function symbol (a relational language in the sequel) and B a Boolean algebra. A B-*valued model* $\mathfrak{M}$ for $\mathcal{L}$ consists of:

1. A non-empty set M
2. The Boolean interpretation of the equality symbol, that is, a function

$$=^{\mathfrak{M}} : M^2 \longrightarrow \mathsf{B}$$

$$\langle \tau, \sigma \rangle \longmapsto [\![\tau = \sigma]\!]_{\mathsf{B}}^{\mathfrak{M}}$$

3. The interpretation of symbols in $\mathcal{L}$, that is:

   - For each n-ary relation symbol $R \in \mathcal{L}$, a function

$$R^{\mathfrak{M}} : M^n \longrightarrow \mathsf{B}$$

$$\langle \tau_1, \ldots, \tau_n \rangle \longmapsto [\![R(\tau_1, \ldots, \tau_n)]\!]_{\mathsf{B}}^{\mathfrak{M}} .$$

   - For each constant symbol $c \in \mathcal{L}$, a name $c^{\mathfrak{M}} \in M$.

We require that the following conditions hold:

1. For all $\tau, \sigma, \pi \in M$,

$$[\![\tau = \tau]\!]_{\mathsf{B}}^{\mathfrak{M}} = \mathbb{1}, \tag{6.1}$$

$$[\![\tau = \sigma]\!]_{\mathsf{B}}^{\mathfrak{M}} = [\![\sigma = \tau]\!]_{\mathsf{B}}^{\mathfrak{M}}, \tag{6.2}$$

$$[\![\tau = \sigma]\!]_{\mathsf{B}}^{\mathfrak{M}} \wedge [\![\sigma = \pi]\!]_{\mathsf{B}}^{\mathfrak{M}} \leq [\![\tau = \pi]\!]_{\mathsf{B}}^{\mathfrak{M}}. \tag{6.3}$$

2. If $R \in \mathcal{L}$ is an n-ary relation symbol, for all $\langle \tau_1, \ldots, \tau_n \rangle, \langle \sigma_1, \ldots, \sigma_n \rangle \in M^n$,

$$\left(\bigwedge_{i=1}^{n} [\![\tau_i = \sigma_i]\!]_{\mathsf{B}}^{\mathfrak{M}} \right) \wedge [\![R(\tau_1, \ldots, \tau_n)]\!]_{\mathsf{B}}^{\mathfrak{M}} \leq [\![R(\sigma_1, \ldots, \sigma_n)]\!]_{\mathsf{B}}^{\mathfrak{M}}. \tag{6.4}$$

We now extend the above semantics for atomic formulae to arbitrary first order formulae. The boolean operations give a natural interpretation for the boolean connectives; however to deal with the semantics of quantifiers, we need to evaluate the formulae in $\mathsf{RO}(\mathsf{B})$ rather than B. We will see that only a certain amount of completeness on B is needed to assign a correct truth value to all formulae.

Following standard patterns for Tarski semantics, we adopt the following strategy to define the boolean valued semantics for first order logic:

Definition 6.1.2 Let $\mathfrak{M} = \langle M, =_M, R_i^{\mathfrak{M}} : i \in I, c_j^{\mathfrak{M}} : j \in J \rangle$ be a B-valued model for the relational language $\mathcal{L} = \{ R_i : i \in I, c_j : j \in J \}$.

Let:

- $\mathcal{L}^*$ be the expansion of $\mathcal{L}$ with constants c_a for all $a \in M \setminus \left\{ c_j^{\mathfrak{M}} : j \in J \right\}$.

- $\mathfrak{M}^*$ be the expansion of $\mathfrak{M}$ which assigns a to each new constant c_a.[1]

We identify B as a dense[2] subalgebra of $\mathsf{RO}(\mathsf{B})$ and evaluate in $\mathsf{RO}(\mathsf{B})$ the $\mathcal{L}_M$-sentences as follows:

- $[\![R_i(c_{j_1}, \ldots, c_{j_k}, c_{a_1}, \ldots, c_{a_n})]\!]_{\mathsf{RO}(\mathsf{B})}^{\mathfrak{M}^*} = R_i^{\mathfrak{M}}(c_{j_1}^{\mathfrak{M}}, \ldots, c_{j_k}^{\mathfrak{M}}, a_1, \ldots, a_n)$.
- $[\![\varphi \wedge \psi]\!]_{\mathsf{RO}(\mathsf{B})}^{\mathfrak{M}^*} = [\![\varphi]\!]_{\mathsf{RO}(\mathsf{B})}^{\mathfrak{M}^*} \wedge_{\mathsf{RO}(\mathsf{B})} [\![\psi]\!]_{\mathsf{B}}^{\mathfrak{M}^*}$.
- $[\![\neg \varphi]\!]_{\mathsf{B}}^{\mathfrak{M}^*} = \neg_{\mathsf{B}} [\![\varphi]\!]_{\mathsf{B}}^{\mathfrak{M}^*}$.
- $[\![\exists x \varphi(x, c_{\bar{a}})]\!]_{\mathsf{RO}(\mathsf{B})}^{\mathfrak{M}^*} = \bigvee_{b \in M} [\![\varphi(c_b, c_{\bar{a}})]\!]_{\mathsf{RO}(\mathsf{B})}^{\mathfrak{M}^*}$.

---

[1] To avoid clashes in interpretations, we may assume J and M are disjoint sets.

[2] B^+—seen as a partial order—is a dense subset of $\mathsf{RO}(\mathsf{B})^+$ (by Corollary 4.2.7, since $(\mathsf{B}^+, \leq)$ is a separative partial order).

If $\phi(x_1, \ldots, x_n)$ is a formula of $\mathcal{L}$ with free variables $x_1, \ldots, x_n$ and ν is an assignment, we let

$$\nu(\phi(x_1, \ldots, x_n)) = [\![\phi(c_{\nu(x_1)}, \ldots, c_{\nu(x_n)})]\!]^{\mathfrak{M}^*}_{\mathsf{RO}(\mathsf{B})} \, .$$

We also denote oftentimes $\nu(\phi(x_1, \ldots, x_n))$ by $[\![\varphi]\!]^{\mathfrak{M},\nu}_{\mathsf{B}}$.

$\mathfrak{M}$ is a *well behaved* B-valued model if $[\![\phi]\!]^{\mathfrak{M}^*}_{\mathsf{RO}(\mathsf{B})} \in \mathsf{B}$ for all $\mathcal{L}_M$-sentences ϕ.

To simplify notation we shall confuse from now on the constant symbol $c_a \in \mathcal{L}_M$ with its intended interpretation $a \in M$; furthermore we may assume that the constant symbols of $\mathcal{L}$ are redundant, for example, by assuming that $\mathcal{L}$ has no constant symbols and that all constants are those appearing in $\mathcal{L}_M$. When working with well behaved B-valued models, we write henceforth $[\![\phi]\!]^{\mathfrak{M}^*}_{\mathsf{B}}$ rather than $[\![\phi]\!]^{\mathfrak{M}^*}_{\mathsf{RO}(\mathsf{B})}$. We also feel free to omit subscripts and superscripts and to freely identify $\mathfrak{M}$ with $\mathfrak{M}^*$ if no confusion on the intended meaning can arise.

Remark 6.1.3 Some comments are:

- If $\mathsf{B} = \{0, 1\}$, the semantics we have just defined is the usual Tarski semantic for first order logic.
- Clearly consistent definitions of $[\![\phi \vee \psi]\!]$ and $[\![\phi \rightarrow \psi]\!]$ can be given in terms of $[\![\neg \varphi]\!]$ and $[\![\varphi \wedge \psi]\!]$. Also $[\![\forall x \varphi(x, \bar{a})]\!]$ can be defined in terms of $[\![\neg \varphi]\!]$ and $[\![\exists x \varphi(x, \bar{y})]\!]$.
- The atomic formulae take values in B; boolean combinations of formulae whose evaluations are in B are also in B.
  On the other hand $[\![\exists x \varphi(x, \bar{a})]\!]^{\mathfrak{M}^*}_{\mathsf{RO}(\mathsf{B})}$ and $[\![\forall x \varphi(x, \bar{a})]\!]^{\mathfrak{M}^*}_{\mathsf{RO}(\mathsf{B})}$ may take values possibly not in B even if $[\![\varphi(c, \bar{a})]\!]^{\mathfrak{M}^*}_{\mathsf{RO}(\mathsf{B})} \in \mathsf{B}$ for all $c \in M$. This motivates the definition of well behaved boolean valued model.

6.1.1   Soundness and Completeness for Boolean Valued Semantics

We show that the semantics we just defined is a natural generalization of Tarski semantics which is sound and complete with respect to first order calculus.

Definition 6.1.4 A sentence φ in the language $\mathcal{L}$ is *valid* in a B-valued model $\mathfrak{M}$ for $\mathcal{L}$ if $[\![\varphi]\!] = 1_{\mathsf{B}}$. A theory T is valid in $\mathfrak{M}$ if every axiom $\varphi \in T$ is valid.

We recall the following deductive system for first order logic which is taken (with slight modifications) from [28, Section 2.6].[3]

Axioms

1. $x = x$.
2. $\varphi(a) \to \exists x \varphi(x)$.
3. $x = y \to [\varphi(x) \to \varphi(y)]$.

Rules

4. $\varphi \vdash \varphi \vee \psi$.
5. $\varphi \vee \varphi \vdash \varphi$.
6. $(\varphi \vee (\psi \vee \chi)) \vdash ((\varphi \vee \psi) \vee \chi)$.
7. $\varphi \vee \psi, \neg\varphi \vee \chi \vdash \psi \vee \chi$.
8. $\forall x(\varphi(x) \to \psi) \vdash (\exists x \varphi(x)) \to \psi$ (with x a variable that does not occur free in ψ).

Recall that a formula ϕ is *syntactically provable* from a theory T, if there is a finite list of formulae such that ϕ is the last appearing in the list and each formula in the list is either an axiom or is a formula in T or is deduced by one of the above rules applied to some of the preceding formulae in the list.

Theorem 6.1.5 (Soundness and Completeness for Boolean Valued Semantics) *Let $\mathcal{L}$ be a relational first order language. The following are equivalent for $\mathcal{L}$-formula φ:*

1. *φ is provable syntactically by an $\mathcal{L}$-theory T.*
2. *For all boolean algebras B, for every assignment $v : FRV(\mathcal{L}) \to M$ on a B-valued model $\mathfrak{M}$ for $\mathcal{L}$, and for every $\psi \in T$,*

$$v(\varphi) \geq v(\psi).$$

The completeness part of the theorem is automatic, since (as we already observed) the Tarski models are a subfamily of the boolean valued models.

On the other hand the soundness part of the theorem requires a proof, which will be given in the rest of this section.

Proof Fix $\mathfrak{M}$ a B-valued model for $\mathcal{L}$ with domain M.

Regarding the axioms, we prove that for all assignments $v : FRV \to M$, and all axioms ϕ in the above list $\llbracket \phi \rrbracket_{\mathsf{B}}^{\mathfrak{M},v} = 1_{\mathsf{B}}$.

---

[3] In listing the axioms below, we adopt the convention that:
- $\phi(x)$ denotes a formula ϕ with displayed free variable x.
- $\phi(a)$ is the uniform substitution of the free variable x with the constant symbol a in the formula ϕ.
- $\phi(y)$ is the uniform substitution of the free variable x with the free variable y in the formula ϕ.

Regarding the rules, we prove for any rule that for all assignments $v : \mathrm{FRV} \to M$

$$\varphi \vdash \psi \Rightarrow [\![\varphi]\!]_{\mathsf{B}}^{\mathfrak{M},v} \le [\![\psi]\!]_{\mathsf{B}}^{\mathfrak{M},v}.$$

The proof is rather straightforward, and we sketch some of its parts. Given $v : \mathrm{FRV} \to M$, we let $v_{x/b}$ be obtained from v by changing $v(x)$ to b and leaving it unchanged on all other free variables.

1. $x = x$. It follows by the definition of boolean valued model that

$$[\![a = a]\!] = 1_{\mathsf{B}}$$

for all $a \in M$. We thus get that $[\![x = x]\!]_{\mathsf{B}}^{\mathfrak{M},v} = 1_{\mathsf{B}}$ for all valuations v and free variables x.

2. $\varphi(a) \to \exists x \varphi(x)$. We have by definition

$$[\![\exists x \varphi(x)]\!] = \bigvee_{b \in M} [\![\varphi(b)]\!] \ge [\![\varphi(a)]\!],$$

so we conclude using Exercise 3.3.3.

3. $x = y \to [\varphi(x) \to \varphi(y)]$. By Exercise 3.3.3 it is sufficient to show that

$$[\![a = b]\!] \le [\![\varphi(x)]\!]_{\mathsf{B}}^{\mathfrak{M},v_{x/a}} \leftrightarrow [\![\varphi(x)]\!]_{\mathsf{B}}^{\mathfrak{M},v_{x/b}}$$

or equivalently

$$[\![a = b]\!] \wedge [\![\varphi(a)]\!] = [\![a = b]\!] \wedge [\![\varphi(b)]\!]$$

for all $a, b \in M$. This is proved by induction on the complexity of φ. Consider in what follows:

- The assignments v and v' defined by

$$v = (x_1 \mapsto a_1, \ldots, x_{i-1} \mapsto a_{i-1}, x_i \mapsto a, x_{i+1} \mapsto a_{i+1} \ldots, x_n \mapsto a_n),$$

$$v' = v_{x_i/b}$$

- The formula $\psi(x_1, \ldots, x_n)$ in displayed free variables

For the sake of notational simplicity, we write $[\![\psi(v)]\!]$ rather than $[\![\psi]\!]_{\mathsf{B}}^{\mathfrak{M}^*,v}$.

Atomic formulae: This immediately follows by the definition of boolean valued model.

Negation: If $\phi \equiv \neg\psi$, by induction we have

$$[\![a = b]\!] \wedge [\![\psi(v)]\!] = [\![a = b]\!] \wedge [\![\psi(v')]\!],$$

which clearly holds if and only if

$$[\![a = b]\!] \wedge [\![\neg\psi(v)]\!] = [\![a = b]\!] \wedge [\![\neg\psi(v')]\!] .$$

Conjunction: If $\phi \equiv \psi \wedge \theta$, we have

$$[\![a = b]\!] \wedge [\![\phi(v)]\!] = ([\![a = b]\!] \wedge [\![\psi(v)]\!]) \wedge ([\![a = b]\!] \wedge [\![\theta(v)]\!])$$

$$= ([\![a = b]\!] \wedge [\![\psi(v')]\!]) \wedge ([\![a = b]\!] \wedge [\![\theta(v')]\!]) = [\![a = b]\!] \wedge [\![\phi(v')]\!] .$$

Existential: If $\phi(x_1, \ldots, x_n) \equiv \exists y \psi(y, x_1, \ldots, x_n)$, we have that

$$[\![a = b]\!] \wedge [\![\phi(v)]\!] = \bigvee_{c \in M} ([\![\psi(y/c, v)]\!] \wedge [\![a = b]\!])$$

$$= \bigvee_{c \in M} ([\![\psi(y/c, v')]\!] \wedge [\![a = b]\!])$$

$$= [\![\phi(v')]\!] \wedge [\![a = b]\!] .$$

4. $\varphi \vdash \varphi \vee \psi$. Immediate since $u \vee v \geq u$ for all $u, v \in \mathsf{B}$.
5. $\varphi \vee \varphi \vdash \varphi$. Immediate since $u \vee u = u$ for all $u \in \mathsf{B}$.
6. $(\varphi \vee (\psi \vee \chi)) \vdash ((\varphi \vee \psi) \vee \chi)$. Immediate since $[\![(\varphi \vee (\psi \vee \chi))]\!] = [\![((\varphi \vee \psi) \vee \chi)]\!]$.
7. $\varphi \vee \psi, \neg\varphi \vee \chi \vdash \psi \vee \chi$.
   This follows easily from Exercise 3.3.4 on boolean algebras:
8. $\forall x (\varphi(x) \rightarrow \psi) \vdash (\exists x \varphi(x)) \rightarrow \psi$.

$$[\![\forall x (\varphi(x) \rightarrow \psi)]\!] = [\![\forall x (\neg\varphi(x) \vee \psi)]\!] = \bigwedge_{b \in M} ([\![\neg\varphi(b)]\!] \vee [\![\psi]\!]) =$$

(using the fact that x is not free in ψ)

$$= (\bigwedge_{b \in M} [\![\neg\varphi(b)]\!]) \vee [\![\psi]\!] = (\neg \bigvee_{b \in M} [\![\varphi(b)]\!]) \vee [\![\psi]\!] = [\![(\exists x \varphi(x)) \rightarrow \psi]\!] .$$

The proof is complete.   □

6.1.2   Boolean Morphisms

For our analysis of boolean valued semantics, it is now convenient to introduce briefly the definition of morphism between B-valued models:

Definition 6.1.6 Fix a relational language $\mathcal{L} = \{R_i : i \in I, c_j : j \in J\}$. Let $k : B \to C$ be a homomorphism of boolean algebras. Let $\mathfrak{M}$ be a well behaved B-valued model for $\mathcal{L}$ with domain M and $\mathfrak{N}$ be a well behaved C-valued model for $\mathcal{L}$ with domain N.

- $\bar{k} : M \to N$ is a k-*morphism* if for all $R \in \mathcal{L}$ of arity n and $a_1, \ldots, a_n \in M$

$$\llbracket R\,(\bar{k}(a_1), \ldots, \bar{k}(a_n)) \rrbracket^{\mathfrak{N}}_C \geq k(\llbracket R\,(a_1, \ldots a_n) \rrbracket^{\mathfrak{M}}_B),$$

and for all $a, b \in M$

$$\llbracket \bar{k}(a) = \bar{k}(b) \rrbracket^{\mathfrak{N}}_C \geq k(\llbracket a = b \rrbracket^{\mathfrak{M}}_B).$$

- $\bar{k} : M \to N$ is a k-*embedding* if all the above inequalities are reinforced to equalities.
- $\bar{k}$ is a k-*isomorphism* if k is an isomorphism, and for all $b \in N$ there is $a \in M$ such that

$$\llbracket \bar{k}(a) = b \rrbracket^{\mathfrak{N}}_C = 1_C.$$

Exercise 6.1.7 Show that an Id_2-morphism (for $Id_2 : 2 \to 2$ the identity map) is a morphism of Tarski models in the classical sense, and similarly for Id_2-embeddings and Id_2-isomorphisms.

6.2   A Discussion on Boolean Valued Semantics

We now introduce informally the main ideas behind the forcing method making an excursion in other areas of mathematics and borrowing our language and terminology from analysis and sheaf theory.

Recall that a function $f : \mathbb{R} \to \mathbb{R}$ is *analytic* in $\mathbb{R}$ if and only if for every $x_0 \in \mathbb{R}$

$$f(x) = \sum_{k=0}^{\infty} \frac{f^{(k)}(x_0)}{k!} (x - x_0)^n.$$

Let $C^\omega(\mathbb{R})$ be the set of all analytic functions over $\mathbb{R}$. Let $B = RO(\mathbb{R})$ the boolean algebra of regular open sets of the real line.

We aim to see $C^\omega(\mathbb{R})$ as a B-*boolean valued extension* of $\mathbb{R}$ which naturally contains $\mathbb{R}$ as a substructure.

The consideration we are going to make will apply equally well for much larger classes of real-valued functions and for other complete boolean algebras than B (e.g., MALG and[4] $L^\infty(\mathbb{R})$ could work equally well in the discussion to follow, mutatis mutandis); however $C^\omega(\mathbb{R})$ and B give a first example of an instructive case in which central concepts of boolean valued semantics display their nontrivial content.

Consider the dense linear order $(\mathbb{R}, <)$. We focus on the analysis of the boolean semantics for boolean expansions of the above structure. This will give an idea of how we can employ boolean valued models to enlarge the domain of certain given first order structures. We will do this with respect to the relational language $\mathcal{L} = \{<\} \cup \mathbb{R}$ with one binary relation symbol and constants for all real numbers.

First of all we need to say what is the boolean value that $C^\omega(\mathbb{R})$ gives to the formula $x < y$ when $x \mapsto f$ and $y \mapsto g$. A natural answer is the following:

$b \in$ B *forces* that $f < g$ if the set of $x \in b$ such that $f(x) < g(x)$ is an open dense subset of b (recall that b is a regular open subset of $\mathbb{R}$).

For example, let $f(x) = \sin(x)$ and $g(x) = -1$ for all x; then $\mathbb{R}$ forces that $g < f$ since the set of points $x \in \mathbb{R}$ on which $f(x) \le g(x)$ is closed and nowhere dense. On the other hand, if $f(x) = \sin(x)$ and $g(x) = \cos(x)$, we have that $f(x) < g(x)$ if and only if $x \in (\pi/4 + 2k \cdot \pi, 5 \cdot \pi/4 + 2k\pi)$ as k ranges in $\mathbb{Z}$. Notice that the above set is open regular and thus

$$a = \bigcup_{k \in \mathbb{Z}} (\pi/4 + 2k\pi, 5/4\pi + 2k\pi)$$

is the largest regular open subset of $\mathbb{R}$ which forces $\sin(x) < \cos(x)$. Notice that the complement $\neg_B a$ in B of this set is exactly the set of points on which $\cos(x) < \sin(x)$ and that what is left out by $a \cup \neg_B a$ is the closed nowhere dense set of points in which $\sin(x) = \cos(x)$ which are the extremes of the intervals defining a.

Guided by this example we can now give an interpretation of the forcing relation and a precise meaning to quantifier free $\mathcal{L}$-formulae in $C^\omega(\mathbb{R})$ as follows:

- $[\![f \ R \ g]\!]$ is the largest regular open set a such that the set

$$\{x \in \mathbb{R} : f(x) \ R \ g(x)\}$$

  has an open dense intersection with a for R any relation among $<, =$.

- $[\![\phi \wedge \psi]\!] = [\![\phi]\!] \cap [\![\psi]\!]$.
- $[\![\neg\phi]\!] = \neg_B [\![\phi]\!]$.

Now we are left to see that $\mathbb{R}$ can be copied inside $C^\omega(\mathbb{R})$. The natural idea is that $\mathbb{R}$ is "represented inside $C^\omega(\mathbb{R})$" by the constant functions $c_a(x) = a$ for all $a \in \mathbb{R}$. Indeed we can check that $a \ R \ b$ holds in $\mathbb{R}$ iff $[\![c_a \ R \ c_b]\!] = \mathbb{R}$ for any binary relation R among $<, =$. So we get that essentially any of the above relations holds on two

---

[4] $L^\infty(\mathbb{R})$ denotes the essentially bounded measurable functions; see also Example 6.4.1 to follow for more details.

real numbers iff $\mathbb{R}$ forces the corresponding relation to hold for the corresponding constant functions. An instructive exercise for the reader is the following:

Exercise 6.2.1 The following holds:

- $\mathfrak{M} = \langle C^\omega(\mathbb{R}), (f, g) \mapsto [\![f < g]\!], (f, g) \mapsto [\![f = g]\!], c_a : a \in \mathbb{R} \rangle$ is a B-valued model.
- The map $a \mapsto c_a$ defines an i^*-morphism of the 2-valued model $(\mathbb{R}, <, a \in \mathbb{R})$ for the language $\mathcal{L}$ in the B-valued model $\mathfrak{M}$, where $i^* : 2 \to B$ is the unique complete homomorphism and has to map $j \mapsto j_B$ for $j = 0, 1$.

Finally we want to show that this boolean expansion of $\mathbb{R}$ is a proper boolean superstructure of $\mathbb{R}$: A natural way to say this is to find some function which is forced by 1_B to be different from all constant functions. It is easily seen that the sinus function or any analytic function which is nowhere locally constant has this property. In particular our $\mathsf{RO}(\mathbb{R})$-boolean expansion $C^\omega(\mathbb{R})$ of $\mathbb{R}$ appears to have added many new elements with respect to $\mathbb{R}$.

This describes the passage from a first order structure M to an associated boolean valued model M^B, which in this case is given by the analytic functions on $\mathbb{R}$.

However there is a disturbing issue of this boolean expansion, i.e., that we are not able to decide many basic facts, for example, is $\sin(x) < \cos(x)$? We have already seen that the boolean values $[\![\sin(x) < \cos(x)]\!]$ and $[\![\sin(x) > \cos(x)]\!]$ are both positive, while $[\![\sin(x) = \cos(x)]\!] = 0_B$.

In particular the boolean expansion already carries enough information to decide whether $\sin(x)$ and $\cos(x)$ represent different objects but is not yet able to decide whether $\sin(x) < \cos(x)$ or the other way round. If we choose to restrict our attention to a small interval like $(\pi/4, 5/4 \cdot \pi)$, this interval will *force* that $\sin(x) < \cos(x)$, but it will not yet be able to decide other basic relations among other functions, for example, whether $\sin(2x) < \cos(2x)$. Making our interval smaller and smaller, we end up "forcing" more and more properties regarding the mutual relationship between functions in $C^\omega(\mathbb{R})$.

It may seem that if we take a decreasing sequence of intervals $\{I_n : n \in \mathbb{N}\}$ with diameter converging to 0 and such that $\mathsf{Cl}(I_{n+1}) \subseteq I_n$, in the limit the unique point $x \in \bigcap_n I_n$ will be able to decide all basic relations among the analytic functions.

This is not yet the case though: For example, no open neighborhood of $\pi/4$ *forces* $\sin(x) < \cos(x)$ and no open neighborhood of $\pi/4$ *forces* $\sin(x) > \cos(x)$. So actually in order to be able to decide all basic relations on the elements of $C^\omega(\mathbb{R})$, it is not enough to select a point and look at the filter given by its regular open neighborhoods.

On the other hand, if we select an ultrafilter G on $\mathsf{RO}(\mathbb{R})$, since for all $f, g \in C^\omega(\mathbb{R})$ we have that

$$[\![f < g]\!] \vee [\![f > g]\!] \vee [\![f = g]\!] = 1_B,$$

we will have that any $G \in \mathsf{St}(B)$ will always pick exactly one among the three values.

We say that a filter G *decides* that ϕ holds when $[\![\phi]\!] \in G$.

According to this terminology, an ultrafilter G decides whether $f < g$, $f = g$, or $f > g$ holds simultaneously for all $f, g \in C^\omega(\mathbb{R})$.

Note that any filter F on $\mathbf{B}$ decides that $\sin(x) \neq c_a$ for any real number a, so indeed $\mathfrak{M}$ is a proper $\mathbf{B}$-extension of $\mathbb{R}$ containing "new" elements.

Clearly any $G \in \mathrm{St}(\mathsf{RO}(\mathbb{R}))$ will always choose a unique point $x_G \in \mathbb{R} \cup \{\pm\infty\}$ which will be the unique point in

$$\bigcap \{\mathsf{Cl}(A) : A \in G\}.$$

Note however that the same point can be associated with incompatible ultrafilters on $\mathsf{RO}(\mathbb{R})$: For example, let G be an ultrafilter extending the regular open neighborhoods of $\pi/4$ with $(-\infty; \pi/4)$ and H extend the same filter of neighborhoods with $(\pi/4; +\infty)$; then $x_G = x_H = \pi/4$.

Note that G decides that $\sin(x) < \cos(x)$ holds, H decides that $\sin(x) > \cos(x)$ holds, while the filter F of regular open neighborhoods of $\pi/4$ does not carry enough information to decide which of the two among $\sin(x) > \cos(x)$ and $\sin(x) < \cos(x)$ should be the case.

Summing up:

1. We have given:
   - A $\mathbf{B}$-valued model for $\mathcal{L}$ which extends the real line with new objects (the nonconstant analytic functions).
   - A forcing relation between properties of analytic functions and regular open intervals: For example, appropriately chosen intervals can decide whether $f < g$ or $f = g$ or $f > g$ holds for given analytic functions f, g.
   - An intuitive notion of what does it mean that a filter on $\mathbf{B}$ decides a basic fact about analytic functions.

2. Intervals (or even real numbers) are not carrying enough information to decide simultaneously all basic facts about analytic functions (e.g., the filter of open neighborhoods of $\pi/4$ does not decide whether $\sin(x) < \cos(x)$ holds or not).
3. Ultrafilters on $\mathsf{RO}(\mathbb{R})$ on the other hand make a coherent selection of decisions on which among $f < g$, $f = g$, $f > g$ holds simultaneously for any pair of analytic functions f, g.

A side observation for those familiar with the dual of $L^\infty(\mathbb{R})$ (see [31, Section 3] for more details): If one made the same example starting with MALG and $L^\infty(\mathbb{R})$, one would have ended up making these same considerations for $G \in \mathrm{St}(\mathsf{MALG})$. Note that (via the identification of MALG with the equivalence classes in $L^\infty(\mathbb{R})$ of the characteristic functions of measurable sets) $\mathrm{St}(\mathsf{MALG})$ is the set of extremal points in the dual of $L^\infty(\mathbb{R})$.

It is time to make a systematic and rigorous development of these considerations.

6.3   Quotients of Boolean Valued Models, Fullness, Łoś Theorem

This section explores the notion of quotient for a boolean valued model and characterizes by means of the fullness property the boolean valued models whose semantics behave properly with respect to quotients.

Definition 6.3.1 Let B be a boolean algebra, and let $\mathcal{L} = \{R_i : i \in I, c_j : j \in J\}$ be a relational language where R_i is an m_i-ary relation symbol for every $i \in I$ and each c_j is a constant symbol. Suppose that $\mathfrak{M} = (M, R_i : i \in I, c_j : j \in J)$ is a B-model for $\mathcal{L}$. Let F be a filter on B. The F-quotient $\mathfrak{M}/F = (M/F, R_i/F : i \in I, [c_j^{\mathfrak{M}}]_F : j \in J)$ is defined as follows:

- $[h]_F = \{f \in M : [\![f = h]\!] \in F\}$ for $h \in M$.
- $M/F = \{[h]_F : h \in M\}$.
- $R_i/F([f_1]_F, \ldots, [f_{m_i}]_F)$ holds if and only if $[\![R_i(f_1, \ldots, f_n)]\!] \in F$ for every $i \in I$.

When G is a ultrafilter on B, we say that $\mathfrak{M}/G$ is the *Tarski quotient of* $\mathfrak{M}$ *by* G.

Exercise 6.3.2 Let F be a filter on a boolean algebra B. Check that the F-quotient of a B-valued model is a well defined B/F-valued model; hence it is a Tarski model when F is an ultrafilter.

6.3.1   Examples of Quotients

The process we described in the previous section is rather flexible and can accommodate many first order structures defined on the domain $\mathbb{R}$ (or even on many other domains, as we shall see below). For example, we could repeat verbatim the same construction for the structure:

$$(\mathbb{R}, \mathbb{Z}, 0, 1, +, \cdot, <)$$

to obtain the $\mathsf{RO}(\mathbb{R})$-boolean expansion:

$$(C^\omega(\mathbb{R}), \mathbb{Z}_{\mathsf{RO}(\mathbb{R})}, c_0, c_1, +_\mathsf{B}, \cdot_\mathsf{B}, <),$$

where $+$ and $\cdot$ are interpreted by the ternary relations of their respective graphs, $+_\mathsf{B}$ is the ternary boolean relation

$$(f, g, h) = \mathsf{Reg}(\{x \in \mathbb{R} : f(x) + g(x) = h(x)\}),$$

and similarly for $\cdot_\mathsf{B}$. $\mathbb{Z}_\mathsf{B}$ is the predicate assigning to each $f \in C^\omega(\mathbb{R})$ the boolean value

$$\llbracket \mathbb{Z}_{\mathsf{RO}(\mathbb{R})}(f) \rrbracket = \mathsf{Reg}(\{x \in \mathbb{R} : f \restriction a \text{ is locally constant with value in } \mathbb{Z} \text{ and } a \ni x\}).$$

The latter predicate has either value $\mathbb{R}$ or $\emptyset$, in any case an open regular subset of $\mathbb{R}$. Here we use a specific property of analytic functions: An analytic function is constant if and only if it is locally constant in some open set of its domain.

We can also check that for all $G \in \mathsf{St}(\mathsf{RO}(\mathbb{R}))$

$$(C^\omega(\mathbb{R})/_G, \mathbb{Z}_\mathsf{B}/_G, [c_0]_G, [c_1]_G, +_\mathsf{B}/_G, \cdot_\mathsf{B}/_G, <_\mathsf{B}/_G)$$

is also an ordered ring with a distinguished predicate $\mathbb{Z}_\mathsf{B}/_G$.

We let the map $i_G^* : \mathbb{R} \to C^\omega(\mathbb{R})/_G$ be defined by $a \mapsto [c_a]_G$. Then it is not hard to check that i_G^* is an injective homomorphism of rings which preserves the order relation and is also such that $i_G^*[\mathbb{Z}] = \mathbb{Z}_\mathsf{B}/_G$.

Exercise 6.3.3 Prove in detail all the above facts about the structures

$$(C^\omega(\mathbb{R}), \mathbb{Z}_\mathsf{B}, 0, 1, +_\mathsf{B}, \cdot_\mathsf{B}, <_\mathsf{B})$$

and

$$(C^\omega(\mathbb{R})/_G, \mathbb{Z}_\mathsf{B}/_G, [0]_G, [1]_G, +_\mathsf{B}/_G, \cdot_\mathsf{B}/_G, <_\mathsf{B}/_G).$$

More precisely let $\mathsf{RO}(\mathbb{R})$ be the complete boolean algebra of regular open subset of $\mathbb{R}$, and show that the map $k : \mathbb{R} \to C^\omega(\mathbb{R})$ sending $a \mapsto c_a$ is an i^*-embedding of B-valued models, where $i^* : 2 \to \mathsf{B}$ is the unique embedding of 2 into B. Show also that $i_G^*(a) = [a]_G$ defines an injective morphism of ordered rings such that $i_G^*[\mathbb{Z}] = \mathbb{Z}_\mathsf{B}/_G$ (HINT: Notice that an analytic function is locally constant with value in $\mathbb{Z}$ iff it is everywhere constant with the same value; the left to right inclusion does not require this property, and the right to left inclusion does).

Exercise 6.3.4 Let $C(\mathbb{R})$ be the family of continuous real-valued functions and B be the cba given by the regular open sets of $\mathbb{R}$ with usual euclidean topology. Show that

$$(C(\mathbb{R}), \mathbb{Z}_\mathsf{B}, 0, 1, +_\mathsf{B}, \cdot_\mathsf{B}, <_\mathsf{B})$$

is a B-valued model (where the definitions of the additional predicates are the same as in the previous exercise but now apply to continuous functions rather than just analytic functions). Show also that for some (actually any) ultrafilter G on B, $i_G[\mathbb{Z}]$ is a proper subset of $\mathbb{Z}_\mathsf{B}/_G$ (HINT: For this strict inclusion note that if one chooses G an ultrafilter which concentrates on $\bigcup_{n \in \mathbb{Z}}(2n; 2n + 1)$ and chooses f to be locally constant on the interval $(2n; 2n + 1)$ with value n and a translate of the identity on the intervals $(2n + 1; 2n + 2)$, then $[f]_G \in \mathbb{Z}_\mathsf{B}/_G$, but f is not in $i_G[\mathbb{R}]$).

6.3.2 Counterexamples

We have no reasons to expect that a formula that is not quantifier free true in a B-valued model $\mathfrak{M}$ is also true in $\mathfrak{M}/_G$, for some $G \in \mathrm{St}(\mathsf{B})$. In general this is false, as the following example shows:

Example 6.3.5 Fix the language $\mathcal{L} = \{<, C\}$ consisting of two relation symbols, where $<$ is binary and C is unary. Let $\mathsf{B} = \mathrm{RO}(\mathbb{R})$, and consider the B-valued model for the language $\mathcal{L}$ given by $\mathfrak{M} = (C^\omega(\mathbb{R}), =, <_\mathsf{B}, C_\mathsf{B})$ with the following interpretation of the atomic formulae:

$$\llbracket f = g \rrbracket = \mathsf{Reg}(\{x \in \mathbb{R} : f(x) = g(x)\}),$$

$$\llbracket f <_\mathsf{B} g \rrbracket = \mathsf{Reg}(\{x \in \mathbb{R} : f(x) < g(x)\}),$$

$$\llbracket C_\mathsf{B}(f) \rrbracket = \mathsf{Reg}(\bigcup \{U : f \upharpoonright_U \text{ is constant}\}).$$

We leave to the reader to check that $(C^\omega(\mathbb{R}), =, <_\mathsf{B}, C_\mathsf{B})$ is a B-valued model. Now, fix any $f \in C^\omega(\mathbb{R})$, and look at the formula $\phi := \exists y \, (f < y \wedge C(y))$.

$$\llbracket \exists y \, (f <_\mathsf{B} y \wedge C_\mathsf{B}(y)) \rrbracket = \bigvee_{g \in C^\omega(\mathbb{R})} \llbracket f <_\mathsf{B} g \wedge C_\mathsf{B}(g) \rrbracket$$

$$\geq \bigvee_{a \in \mathbb{R}} \llbracket f <_\mathsf{B} c_a \rrbracket \wedge \llbracket C_\mathsf{B}(c_a) \rrbracket$$

$$\text{where } c_a \text{ is the constant function } c_a(x) = a$$

$$= \bigvee_{a \in \mathbb{R}} \llbracket f <_\mathsf{B} c_a \rrbracket$$

$$\geq \bigvee_{n \in \mathbb{Z}} \llbracket f < c_{a_n} \rrbracket \quad \text{where } a_n = \max(f \upharpoonright [n-1; n]) + 1$$

$$\geq \mathsf{Reg}(\bigcup_{n \in \mathbb{Z}} (n-1; n)) = \mathbb{R}.$$

Therefore, we have that $\mathfrak{M} \models \phi(f)$ and in particular

$$\mathfrak{M} \models \exists y \, (id_\mathbb{R} < y \wedge C(y)),$$

where $id_\mathbb{R}$ is the identity function $x \mapsto x$.

Now, consider $F = \{(a; +\infty) : a \in \mathbb{R}\} \subseteq \mathrm{RO}(\mathbb{R})$. Since F satisfies the finite intersection property (i.e., F is closed under intersection of finite subsets), we can extend F to some $G \in \mathrm{St}(\mathrm{RO}(\mathbb{R}))$. Consider the quotient $\mathfrak{M}/_G$. The identity function $id_{\mathbb{R}}$ has the property that for any $a \in \mathbb{R}$

$$[\![\neg(id_{\mathbb{R}} <_B c_a)]\!] = (a, +\infty) \in G.$$

It follows that $\mathfrak{M}/G \models \neg \exists y \, ([id_{\mathbb{R}}]_G < y \wedge C(y))$.

6.3.3 Łoś Theorem for Full Boolean Valued Models

Example 6.3.5 shows that quotients of boolean valued models may not preserve validity of formulae with quantifiers. To overcome this issue, we are led to the definition of full boolean valued models as those boolean valued models for which the above problem does not occur. We show that fullness characterizes the preservation of satisfiability in any quotient, and we give several examples of full boolean valued models.

Definition 6.3.6 Fix a language $\mathcal{L}$ and a boolean algebra B; a well behaved B-valued model $\mathfrak{M}$ for $\mathcal{L}$ is *full* if for every formula $\phi(x, \bar{y})$ and $\bar{f} \in M^{|y|}$ there are $h_1, \ldots, h_k \in M$

$$[\![\exists x \phi(x, \bar{f})]\!]_B^{\mathfrak{M}} = \bigvee_{i=1}^{k} [\![\phi(h_i, \bar{f})]\!]_B^{\mathfrak{M}} .$$

Theorem 6.3.7 (Łoś Theorem) *Let* B *be a boolean algebra. Assume* $\mathfrak{M}$ *is a full* B-*valued model for the language* $\mathcal{L}$. *For any* $G \in \mathrm{St}(B)$, $f_1, \ldots, f_n \in M$, *and for all formulae* $\phi(f_1, \ldots, f_n)$,[5]

$$\mathfrak{M}/G \models \phi([f_1]_G, \ldots, [f_n]_G) \quad iff \quad [\![\phi(f_1, \ldots, f_n)]\!]_B^{\mathfrak{M}} \in G.$$

Proof By induction on the complexity of $\phi(f_1, \ldots, f_n)$:

- If $\phi(f_1, \ldots, f_n) = R(f_1, \ldots, f_n)$ for some relational symbol R, then

$$\mathfrak{M}/G \models R([f_1]_G, \ldots, [f_n]_G) \quad \text{iff} \quad [\![R(f_1, \ldots, f_n)]\!]_B^{\mathfrak{M}} \in G$$

by definition.
- If $\phi(f_1, \ldots, f_n) = \psi(f_1, \ldots, f_n) \wedge \chi(f_1, \ldots, f_n)$, then

$$\mathfrak{M}/G \models \phi([f_1]_G, \ldots, [f_n]_G)$$
$$\text{iff} \quad \mathfrak{M}/G \models \psi([f_1]_G, \ldots, [f_n]_G), \chi([f_1]_G, \ldots, [f_n]_G)$$

---

[5] Since $\mathfrak{M}/G$ is a 2-valued model, i.e., an $\mathcal{L}$-structure for Tarski semantics, $\models$ refers to its usual meaning for $\mathfrak{M}/G$ seen as a Tarski structure.

$$\text{iff} \quad [\![\psi(f_1, \ldots, f_n)]\!]_{\mathsf{B}}^{\mathfrak{M}} \in G \text{ and } [\![\chi(f_1, \ldots, f_n)]\!]_{\mathsf{B}}^{\mathfrak{M}} \in G$$

$$\text{iff} \quad [\![\psi(f_1, \ldots, f_n) \wedge \chi(f_1, \ldots, f_n)]\!]_{\mathsf{B}}^{\mathfrak{M}} \in G.$$

- If $\phi(f_1, \ldots, f_n) = \neg\psi(f_1, \ldots, f_n)$, then

$$\mathfrak{M}/_G \models \phi([f_1]_G, \ldots, [f_n]_G) \quad \text{iff} \quad \mathfrak{M}/_G \models \neg\psi([f_1]_G, \ldots, [f_n]_G)$$

$$\text{iff} \quad \mathfrak{M}/_G \not\models \psi([f_1]_G, \ldots, [f_n]_G)$$

$$\text{iff} \quad [\![\psi(f_1, \ldots, f_n)]\!]_{\mathsf{B}}^{\mathfrak{M}} \notin G$$

$$\text{iff} \quad [\![\neg\psi(f_1, \ldots, f_n)]\!]_{\mathsf{B}}^{\mathfrak{M}} \in G.$$

- If $\phi(f_1, \ldots, f_n) = \exists x \psi(x, f_1, \ldots, f_n)$, then

$$\mathfrak{M}/_G \models \exists x \psi(x, [f_1]_G, \ldots, [f_n]_G) \quad \text{iff} \quad \mathfrak{M}/_G \models \psi([h]_G, [f_1]_G, \ldots, [f_n]_G)$$
$$\text{for some } h \in M$$

$$\text{iff} \quad [\![\psi(h, f_1, \ldots, f_n)]\!]_{\mathsf{B}}^{\mathfrak{M}} \in G$$
$$\text{for some } h \in M,$$

$$\text{which implies} \quad [\![\exists x \psi(x, f_1 \ldots, f_n)]\!]_{\mathsf{B}}^{\mathfrak{M}} \in G.$$

This vice versa also holds, by fullness: Let $h_1 \ldots, h_k \in M$ be such that

$$[\![\exists x \psi(x, f_1, \ldots, f_n)]\!]_{\mathsf{B}}^{\mathfrak{M}} = \bigvee_{i=1}^{k} [\![\phi(h_i, f_1, \ldots, f_n)]\!]_{\mathsf{B}}^{\mathfrak{M}}.$$

Assuming $[\![\exists x \psi(x, f_1, \ldots, f_n)]\!]_{\mathsf{B}}^{\mathfrak{M}} \in G$, since G is an ultrafilter, there is some $j \le k$ such that $[\![\phi(h_j, f_1, \ldots, f_n)]\!]_{\mathsf{B}}^{\mathfrak{M}} \in G$. Then by inductive assumptions,

$$\mathfrak{M}/_G \models \psi([h_j]_G, [f_1]_G, \ldots, [f_n]_G),$$

which yields that

$$\mathfrak{M}/_G \models \exists x \, \psi(x, [f_1]_G, \ldots, [f_n]_G),$$

as was to be shown.

□

A key property of fullness is its preservation by quotients:

Lemma 6.3.8 *Assume $\mathfrak{M}$ is a full B-valued model and F is a filter on B. Then $\mathfrak{M}/_F$ is a full $\mathsf{B}/_F$-valued model.*

Proof It suffices to show by induction on $\phi(y_1, \ldots, y_n)$ that for all $f_1 \ldots f_n \in \mathcal{M}$

$$[\![\phi([f_1]_F, \ldots, [f_n]_F)]\!]_{\mathsf{B}/F}^{\mathfrak{M}/F} = [[\![\phi(f_1, \ldots, f_n)]\!]_{\mathsf{B}}^{\mathfrak{M}}]_F. \tag{6.5}$$

If this is the case, then for any $\phi(x, y_1, \ldots, y_n)$ we have that for all $f_1, \ldots, f_n \in \mathfrak{M}$ there are $h_1, \ldots, h_k$ such that

$$[\![\exists x \phi(x, [f_1]_F, \ldots, [f_n]_F)]\!]_{\mathsf{B}/F}^{\mathfrak{M}/F} = [[\![\exists x \phi(x, f_1, \ldots, f_n)]\!]_{\mathsf{B}}^{\mathfrak{M}}]_F$$

$$= [\bigvee_{i=1}^{k} [\![\phi(h_i, f_1, \ldots, f_n)]\!]_{\mathsf{B}}^{\mathfrak{M}}]_F$$

$$= \bigvee_{i=1}^{k} [[\![\phi(h_i, f_1, \ldots, f_n)]\!]_{\mathsf{B}}^{\mathfrak{M}}]_F$$

$$= \bigvee_{i=1}^{k} [\![\phi([h_i]_F, [f_1]_F, \ldots, [f_n]_F)]\!]_{\mathsf{B}/F}^{\mathfrak{M}/F}.$$

Note that 6.5 holds true for atomic formulae by definition and is easily shown to be the case for boolean combinations of formulae for which 6.5 holds.

Now assume 6.5 holds for $\phi(x, y_1, \ldots, y_n)$ for all $h, f_1, \ldots, f_n \in \mathfrak{M}$. Then

$$[\![\exists x \phi(x, [f_1]_F, \ldots, [f_n]_F)]\!]_{\mathsf{B}/F}^{\mathfrak{M}/F} = \bigvee_{h \in \mathfrak{M}} [\![\phi([h]_F, [f_1]_F, \ldots, [f_n]_F)]\!]_{\mathsf{B}/F}^{\mathfrak{M}/F}$$

$$= \bigvee_{h \in \mathfrak{M}} [[\![\phi(h, f_1, \ldots, f_n)]\!]_{\mathsf{B}}^{\mathfrak{M}}]_F$$

$$\leq [\bigvee_{h \in \mathfrak{M}} [\![\phi(h, f_1, \ldots, f_n)]\!]_{\mathsf{B}}^{\mathfrak{M}}]_F$$

$$= [[\![\exists x \phi(x, f_1, \ldots, f_n)]\!]_{\mathsf{B}}^{\mathfrak{M}}]_F$$

holds trivially. On the other hand, for suitably chosen $h_1, \ldots, h_k \in \mathfrak{M}$,

$$[\![\exists x \phi(x, [f_1]_F, \ldots, [f_n]_F)]\!]_{\mathsf{B}/F}^{\mathfrak{M}/F} = \bigvee_{h \in \mathfrak{M}} [\![\phi([h]_F, [f_1]_F, \ldots, [f_n]_F)]\!]_{\mathsf{B}/F}^{\mathfrak{M}/F}$$

$$= \bigvee_{h \in \mathfrak{M}} [[\![\phi(h, f_1, \ldots, f_n)]\!]_{\mathsf{B}}^{\mathfrak{M}}]_F$$

$$\geq [\bigvee_{i=1}^{k} [\![\phi(h_i, f_1, \ldots, f_n)]\!]_{\mathsf{B}}^{\mathfrak{M}}]_F$$

$$= [[\![\exists x \phi(x, f_1, \ldots, f_n)]\!]_{\mathsf{B}}^{\mathfrak{M}}]_F$$

follows trivially from the properties of sups and the fullness of $\mathfrak{M}$ as a B-valued model.

This establishes 6.5 for all formulae.

□

6.3.4 Forcing and Fullness

The following section introduces the forcing relation, analyzes its basic properties, and outlines a fundamental link between full B-valued models and the topological properties of St(B).

Definition 6.3.9 Let B be a boolean algebra. Given a well behaved B-valued model $\mathfrak{M}$ for $\mathcal{L}$, $\phi(x_0, \ldots, x_n)$ a formula of the language $\mathcal{L}$, $a_0, \ldots, a_n \in M$, define

$$b \Vdash \phi(a_0, \ldots, a_n) \text{ (to be read as } b \text{ forces } \phi(a_0, \ldots, a_n))$$

iff $b \leq [\![\phi(a_0, \ldots, a_n)]\!]$.

Lemma 6.3.10 (Forcing Lemma I) *Let* B *be a boolean algebra,* $\mathfrak{M}$ *be a full* B-*model, and* $G \in$ St(B). *Then, for any formula* ϕ *and any* $b \in$ B, *the following are equivalent:*

1. $b \Vdash [\![\phi]\!]$.
2. $D_\phi = \{G \in \text{St(B)} : M/G \models \phi\}$ *is dense in* N_b.
3. $D_\phi \supseteq N_b$.

Proof

$1 \Leftrightarrow 3$ This is a straightforward consequence of Łoś' theorem.

$1 \Rightarrow 2$ Fix some formula ϕ. Let $b \in$ B such that $[\![\phi]\!] \geq b$, and assume that D_ϕ is not dense in N_b, aiming for a contradiction. Then, by our assumption, there exists some $c \in$ B$^+$, $c \leq b$, such that $N_c \neq \emptyset$ and N_c is disjoint from D_ϕ. Now for every $G \in N_c$, $\mathfrak{M}/G \not\models \phi$. By Łoś theorem, it follows that for all $G \in N_c$, $[\![\neg\phi]\!] \in G$, and thus $[\![\neg\phi]\!] \wedge c \in G$. Therefore, we have that

$$0_\text{B} < [\![\neg\phi]\!] \wedge c \leq [\![\neg\phi]\!] \wedge b \leq [\![\neg\phi]\!] \wedge [\![\phi]\!] = 0_\text{B},$$

which is a contradiction.

$2 \Rightarrow 1$ Assume that $[\![\phi]\!] \not\geq b$, that is, $[\![\neg\phi]\!] \wedge b > 0$. So, let $d = [\![\neg\phi]\!] \wedge b$, and look at N_d, which is contained in N_b. Now, observe that if $d \in G$, then also $[\![\neg\phi]\!] \in G$, which implies that $\mathfrak{M}/G \models \neg\phi$. It follows that for every $G \in N_d$, $\mathfrak{M}/G \models \neg\phi$, and therefore $D_\phi \cap N_d = \emptyset$. This proves that D_ϕ is not dense in N_b.

□

According to Lemma 6.3.10, if $\mathfrak{M}$ is full, to check that a formula ϕ is valid in $\mathfrak{M}$ it suffices to show that it is valid in $\mathfrak{M}/_G$ for densely many $G \in \mathrm{St}(\mathsf{B})$.

Lemma 6.3.11 (Forcing Lemma II) *Let* B *be a boolean algebra and* $\mathfrak{M}$ *be a well behaved* B-*valued model for* $\mathcal{L}$ *with domain* M, $\phi(x_0, \ldots, x_n)$ *a formula of the language* $\mathcal{L}$, $a_0, \ldots, a_n \in M$.

The following hold for all $b \in \mathsf{B}$:

1. $b \Vdash \phi$ *iff the set of* $G \in \mathrm{St}(\mathsf{B})$ *such that* $\mathfrak{M}/_G \models \phi$ *is dense in* N_b.
2. $b \Vdash \phi \wedge \psi$ *iff* $b \Vdash \phi$ *and* $b \Vdash \psi$.
3. $b \Vdash \neg\phi$ *iff* $c \nVdash \phi$ *for any* $c \leq b$.
4. $b \Vdash \phi \vee \psi$ *iff the set of* $c \leq b$ *such that* $c \Vdash \phi$ *or* $c \Vdash \psi$ *is dense below* b *in* B^+.
5. $b \Vdash \exists x\phi(x)$ *iff the set of* $c \leq b$ *such that* $c \Vdash \phi(\sigma)$ *for some* $\sigma \in M$ *is dense below* b.

Furthermore the following hold for all $G \in \mathrm{St}(\mathsf{B})$ *and all* ϕ *formulae with parameters in* M *and no free variable:*

(i) $\mathfrak{M}/_G \models \phi$ *if and only if* $b \Vdash \phi$ *for some* $b \in G$.
(ii) $[\![\phi]\!] = \bigvee \{b \in \mathsf{B} : b \Vdash \phi\}$.

Proof Left to the reader. $\square$

Exercise 6.3.12 Show that a well behaved B-valued model $\mathfrak{M}$ is full if and only if the conclusion of Łoś theorem holds for $\mathfrak{M}$. (HINT: It is not so trivial; there is a compactness argument to infer that $\mathfrak{M}/_G \models \exists x\psi$ for all $G \in N_{[\![\exists x\psi]\!]}$ if and only if there are $h_1, \ldots, h_k \in M$ such that $N_{[\![\exists x\psi]\!]} = \bigcup_{i=1}^{k} N_{[\![\psi[h_i/x]]\!]}$).

6.3.5 The Mixing Property and Fullness

Fullness for a boolean valued model $\mathfrak{M}$ for $\mathcal{L}$ is a desirable feature of $\mathfrak{M}$ but needs to be checked on the infinitely many formulae of $\mathcal{L}$. It is oftentimes simpler and convenient to establish a property of the model which implies a strong form of fullness. The mixing property gives a sufficient condition for having the fullness property which is, usually, easier to check.

Definition 6.3.13 Let κ be a cardinal, $\mathcal{L}$ a first order language, B a κ-complete boolean algebra, and $\mathfrak{M}$ a B-valued model for $\mathcal{L}$.

- $\mathfrak{M}$ satisfies the κ-*mixing property* if for every antichain $A \subset \mathsf{B}$ of size at most κ and for every subset $\{\tau_a : a \in A\} \subseteq M$, there exists $\tau \in M$ such that $a \leq [\![\tau = \tau_a]\!]$ for every $a \in A$.
- $\mathfrak{M}$ satisfies the $< \kappa$-*mixing property* if it satisfies the λ-mixing property for all cardinals $\lambda < \kappa$.
- $\mathfrak{M}$ satisfies the *mixing property* if it satisfies the $|\mathsf{B}|$-mixing property.

In [14] models with the $< \omega$-mixing property are called models which *admit gluing*.

Whether a B-valued model $\mathfrak{M}$ for some signature $\mathcal{L}$ has the mixing property depends only on the interpretation of the equality symbol by $[\![\cdot = \cdot]\!]_B^{\mathfrak{M}}$.

Proposition 6.3.14 *Let* B *be a complete boolean algebra, and let* $\mathfrak{M}$ *be a* B-*valued model for* $\mathcal{L}$. *Assume that* $\mathfrak{M}$ *satisfies the* κ-*mixing property for some* $\kappa \geq \min \{|B|, |M|\}$. *Then* $\mathfrak{M}$ *is full.*

Proof Fix a formula $\phi(x, y_1, \ldots, y_n)$ in $\mathcal{L}$ and $\sigma_1, \ldots, \sigma_n \in M$. Fix moreover an enumeration $\langle \tau_i : i \in \gamma \rangle$ of M. Since $[\![\exists x \phi(x, \sigma_1, \ldots, \sigma_n)]\!] = \bigvee_{i \in I} [\![\phi(\tau_i, \sigma_1, \ldots, \sigma_n)]\!]$, we can refine the family $\{ [\![(\phi(\tau_i, \sigma_1, \ldots, \sigma_n)]\!] : i \in \gamma \}$ to an antichain $\{ a_j : j \in J \}$ as follows: Let

$$J := \gamma \setminus \left\{ i \in I : [\![\phi(\tau_i, \sigma_1, \ldots, \sigma_n)]\!] \setminus \bigvee_{j<i} [\![\phi(\tau_j, \sigma_1, \ldots, \sigma_n)]\!] = 0_B \right\}.$$

In particular, J is well ordered with the order induced by γ, and we have that $\min J = 0$. Define

$$a_0 := [\![\phi(\tau_0, \sigma_1, \ldots, \sigma_n)]\!]$$

and, for $J \ni i > 0$,

$$a_i := [\![\phi(\tau_i, \sigma_1, \ldots, \sigma_n)]\!] \setminus \bigvee_{J \ni j < i} [\![\phi(\tau_j, \sigma_1, \ldots, \sigma_n)]\!].$$

If $A := \{ a_j : j \in J \}$, it is clear that $\bigvee A = [\![\exists x \phi(x, \sigma_1, \ldots, \sigma_n)]\!]$ and $|A| \leq |M|, |B| \leq \kappa$. Since $\mathfrak{M}$ satisfies the κ-mixing property, there exists $\tau \in M$ such that

$$a_i \leq [\![\tau = \tau_i]\!]$$

for every $i \in J$. In particular, since $a_i \leq [\![\phi(\tau_i, \sigma_1, \ldots, \sigma_n)]\!]$, we have that

$$a_i = a_i \wedge [\![\tau = \tau_i]\!] \leq [\![\phi(\tau_i, \sigma_1, \ldots, \sigma_n)]\!] \wedge [\![\tau = \tau_i]\!] \leq [\![\phi(\tau, \sigma_1, \ldots, \sigma_n)]\!]$$

for every $i \in J$. Hence $[\![\phi(\tau, \sigma_1, \ldots, \sigma_n)]\!] \geq \bigvee_{i \in J} a_i = \bigvee A = [\![\exists x \phi(x, \sigma_1, \ldots, \sigma_n)]\!]$ and so $\mathfrak{M}$ is full.   □

Remark 6.3.15 The result just proven actually shows that the mixing property implies a strong version of fullness, that is, for every formula $\phi(x_0, x_1, \ldots, x_n)$ and for every $\tau_1, \ldots, \tau_n \in \mathfrak{M}$, there exists an element τ_0 such that

$$\bigvee_{\sigma \in \mathfrak{M}} [\![\phi(\sigma, \tau_1, \ldots, \tau_n)]\!] = [\![\phi(\tau_0, \tau_1, \ldots, \tau_n)]\!].$$

This is actually the definition of fullness one can find, for instance, in [16]. It is easy to see that this property is true in every full model $\mathfrak{M}$ satisfying the $< \omega$-mixing property.

6.4 Examples of Boolean Valued Models with the Mixing Property

6.4.1 Example I: Spaces of Measurable Functions

Let

$L^{\infty+}([0, 1])$

$$= \{f : [0, 1] \to \mathbb{R} \cup \{\infty\} : f \text{ is Lebesgue measurable and } \mu(f^{-1}[\{\infty\}]) = 0\}.$$

Recall that $\mathsf{MALG} = \mathrm{Bor}([0, 1])/\mathrm{NULL}$ is a complete boolean algebra (cf. Corollary 4.3.9). $L^{\infty+}([0, 1])$ is a natural enlargement of $L^{\infty}([0, 1])$, the space of essentially bounded measurable functions (i.e., those measurable $f : [0, 1] \to \mathbb{R}$ such that $\mu(\{x : |f(x)| > C\}) = 0$ for some $C > 0$).

Proposition 6.4.1 $(L^{\infty+}([0, 1]), <_{\mathsf{B}}, C_{\mathsf{B}})$ *is a* MALG*-valued model for* $\mathcal{L} = \{<, C\}$ *with the mixing property.*

Proof Let $\mathsf{B} = \mathsf{MALG}$ in what follows. Assume $\{[X_i]_{\mathrm{NULL}} : i < \omega\}$ is a maximal antichain (recall that MALG has the CCC).

Let $\{f_i : i \in \omega\}$ be a countable family of functions in $L^{\infty+}([0, 1])$. Without loss of generality we can suppose that $\{X_i : i \in \omega\}$ consists of a partition of $\mathbb{R}$ in measurable pieces.

Set $g = \bigcup_{n \in \omega} f_n$. Then $g : [0; 1] \to \mathbb{R} \cup \{\infty\}$ is measurable, and $g^{-1}[\{\infty\}] = \bigcup_{n \in \omega} f_n^{-1}[\{\infty\}]$ has measure 0; hence $g \in L^{\infty+}([0, 1])$. Clearly $[\![g = f_n]\!] \geq [X_n]$ for all $n \in \omega$. □

Exercise 6.4.2 Explain what goes wrong if you try to prove the same for $L^{\infty}([0, 1])$ (HINT: In this case we can only guarantee that g is measurable and $g \upharpoonright X_i \in L^{\infty}(X_i)$, however note that $g \upharpoonright \bigcup_{i \in \omega} X_i$ may not be essentially bounded in $\bigcup_{i \in \omega} X_i$, for example you can use the counterexample to the fullness of $C^{\omega}(\mathbb{R})$ replacing Id with the function $x \mapsto 1/x$).

Proposition 6.4.3 $(L^{\infty+}([0; 1]), +_{\mathsf{B}}, \cdot_{\mathsf{B}}, 0, 1)$ *is a* MALG*-model for the theory of fields.*

Proof We just prove that $(L^{\infty+}([0;1]), +, \cdot, 0, 1)$ satisfies the existence of the inverse for every nonzero element. We must show that $[\![\forall x(x \neq 0 \to \exists y(x \cdot y = 1))]\!] = 1_{\text{MALG}}$. Since

$$[\![\forall x(x \neq 0 \to \exists y(x \cdot y = 1))]\!] = \bigwedge_{g \in L^{\infty+}([0,1])} [\![(g \neq 0) \to \exists y(g \cdot y = 1))]\!],$$

it suffices to prove that $[\![(g \neq 0) \to \exists y(g \cdot y = 1))]\!] = 1_{\text{MALG}}$ for all $g \in L^{\infty+}([0,1])$. Fix $g \in L^{\infty+}([0,1])$, and define

$$A_0 := \{x : g(x) = 0\}$$
$$A_1 := [0,1] \smallsetminus A_0.$$

Consider the following function:

$$g^{-1}(x) = \begin{cases} 0 & \text{if } x \in A_0, \\ \frac{1}{g(x)} & \text{otherwise.} \end{cases}$$

Now, observe that $[\![g = 0]\!] = [A_0]_{\text{NULL}}$, and therefore

$$[\![g = 0 \vee \exists y(g \cdot y = 1)]\!] = [A_0]_{\text{NULL}} \vee [\![\exists y(g \cdot y = 1)]\!]$$

$$= [A_0]_{\text{NULL}} \vee \bigvee_{h \in L^{\infty+}} [\![g \cdot h = 1)]\!]$$

$$\geq [A_0]_{\text{NULL}} \vee \left[\!\!\left[g \cdot g^{-1} = 1 \right]\!\!\right]$$

$$= [A_0]_{\text{NULL}} \vee [A_1]_{\text{NULL}} = 1_{\text{MALG}}.$$

$\square$

Exercise 6.4.4 $(L^{\infty}([0,1]), +_{\text{MALG}}, \cdot_{\text{MALG}}, 0, 1)$ is also a **MALG**-model for the theory of fields. (HINT: Show that locally any function is invertible, i.e., given f show that one can split $\{x : f(x) \neq 0\}$ in countably many pairwise disjoint sets A_n for $n \in \omega$ such that $f \upharpoonright A_n$ has an inverse g_n which is in $L^{\infty}(A_n)$. The problem is that $g = \bigcup_{n \in \omega} g_n$ is measurable but possibly not in $L^{\infty}(\bigcup_n A_n)$. On the other hand, g will always be in $L^{\infty+}(\bigcup_n A_n)$....).

Exercise 6.4.5 Prove that the model of Exercise 6.4.4 is not full (HINT: Show that the quotients $L^{\infty}([0,1])/_G$ by any ultrafilter G are not fields, since the germ in G of some $f \in L^{\infty}([0,1])$ has an inverse in $L^{\infty+}([0,1])/_G \setminus L^{\infty}([0,1])/_G$).

6.4.2 Example II: Standard Ultraproducts

We now sketch an argument to show that the familiar notion of ultraproduct of Tarski models is a special case of a quotient of a boolean valued model with the mixing property.

Let X be a set. Then $\mathcal{P}(X)$ is an atomic complete boolean algebra. Notice that all theorems proved so far apply equally well to *atomic, complete* boolean algebras even if in the examples we focused on *atomless, complete* boolean algebras. A key observation is that $\{\{x\} : x \in X\}$ is a maximal antichain and a dense open set in $\mathcal{P}(X)^+$. Now observe that $\mathrm{St}(\mathcal{P}(X))$ is the space of ultrafilters on X and X can be identified inside $\mathrm{St}(\mathcal{P}(X))$ as the open dense set $\{G_x : x \in X\}$, where G_x is the principal ultrafilter on $\mathcal{P}(X)$ given by all supersets of $\{x\}$. Another key observation is the following:

Fact 6.4.6 *Let $(M_x : x \in X)$ be a family of Tarski models in the first order relational language $\mathcal{L}$. Then $N = \prod_{x \in X} M_x$ is a $\mathcal{P}(X)$-model with the mixing property (letting for each n-ary relation symbol $R \in \mathcal{L}$, $[\![R(f_1, \ldots, f_n)]\!]_{\mathcal{P}(X)} = \{x \in X : M_x \models R(f_1(x), \ldots, f_n(x))\}$).*

Proof We leave the proof as an instructive exercise for the reader. $\quad\square$

Let G be any nonprincipal ultrafilter on X. Then, using the notation of the previous fact, N/G is the familiar ultraproduct of the family $(M_x : x \in X)$ by G, and the usual Łoś theorem for ultraproducts of Tarski models is the specialization to the case of the full $\mathcal{P}(X)$-valued model N of Theorem 6.3.7. Notice that in this special case, if the ultraproduct is an ultrapower of a model M, the embedding $a \mapsto [c_a]_G$ (where $c_a(x) = a$ for all $x \in X$ and $a \in M$) is elementary. This is not always the case for all the other examples of full B-valued models we are giving in these notes.

6.4.3 Example III: $C(St(\mathbf{B}), 2^\omega)$

We introduce a last example of full boolean valued model, which is more in the spirit of what we are aiming for, since it can give an approach to forcing completely equivalent to the one we pursue in the next chapter.

Exercise 6.4.7 Let $\mathbf{B}$ be an arbitrary (complete) boolean algebra. Let $M = C(St(\mathbf{B}), 2^\omega)$ be the family of continuous functions from $\mathrm{St}(\mathbf{B})$ into 2^ω. Fix R a binary Borel relation on 2^ω. The continuity of f, g grants that the set

$$\{G : f(G) \ R \ g(G)\} = (f \times g)^{-1}[R]$$

is Borel (and therefore has the Baire property in $St(B)$, where $f \times g(G) = (f(G), g(G))$). So we can define

$$R^M(f, g) = [\![f \ R \ g]\!] = \mathsf{Reg}(\{G : f(G) \ R \ g(G)\}).$$

Also since the diagonal is closed (hence has the Baire property) in $(2^\omega)^2$,

$$=^M (f, g) = [\![f = g]\!] = \mathsf{Reg}(\{G : f(G) = g(G)\})$$

is well defined.

- Check that $(C(St(B), 2^\omega), =^M, R^M)$ is a full B-valued extension of the structure $(2^\omega, =, R)$ (where 2^ω is copied inside $C(St(B), 2^\omega)$ as the set of constant functions). (HINT: The Baire property ensures that

$$\mathsf{Reg}(\{G : f(G) \ P \ g(G)\}) = X \setminus \mathsf{Cl}(\{G : \neg(f(G) \ P \ g(G))\}),$$

  and

$$\mathsf{Reg}(\{G : f(G) \ P \ g(G)\}) \cap \mathsf{Reg}(\{G : h(G) \ Q \ k(G)\})$$
$$= \mathsf{Reg}(\{G : f(G) \ P \ g(G)\} \cap \{G : h(G) \ Q \ k(G)\})$$

  holds for P, Q among $R, =$ and $f, g, h, k \in C(St(B), 2^\omega)$).
- Check also that whenever G is an ultrafilter on $St(B)$, the map $i_G : 2^\omega \rightarrow C(St(B), 2^\omega)/G$ given by $x \mapsto [c_x]_G$ (the constant function with value x) defines an injective morphism of the 2-valued structure $(2^\omega, =, R)$ into the 2-valued structure $(C(St(B), 2^\omega)/G, =^M /G, R^M /G)$.

The above exercise outlines a general strategy to expand many first order structures on 2^ω to extensions $C(St(B), 2^\omega)/G$ indexed by $G \in St(B)$ in such a way that the first properties of the structure $C(St(B), 2^\omega)/G$ are finely controlled by the topological properties of $St(B)$ and the algebraic properties of B via the embeddings i_G. Note that many interesting binary relations on 2^ω have a Borel graph (e.g., $f =^* g$ if $\{n : f(n) \neq g(n)\}$ is finite).

The general idea of forcing is to develop this technique in order to be able to replace first order structures with domain 2^ω by any first order model (M, E) of ZFC. For the sake of simplicity, we assume now that M is transitive and E is $\in \cap M^2$. $(M, \in, \subseteq, =)$ is expanded to a boolean extension $(M^B, \in^B, \subseteq^B, =^B)$ defined by means of a boolean algebra $B \in M$ in such a way to define a forcing relation which ties the logical property of the boolean structure $(M^B, \in^B, \subseteq^B, =^B)$ to the topological properties of the space $St(B)$ or, equivalently (via Stone's Duality), to the algebraic properties of B. We can then pass to a natural quotient structure M^B/G, which is now a Tarski model for the language of M and which naturally contains an isomorphic copy of M. The definition of M^B will be done reversing the arrows and exploiting Stone's Duality between zero-dimensional compact Hausdorff spaces and boolean algebras. We will develop the theory of boolean valued models for set theory defining M^{B} as an appropriate bunch of functions from M to B, rather than as a set of continuous functions from $\mathrm{St}(\mathsf{B})$ to (what should be) some compactification of M.

Chapter 7
Forcing

In this chapter we present the technique of forcing. Forcing has a crucial role in the development of modern set theory, and it has had and has an immense number of applications in this field of research. For example, forcing is the standard tool to prove the consistency with the standard axioms of ZFC of a mathematical statement which can be formulated as a first order $\in$-formula ϕ in the language of set theory. Many mathematical theories can get a natural interpretation as subtheories of ZFC. In this way forcing provides an extremely powerful tool to investigate the undecidability of a given mathematical problem,[1] since this problem can in most cases be formulated as a first order statement in the theory ZFC. The first and most celebrated example of an unexpected undecidability result is the proof of the independence of the continuum hypothesis CH from ZFC, and the aim of this chapter is to develop forcing far enough in order to be able to give a complete proof of this result. The general idea of forcing is the following: We want to get a model of some first order statement ϕ in the language of set theory which we aim to show to be consistent with ZFC. To do so we enlarge the universe V which is the "standard" model of set theory to another universe of sets $N \supset V$ which is still a model of ZFC so that we are able to *force* N to be a model of ϕ. If we adopt a platonistic stance toward set theory, the statement "$N \supset V$" is nonsense since all possible sets are already elements of V, and so there cannot be a proper superuniverse N of V. To overcome this difficulty we assume that in V there is a *countable transitive model* $M \in V$ of ZFC, and we extend M to a *generic extension* $N \supset M$ which is also in V and which is a model of ZFC $+ \phi$. This approach requires us to work in a theory which is slightly stronger than ZFC, since by Gödel's incompleteness theorem, ZFC cannot (unless ZFC is not consistent) derive the statement *there is a countable transitive model of* ZFC, while we will assume that V models ZFC$^+$, where ZFC$^+$ stands for ZFC plus the latter statement. Nonetheless in our eyes

---

[1] More precisely, rather than undecidability one could say: independence from the axiom system ZFC of any of its possible solutions.

© The Author(s), under exclusive license to Springer Nature Switzerland AG 2024   107
M. Viale, *The Forcing Method in Set Theory*, La Matematica per il 3+2 168,
https://doi.org/10.1007/978-3-031-71660-7_7

requiring V to be a model of ZFC^+ allows for a simpler exposition of the semantics of the forcing method and does not weaken substantially the undecidability results we are able to obtain with respect to ZFC^+ (the statement *there is a countable transitive model of* ZFC follows from the theory $\mathsf{ZFC}+$*there exists an inaccessible cardinal*—see Sect. A.5.1—and is equiconsistent with ZFC).[2]

The general strategy to prove the undecidability of ϕ by means of forcing is to start from given known countable transitive models of ZFC M_0, M_1 and to produce by means of forcing generic extensions $N_i \supseteq M_i$ such that N_0 models $\mathsf{ZFC} + \phi$ and N_1 models $\mathsf{ZFC} + \neg\phi$. In this chapter we just assume that there is one given countable transitive model of ZFC M, and we will build all our generic extensions over this M.

We also need to give some intuition on the reasons why enlarging V to a larger N we can hope to be able to show that N is a model of ϕ. The strategy we will follow is that leading from a two valued logic where all statements are either true or false to a boolean valued logic where statements ϕ get evaluated as elements $[\![\phi]\!] \in \mathsf{B}$ for some boolean algebra B, we consider ϕ true if $[\![\phi]\!] = 1_\mathsf{B}$, false if $[\![\phi]\!] = 0_\mathsf{B}$, undetermined otherwise. Now observe that B corresponds to the clopen sets of $St(\mathsf{B})$ its space of ultrafilters and that selecting a point $G \in St(\mathsf{B})$ allows us to decide which as yet undecided statements ϕ are true or false according to G: They will be considered true by G if and only if $[\![\phi]\!] \in G$. So we are led to the consideration that boolean algebras B allow to define a B-valued logic in which we have not yet compromised ourselves on the truth values of certain first order statements ϕ (those for which $0_\mathsf{B} < [\![\phi]\!] < 1_\mathsf{B}$), and that the points $G \in St(\mathsf{B})$ will *force* us to accept ϕ as true iff $[\![\phi]\!] \in G$. Now the idea of the forcing method is to employ these boolean valued logics as follows: We start from a *transitive* model M of ZFC (where we are not able to compute ϕ), and we choose in M a boolean algebra B for which we are able to calculate its combinatorial properties. We extend M to a boolean valued model M^B which is a definable class in M and contains an isomorphic copy of M as a B-valued substructure. M^B is such that for all formulae ϕ we are able to define an evaluation map $[\![\phi]\!]$ which links (in a manner which is possible to compute inside M) the B-valued semantics of M^B to the combinatorial properties of B in M. We can pick any $G \in St(\mathsf{B})$, and we get that G decides that ϕ holds iff $[\![\phi]\!] \in G$. Things can be done so properly that our heuristic assertion "G decides that ϕ holds" can be expanded in the precise statement: "The Tarski structure $M^\mathsf{B}/_G \supset M$ is a first order model of ZFC in which ϕ holds and which properly contains M."

There are different approaches to the technique of forcing. We follow the one through boolean algebras and boolean valued models, as in [16]. To this aim it

---

[2] The interested reader can find in the first pages of [19, Chapter 7] several arguments which translate the undecidability results obtained by means of forcing over the theory ZFC^+, to undecidability results obtained over ZFC. Another possibility is to explore Appendix B and the last part of Appendix C to see how the reflection theorem can be used to remove the unnecessary assumption present in ZFC^+.

is fundamental to exploit the theory of boolean valued models we developed in Chap. 6.

The remainder of this chapter consists of four sections:

1. In the first section we define a procedure which given any (transitive) first order structure M which is a model of ZFC and a boolean algebra $B \in M$ (which M models to be complete) produces a *full* boolean valued model of set theory M^B, i.e., a boolean valued model of the language of set theory according to the boolean semantics we defined in the previous chapter. We also try to give an heuristic for forcing following Cohen's original argument to add a new element of 2^ω to M by describing it inside M using the poset of finite strings of $0, 1$. Moreover we explain how the semantics of M^B is governed by means of the notion of M-genericity introduced by Cohen.

2. In the second section we develop in full detail the key ideas in Cohen's development of the forcing method, which allows one to start from a *countable transitive model* M of ZFC, and to pass to the quotient $M^B/_G$ obtained by selecting some $G \in \mathrm{St}(B)$. Under the key assumption that G is an M-generic filter for B, we can prove that $M^B/_G$ is well founded and show that its transitive collapse, the *generic extension $M[G]$*, is countable, transitive, and contains $M \cup \{G\}$. Cohen's forcing theorem shows that there is a fine-tuning between the first order properties of the structure $(M[G], \in)$ and the combinatorial properties of B.

3. In the third section we define a boolean algebra $B \in M$ in such a way that M^B models $\neg$CH and another boolean algebra $C \in M$ in such a way that M^C models CH.

4. In the fourth section we show that no matter how we select M countable transitive model of ZFC and $B \in M$, the axioms of ZFC all get boolean value 1_B in the boolean valued model M^B and thus, by means of the forcing theorem, also hold in $M[G]$ for any M-generic filter G for B. This will show that M[G] is the smallest transitive model of ZFC containing $M \cup \{G\}$.

All over this chapter we assume the reader familiar with the basics of set theory as exposed in an undergraduate level course on the topics (familiarity with most of the arguments covered in [15] is more than sufficient). We also assume the reader is familiar with standard set theoretic notation such as α, β, γ to denote specific von Neumann ordinals, Ord, to denote the proper class of von Neumann ordinals, etc. Finally we assume the reader is familiar with absoluteness arguments in set theory, for example: which fundamental set theoretic concepts are provably Δ_1 in models of set theory (all those defined by transfinite recursion in terms of Δ_0-properties) and which are not (cardinality, powerset). This part of set theoretic background is covered in detail in Appendix A.

The following piece of notation is essential (and has already been implicitly adopted in the preceding discussion):

Notation 7.0.1 Let M be a transitive set or class and $\vec{a}$ be a tuple of elements in the relevant set or class. We write $M \models \phi(\vec{a})$ as a shorthand for

$$(M, \in \cap M^2, \subseteq \cap M^2, =) \models \phi(\vec{a}).$$

More generally any transitive set M which models ZF^- admits a unique extension to an $\in_{\Delta_0}$-structure $(M, \in^M_{\Delta_0})$ which is a model of T_{Δ_0} (as defined in Notation A.2.3) and is an $\in_{\Delta_0}$-substructure of V. When ϕ is an $\in_{\Delta_0}$-formula and $\vec{a}$ a tuple in M, again we write $M \models \phi(\vec{a})$ as a shorthand for

$$(M, \in^M_{\Delta_0}) \models \phi(\vec{a}).$$

7.1   Boolean Valued Models for Set Theory

We have seen the definition of a boolean valued model for the language $\mathcal{L}$ consisting of two relation symbols $\in, \subseteq$ other than the equality symbol $=$. Now we want something more adherent to the intended meaning we have in mind for the symbols $\in, \subseteq$. To achieve this, we need to add some requests to Definition 6.1.1.

Definition 7.1.1 Given a boolean algebra B a B-*valued model for set theory* is a model $\mathfrak{M}$ (with domain M) for $\mathcal{L} = \{\in, \subseteq, =\}$, where, for any $\sigma, \tau, \eta \in M$:

1. $[\![\tau \subseteq \sigma]\!] \wedge [\![\sigma \subseteq \tau]\!] = [\![\tau = \sigma]\!]$.
2. $[\![\tau \in \sigma]\!] \wedge [\![\sigma \subseteq \eta]\!] \leq [\![\tau \in \eta]\!]$.

We now define the boolean valued models for set theory we will be working with in the sequel of this chapter.

Definition 7.1.2 Let M be any *transitive*[3] first order model of ZFC and $\mathsf{B} \in M$ be such that M models B to be a complete boolean algebra. We let for $\alpha \in \mathrm{Ord} \cap M$:

$$M^\mathsf{B}_0 = \emptyset.$$

$$M^\mathsf{B}_{\alpha+1} = \{f : M^\mathsf{B}_\alpha \to \mathsf{B} : f \text{ is partial}\}.$$

$$M^\mathsf{B}_\beta = \bigcup_{\alpha < \beta} M^\mathsf{B}_\alpha, \text{ where } \beta \text{ is limit.}$$

---

[3] As we shall see below the requirement that M is transitive is redundant and we put it here just to give a clearer intuition of how the class M^B is generated inside M.

Finally:

$$M^B = \bigcup_{\alpha \in \mathrm{Ord}^M} M_\alpha^B.$$

The elements of V^B form the family of B-*names*.

Remark 7.1.3

- It is an instructive exercise (see below) to show that the class M^B is definable in M using the transfinite recursion theorem (applied in M) to obtain it as the extension in M of a formula in the parameter B. Formally, this argument can be carried in any first order model of ZFC, and thus the requirement that M is transitive in the above definition is redundant (even though if M is ill-founded, it is not at all transparent what is the correct interpretation of the objects of M defined by means of the transfinite recursion theorem). In this chapter we are interested just in *transitive* well-founded models of ZFC.
  For the sake of completeness, here is how the class V^B can be defined as the extension of a formula in the parameter B inside V: Let $F : V \to V$ be defined as follows:

$$
\begin{cases}
F(g) = \{f : X \to B : f \text{ is a partial function}\} \\
\qquad \text{if for some ordinal } \alpha, \, g : \alpha + 1 \to V \text{ is a function and } X = g(\alpha), \\
\\
F(g) = \bigcup \mathrm{ran}(g) \text{ otherwise.}
\end{cases}
$$

  Then $G : \mathrm{Ord} \to V$ defined by $G(\alpha) = F(G \restriction \alpha)$ enumerates the V_α^B and $V^B = \bigcup_{\alpha \in \mathrm{Ord}} G(\alpha)$.
- If V is the standard model of set theory, the definition of $V_{\alpha+1}^B$ gives the "boolean powerset" of V_α^B for the boolean algebra B much in the same way as $V_{\alpha+1} = \mathcal{P}(V_\alpha)$ is the powerset of V_α for the boolean algebra $\{0, 1\}$: Indeed, given a set X, $\mathcal{P}(X)$ can be identified as the set of the characteristic functions of its elements. We generalize the notion of powerset using the identification of a "boolean" subset of X with its "characteristic" (partial) function $f : X \to B$, with the further hidden complication that for any partial function $f : X \to B$, the evaluation of how much an element of X on which f is undefined belongs to f is postponed to a later stage (i.e., Definition 7.1.25). We need to consider the family of partial functions from X to B to define the boolean powerset of X for technical reasons which will become transparent in the sequel.

Every element $X \in V$ can be identified by a partial characteristic function which takes as domain X and has value constantly 1; we use these types of functions to define canonical B-names for elements of V inside V^B.

Definition 7.1.4 Given any boolean algebra $B \in V$, for every element u of V we define by induction on its rank:

$$\check{u} = \{\langle \check{v}, 1_B \rangle : v \in u\}.$$

For example, if $a = \{\emptyset, \{\emptyset\}\}$, then

$$\check{\emptyset} = \emptyset,$$

$$\{\check{\emptyset}\} = \{\langle \check{\emptyset}, 1_B \rangle\} = \{\langle \emptyset, 1_B \rangle\},$$

$$\check{a} = \{\langle \check{\emptyset}, 1_B \rangle, \langle \{\check{\emptyset}\}, 1_B \rangle\} = \{\langle \emptyset, 1_B \rangle, \langle \{\langle \emptyset, 1_B \rangle\}, 1_B \rangle\}.$$

The above family denotes the canonical B-names for elements of V and is definable in parameter 1_B:

Definition 7.1.5 Let $B \in V$ be a complete boolean algebra.

$$\check{V} = \left\{ \check{a} : a \in V \right\}.$$

There is also another way to define V^B which is more convenient since it will give a Δ_1-definition in the parameter B of the class V^B inside V.

In the sequel $\mathrm{trcl}(x)$ denotes the smallest transitive set containing x.

Definition 7.1.6 Given a set X of partial functions in V, we define

$$\overset{*}{\bigcup} X = \bigcup \{\mathrm{dom}(z) : z \in X\},$$

and for f a partial function

$$\mathrm{trcl}^*(f) = f \cup \bigcup \{ (\overset{*}{\bigcup})^n f : n \in \omega, n > 0\}.$$

Exercise 7.1.7 Show that the operation $f \mapsto \mathrm{trcl}^*(f)$ is Δ_1-definable without parameters in any model of ZFC and thus is absolute between transitive structures which model the relevant fragment of ZFC.

We leave as an instructive exercise to check the following:

Fact 7.1.8 $\tau \in V^B$ *if and only if*

$$\tau \text{ is a function } \wedge \mathrm{ran}(\tau) \subseteq B \wedge \forall \sigma \in \mathrm{trcl}^*(\tau)[\sigma \text{ is a function } \wedge \mathrm{ran}(\sigma) \subseteq B].$$

The latter property is Δ_1-definable in the parameter B in any model of ZFC.

Remark 7.1.9 $\tau \in \check{V}$ if and only if $\tau \in V^{\mathsf{B}}$ and

$$\forall \sigma_0, \sigma_1 \in \mathrm{trcl}^*(\tau)(\sigma_0 \in \mathrm{dom}(\sigma_1) \Rightarrow \sigma_1(\sigma_0) = 1_{\mathsf{B}}).$$

Hence

$$\check{V} \subseteq V^{\mathsf{B}}.$$

7.1.1   External Definition of Forcing

This section is of a rather peculiar nature: We try to give some more intuition on forcing. This forces us to mix some precise mathematical definitions with rather general considerations. In order to keep a straight division between the different levels of our discourse, we adopt the following typographical convention:

- The parts which introduce definitions and prove facts which will be needed also in the remainder of this chapter will maintain the usual font.
- The parts which are in our eyes of central interest to understand the basic ideas of forcing but are not introducing mathematical definitions and results which will be needed in the remainder of this chapter will be put in font UTOPIA.

Assume we want to construct a new element r of 2^ω not in V. This is clearly not possible since all sets are in V and r is a set. Let us sidestep this problem, assuming that there is $M \in V$ countable and transitive such that $M \models \mathsf{ZFC}$ (this is the case if there is an inaccessible cardinal in V, by the results of Sect. A.5.1). So we can assume $r \in V \setminus M$ is a "new" element r of 2^ω with respect to the ZFC-model M.

The fact that M is a transitive model of ZFC simplifies enormously the comparison of M with V: All the standard absoluteness properties established in Appendix A hold between the transitive ZFC-models $M \subseteq V$, giving that most computations yield the same results when carried inside M or in V.

Exercise 7.1.10 Assume M is a transitive model of ZFC. The following notions are absolute for M and V:

- $(\mathsf{B}, \wedge, \neg, 0_{\mathsf{B}}, 1_{\mathsf{B}}) \in M$ is a boolean algebra. (HINT: The property of being a boolean algebra can be formalized as a Σ_0-property of the tuple $(\mathsf{B}, \wedge, \vee, \neg, 0_{\mathsf{B}}, 1_{\mathsf{B}})$, which requires just to quantify over B^n for a large enough n and over $\wedge, \vee \subseteq \mathsf{B}^3$ and $\neg \subseteq \mathsf{B}^2$, and each of these sets is an element of M).
- $(P, \leq_P)$ is a partial order (with $P, \leq_P \in M$).

- $i : P \to Q$ is an order and incompatibility preserving map with a dense image between the partial orders $(P, \leq_P)$ and $(Q, \leq_Q)$ (with $i, P, \leq_P, Q, \leq_Q \in M$).
- $a \in P$ is an atom of the partial order $(P, \leq_P)$ (with $a, P, \leq_P \in M$).
- $(P, \leq_P) \in M$ is an atomless partial order.
- $(P, \leq_P) \in M$ is a separative partial order.
- $G \in M$ is a filter on a partial order $(P, \leq_P) \in M$ (or a ultrafilter on a boolean algebra B).
- $b = \bigvee A$ for $A \in \mathcal{P}(\mathsf{B}) \cap M$.
- $X \subseteq P$ is predense in the partial order $(P, \leq_P)$ (with $X, P, \leq_P \in M$).
- $X \subseteq P$ is predense in the partial order P below the condition p (with $X, P, \leq_P \in M$).
- $A \subseteq P$ is an (maximal) antichain in the partial order $(P, \leq_P)$ (with $A, P, \leq_P \in M$).
- $O \subseteq P$ is open in the partial order $(P, \leq_P)$ (with $O, P, \leq_P \in M$).

Hence for $\mathsf{B} \in M$ a boolean algebra $\mathrm{St}(\mathsf{B})^M = \mathrm{St}(\mathsf{B}) \cap M$, since:

$$\mathrm{St}(\mathsf{B})^M = \left\{ G \in \mathcal{P}(\mathsf{B})^M : M \models G \text{ is a ultrafilter on } \mathsf{B} \right\}$$

$$= \{ G \in \mathcal{P}(\mathsf{B}) \cap M : V \models G \text{ is a ultrafilter on } \mathsf{B} \} = \mathrm{St}(\mathsf{B}) \cap M,$$

where in the second equality, we used that $\mathcal{P}(X)^M = \mathcal{P}(X) \cap M$ for all $X \in M$.

Take the poset $(2^{<\omega}, \supseteq)$. By the first item in Remark 5.2.2, its boolean completion is isomorphic to the regular open sets of 2^ω endowed with product topology (see Remark 4.2.10). Notice that, by absoluteness arguments, since $M \subseteq V$ are both transitive models of ZFC, $2^{<\omega} = (2^{<\omega})^M = (2^{<\omega})^V$.

Consider in M the sets

$$D_f = \left\{ s \in 2^{<\omega} : s \not\subseteq f \right\} \in M$$

for each $f \in (2^\omega)^M$ and the sets

$$E_n = \left\{ s \in 2^{<\omega} : n \in \mathrm{dom}(s) \right\} \in M.$$

These are easily seen to be dense subsets of $2^{<\omega}$ (see Exercise 5.1.6) which belong to M applying the comprehension axiom in M to the formula defining them.

On the other hand in V, given $r \in 2^\omega \setminus M$ define

$$G = \left\{ s \in 2^{<\omega} : s \subseteq r \right\}.$$

Then $G \in V$ and $r = \cup G$, thus $G \notin M$, else $r = \cup G \in M$ as well. G is a filter on $2^{<\omega}$, and G meets all the dense sets D_f for all $f \in (2^\omega)^M$: Pick

$f \in (2^\omega)^M$, since $r = \cup G \neq f$, we can find n such that $r(n) \neq f(n)$, and thus $r \upharpoonright n + 1 \in G \cap D_f$. G meets also the dense sets E_n for $n \in \omega$, since $r \upharpoonright n + 1 \in G \cap E_n$ for all $n \in \omega$.

Now assume $H \in M$ is a filter on $2^{<\omega}$ meeting all the dense set E_n for all $n \in \omega$ (H exists by Lemma 5.1.2 applied in M which is a model of ZFC with $\{E_n : n \in \omega\} \in M$ which is a countable set also according to M). Then $h = \cup H \in 2^\omega$ and $H \cap D_h$ is empty, else for some $s \in H$ and $n \in \mathrm{dom}(s)$, $s(n) \neq h(n)$ with $h \supseteq s$, which is a contradiction.

Hence $G \in V \setminus M$ is a filter on $2^{<\omega}$ which meets the family $\{E_n : n \in \omega\} \cup \{D_f : f \in (2^\omega)^M\}$ of dense subsets of $2^{<\omega}$. This family is contained in M and cannot be simultaneously met by any filter $H \in M$ on $2^{<\omega}$ meeting all the dense sets E_n.[4]

So let us make a step further, and let us assume that $r \in 2^\omega \setminus M$ is such that $G = \{s : s \subseteq r\}$ is a filter on $2^{<\omega}$ meeting *all* the dense subsets of $2^{<\omega}$ which belong to M (this is possible since M is countable in V). Notice that $r = \cup G$.

Before proceeding in our analysis of this specific example, we need to introduce some general concepts.

M-Generic Ultrafilters, and the Induced Valuation Map

Let us recall Definition 5.1.1 and specify it to the context we are interested.

Definition 7.1.11 Let M be a transitive model of ZFC and $P \in M$ be a partial order. A filter $G \subseteq P$ is M-generic for P if $G \cap D \neq \emptyset$ for all $D \in M$ predense subset of P.

For a boolean algebra $\mathsf{C} \in M$, $G \in \mathrm{St}(\mathsf{C})$ is M-generic for C if it is M-generic for C^+.

Exercise 7.1.12 Let M be a countable transitive model of ZFC and $\mathsf{B} \in M$ a boolean algebra. Given $\mathcal{D} \in M$ family of subsets of B, show that

$$(\{G \in \mathrm{St}(\mathsf{B}) : \forall X \in \mathcal{D}\, (G \cap X \neq \emptyset)\})^M$$
$$= \{G \in \mathrm{St}(\mathsf{B}) : \forall X \in \mathcal{D}\, (G \cap X \neq \emptyset)\} \cap M.$$

Letting

$$\mathcal{D} = \{X \in M \cap \mathcal{P}(\mathsf{B}) : M \models X \text{ is a predense subset of } \mathsf{B}\} \in M,$$

---

[4] Note that this family belongs to M but has size continuum in M: The map $f \mapsto D_f$ from $(2^\omega)^M$ into the above family is also an element of M and is an injection.

we have that $G \in \mathrm{St}(\mathsf{B})$ is M-generic if $G \cap X \neq \emptyset$ for all $X \in \mathcal{D}$ and

$$\{G \in \mathrm{St}(\mathsf{B}) : G \text{ is } M\text{-generic for } \mathsf{B}\} \cap M$$
$$= \{G \in \mathrm{St}(\mathsf{B}) : M \models \forall X \in \mathcal{D}\, (G \cap X \neq \emptyset)\}.$$

Using Exercise 5.1.5, show that this latter set is empty if B is atomless, and consists just of the principal ultrafilters on B of the form G_a (each of which belongs to M) for a atom of B, if B is atomic. Show also that in any case the set of $G \in \mathrm{St}(\mathsf{B})$ which are M-generic is a dense subset of $\mathrm{St}(\mathsf{B})$ in V, with the property that none of the M-generic filters is in M if B is atomless.

A basic intuition on M-genericity is that dense open subsets of $\mathrm{St}(\mathsf{B})$ are the large sets and M-generic filters denote the points of $\mathrm{St}(\mathsf{B})$ which are in all large subsets of $\mathrm{St}(\mathsf{B})$ which M knows of.

Recall that any complete boolean algebra splits in the disjoint sum of an atomless cba and of an atomic cba (Lemma 4.0.3). We will see that the forcing method gains traction (i.e., it can be used to produce new interesting model of ZFC) just when it is applied to atomless cbas.

Remark 7.1.13 In the remainder of this chapter, we focus on the notion of M-genericity for *atomless* cbas in M. In this case the notion of M-genericity can be used to describe inside M (by means of M^{B}) enlargements of M obtained by adding to M some $G \in \mathrm{St}(\mathsf{B})$ which is M-generic.

The following exercise briefly explains what happens if B is atomic:

Exercise 7.1.14 Assume $\mathsf{C} \in M$ is such that M models C is an *atomic* cba:

- Let

$$A_{\mathsf{C}} = \big\{a \in \mathsf{C} : a \text{ is an atom}\big\}.$$

Then

$$A_{\mathsf{C}} = \bigcap \big\{D \subseteq \mathsf{C}^+ : D \text{ is dense}\big\}$$

is open dense in C^+.
- Show that for any atom a of C

$$G_a = \big\{b \in \mathsf{C} : a \leq b\big\} \in \mathrm{St}(\mathsf{C}) \cap M$$

is M-generic for C.
- Any $G \in \mathrm{St}(\mathsf{C})$ which belongs to V and which is M-generic for C is of the form G_a for some $a \in \mathsf{C}$ atom of C.

The following exercise shows that the notion of being a complete cba is not absolute between M and V.

Exercise 7.1.15 Assume $\mathsf{B} \in M$ is an infinite boolean algebra. Then B is not complete in V. (HINT: Assume the set $A \in M$ of atoms of B is infinite. Pick $Y \subseteq A$ with $Y \notin M$. Then Y cannot have a suprema $b \in \mathsf{B} \subseteq M$, else $Y = \{a \in A : a \leq b\} \in M$. Assume B has a finite set A of atoms. Then $c = \bigvee A \in M$ and $\mathsf{B} \upharpoonright \neg c \in M$ is atomless. Hence we can assume that B is atomless. Enumerate in M an infinite antichain $A = \{a_n : n \in \omega\} \in M$ of B (which exists since B is atomless and infinite in M). Find $X \subseteq \omega$ such that $X \notin M$. Then $\bigvee \{a_n : n \in X\}$ does not exist: Assume $a = \bigvee \{a_n : n \in X\}$. Then $a \in \mathsf{B} \subseteq M$ and $\{a_n : n \in X\} = \{c \in A : c \leq a\} \in M$, giving that $X \in M$ as well.)

Definition 7.1.16 Let M be a countable transitive model of ZFC and C be a complete boolean algebra in M.

For $\sigma, \tau \in M^{\mathsf{C}}$, define $\sigma \, E^{\mathsf{C}} \, \tau$ iff $\sigma \in \text{dom}(\tau)$ and

$$\text{rk}_{\mathsf{C}}(\tau) = \sup \left\{ \text{rk}_{\mathsf{C}}(\sigma) + 1 : \sigma \, E^{\mathsf{C}} \, \tau \right\}.$$

Fact 7.1.17 *Let M be a countable transitive model of ZFC and C be a complete boolean algebra in M. Then $E^{\mathsf{C}} \subseteq (M^{\mathsf{C}})^2$ is definable in M and well founded in M and V as witnessed by the rank function*

$$\text{rk}_{\mathsf{C}} : M^{\mathsf{C}} \to \text{Ord} \cap M,$$

which is as well a definable class function in M.

Proof Clearly E^{C} is definable by the Σ_0-property $\phi(x, y) \equiv x \in \text{dom}(y)$. The map rk_{C} is defined by transfinite recursion inside M using the absolute function $F^M : M^{\mathsf{C}} \times M \to M$ given by

$$F^M(\sigma, h) = \cup \{ h(z) + 1 : z \in \text{dom}(\sigma) \}$$

and setting

$$\text{rk}_{\mathsf{C}}(\sigma) = F(\sigma, \text{rk}_{\mathsf{C}} \upharpoonright \text{pred}_{E^{\mathsf{C}}}(\sigma)) =$$

$$F^M(\sigma, \text{rk}_{\mathsf{C}} \upharpoonright \text{dom}(\sigma)) = \cup \{ \text{rk}_{\mathsf{C}}(z) + 1 : z \in \text{dom}(\sigma) \}.$$

$\square$

Definition 7.1.18 Let M be a countable transitive model of ZFC and C be a complete boolean algebra in M. Let $G \in \text{St}(\mathsf{C})$.

$$\text{val}_G : M^{\mathsf{C}} \to V$$

$$\sigma \mapsto \text{val}_G(\sigma) = \sigma_G$$

is defined in V by recursion on E^C for $\tau \in M^C$ by the rule

$$\tau_G = \{\sigma_G : \tau(\sigma) \in G\}$$

for any given $\tau \in M^C$.

For any $G \in \mathrm{St}(C)$

$$M[G] = \{\tau_G : \tau \in M^C\}.$$

Remark 7.1.19 The definition of $M[G]$ is by recursion with parameters M, C, G, and can be carried in any model of ZFC to which all the relevant parameters belong.

val_G is defined by recursion on E^C using the function $F : M^B \times V \to V$ given by

$$(f, g) \mapsto \{g(z) : z \in \mathrm{dom}(f) \text{ and } f(z) \in G\}$$

by the rule[5]

$$\mathrm{val}_G(\tau) = F(\tau, \mathrm{val}_G \upharpoonright \mathrm{pred}_E(\tau)) = F(\tau, \mathrm{val}_G \upharpoonright \mathrm{dom}(\tau)).$$

In particular $M[G]$ can be defined in V (but a priori not in M whenever $G \notin M$).

Moreover:

Fact 7.1.20 *Sticking to the above assumptions on M, C, $M[G]$ is transitive for any ultrafilter $G \in \mathrm{St}(C)$ and*

$$M = \left\{ \mathrm{val}_G(\check{a}) : a \in M \right\}.$$

Proof $a \in M[G]$ entails

$$a = \sigma_G \subseteq \mathrm{val}_G[\mathrm{dom}(\sigma)] \subseteq M[G].$$

For the second fact see below (Eq. 7.1).   □

The following basic exercise shows that $M[G]$ has nice closure properties and introduces the canonical operations up, op denoting respectively the *unordered pair* and the *ordered pair* of two elements (we adopt Kuratowski's trick to denote ordered pairs):

Exercise 7.1.21 Let M be a transitive model of ZFC and $G \in \mathrm{St}(B)$ for some $B \in M$ which M models to be a complete boolean algebra. Define up : $(M^B)^2 \to M^B$

---

[5] For a binary relation R on A $\mathrm{pred}_R(b) = \{a \in A : a \, R \, b\}$.

and op $: (M^B)^2 \to M^B$ by the rules

$$\text{up}(\sigma, \tau) = \{\langle \sigma, 1_B \rangle, \langle \tau, 1_B \rangle\},$$

$$\text{op}(\sigma, \tau) = \{\langle \text{up}(\sigma, \tau), 1_B \rangle, \langle \text{up}(\sigma, \sigma), 1_B \rangle\}.$$

Show that up, op are definable class functions in M and that $\text{up}(\sigma, \tau)_G = \{\sigma_G, \tau_G\}$, $\text{op}(\sigma, \tau)_G = \langle \sigma_G, \tau_G \rangle$ for all $\sigma, \tau \in M^B$.

How to Describe an M-generic Filter G for $2^{<\omega}$ Inside M

Is it conceivable to describe the properties of an $r \notin M$ defining an M-generic filter $G = \{s : s \subseteq r\}$ for $2^{<\omega}$ reasoning just about what M can say using first order logic about itself?

The (may be surprising) answer is yes. This is what Cohen has shown with the invention of the forcing method. How can we hope to describe inside M this $r \notin M$?

First of all we can develop the notion of boolean valued model relative to M for $B = RO(2^\omega)^M$ (which is a complete atomless boolean algebra in M). To this aim let

$$O_s = \left\{ f \in (2^\omega)^M : s \subseteq f \right\} = N_s \cap M$$

(i.e., O_s is what M thinks is the basic open set of $(2^\omega)^M = 2^\omega \cap M$ induced by functions extending the finite string s of $0, 1$).

Then $E = \{O_s : s \in 2^{<\omega}\} \in M$ (since $M \models ZFC$ and E is defined inside M as a subset of $RO(2^\omega)^M = B$ obtained by applying the comprehension axiom inside M). Moreover M models that E is a dense subset of B^+, since M (being a transitive model of ZFC) models also that:

$$\left\{ O_s = \left\{ f \in (2^\omega)^M : s \subseteq f \right\} : s \in 2^{<\omega} \right\}$$

is a basis consisting of clopen sets for the complete boolean algebra $RO(2^\omega)^M = B$.

Also (once again because M is a transitive model of ZFC) the map

$$k : 2^{<\omega} \to B \qquad\qquad s \mapsto O_s$$

belongs to M, since it is obtained as a subset of $2^{<\omega} \times RO(2^\omega)^M$ (a set in M) applying an instance of the comprehension axiom in M to a formula with parameters in M. Moreover, M models that the map k implements an isomorphism of $(2^{<\omega}, \supseteq)$ with $(E, \subseteq)$ since this is a Σ_0-property of this map which is true in V and thus also in M.

We can also check the following:

Fact 7.1.22 *Assume G is M-generic for $(2^{<\omega}, \supseteq)$. Then $\bar{G} =\uparrow \{O_s : s \in G\}$ is a ultrafilter on B which is M-generic for B^+.*

Proof By assumption G meets *all* dense subsets of $2^{<\omega}$ in M. Assume we are given $D \subseteq \mathsf{B}^+$ dense open subset of B^+ and in M, we get that $D \cap E \in M$ is also a dense subset of $(E, \leq_{\mathsf{B}})$, and thus

$$\left\{ s \in 2^{<\omega} : O_s \in D \cap E \right\} \in M$$

is a dense subset of $2^{<\omega}$ in M. Thus G meets this dense set, so there is some $s \in G$ such that $O_s \in D \cap E \cap \bar{G}$. $\square$

In V, let us fix some $\bar{G}$ M-generic for B^+, and let us consider the map

$$\mathrm{val}_{\bar{G}} : M^{\mathsf{B}} \to V$$

$$\sigma \mapsto \mathrm{val}_{\bar{G}}(\sigma) = \sigma_{\bar{G}}$$

given by the rule $\tau \mapsto \left\{ \sigma_{\bar{G}} : \tau(\sigma) \in \bar{G} \right\}$. Let us check what this map does on the elements of $\check{M} = \check{V}^M = \check{V} \cap M$: By definition:

$$\mathrm{val}_{\bar{G}}(\check{\emptyset}) = \left\{ \sigma : \sigma \in \mathrm{dom}(\check{\emptyset}) \text{ and } \check{\emptyset}(\sigma) \in \bar{G} \right\},$$

but $\check{\emptyset} = \emptyset$ is the empty function, in particular it has empty domain. We get that

$$\mathrm{val}_{\bar{G}}(\check{\emptyset}) = \left\{ \sigma_{\bar{G}} : \sigma \in \mathrm{dom}(\check{\emptyset}) \text{ and } \check{\emptyset}(\sigma) \in \bar{G} \right\} = \emptyset.$$

Next:

$$\mathrm{val}_{\bar{G}}(\{\check{\emptyset}\}) = \left\{ \sigma : \sigma \in \mathrm{dom}(\{\check{\emptyset}\}) \text{ and } \{\check{\emptyset}\}(\sigma) \in \bar{G} \right\},$$

but $\{\check{\emptyset}\} = \left\{ \langle \check{\emptyset}, 1_{\mathsf{B}} \rangle \right\}$ and $1_{\mathsf{B}} \in \bar{G}$, and thus

$$\mathrm{val}_{\bar{G}}(\{\check{\emptyset}\}) = \left\{ \sigma_{\bar{G}} : \sigma \in \mathrm{dom}(\{\check{\emptyset}\}) \text{ and } \{\check{\emptyset}\}(\sigma) \in \bar{G} \right\} = \left\{ \check{\emptyset}_{\bar{G}} \right\} = \{\emptyset\}.$$

Now by induction on the ranks, assuming $\check{y}_{\bar{G}} = y$ for all $y \in x$, we get that

$$\check{x}_{\bar{G}} = \left\{ \sigma_{\bar{G}} : \sigma \in \mathrm{dom}(\check{x}) \text{ and } \check{x}(\sigma) \in \bar{G} \right\} =$$

$$= \left\{ \check{y}_{\bar{G}} : \check{y} \in \mathrm{dom}(\check{x}) \text{ and } \check{x}(\check{y}) = 1_{\mathsf{B}} \in \bar{G} \right\} = \left\{ \check{y}_{\bar{G}} : y \in x \right\} = x. \qquad (7.1)$$

Thus $\check{M} = (\check{V})^M \subseteq M^{\mathsf{B}}$ is giving B-names for the objects of M.

Now let us take the following B-name:

$$\dot{G} = \{\langle \check{s}, O_s \rangle : s \in 2^{<\omega}\} \in M^{\mathsf{B}}.$$

$\dot{G} \in M^{\mathsf{B}}$ since it is obtained by applying the replacement axiom to the function $2^{<\omega} \to M$ given by $s \mapsto \langle \check{s}, O_s \rangle$. Such a function is a definable class in M and thus its image belongs to M. It is immediate to check that $\dot{G}$ satisfies the clause for the definition of B-names in M.

We obtain

$$\dot{G}_{\bar{G}} = \{\check{s}_{\bar{G}} : \dot{G}(\check{s}) = O_s \in \bar{G}\} = \{s : O_s \in \bar{G}\} = G.$$

We have the surprising fact that the object $G \in V \setminus M$ is described by an element of M^{B}! Similarly

$$\dot{H} = \left\{ \langle \check{b}, b \rangle : b \in \mathsf{B}^+ \right\} \in M^{\mathsf{B}}$$

is such that

$$\dot{H}_{\bar{G}} = \left\{ \check{b}_{\bar{G}} : \dot{H}(\check{b}) = b \in \bar{G} \right\} = \{b : b \in \bar{G}\} = \bar{G}.$$

To get another example, let $r = \cup G$, then $G = \{s \in 2^{<\omega} : s \subseteq r\}$, and we will exhibit a B-name for r in M^{B}. Recall the operations on M^{B}-names defined by

$$\mathrm{up}(\sigma, \tau) = \{\langle \sigma, 1_{\mathsf{B}} \rangle, \langle \tau, 1_{\mathsf{B}} \rangle\}$$

and

$$\mathrm{op}(\sigma, \tau) = \{\langle \mathrm{up}(\sigma, \tau), 1_{\mathsf{B}} \rangle, \langle \mathrm{up}(\sigma, \sigma), 1_{\mathsf{B}} \rangle\}$$

introduced in Exercise 7.1.21. Now let

$$\dot{r} = \left\{ \langle \mathrm{op}(\check{n}, \check{i}), O_{\langle n, i \rangle} \rangle : n < \omega, i < 2 \right\} \in M^{\mathsf{B}},$$

where $O_{\langle n, i \rangle} = \left\{ f \in (2^{\omega})^M : f(n) = i \right\}$. Then (by Exercise 7.1.21)

$$\dot{r}_{\bar{G}} = \left\{ \mathrm{op}(\check{n}, \check{i})_{\bar{G}} : O_{\langle n, i \rangle} \in \bar{G}, n < \omega, i < 2 \right\}$$

$$= \left\{ \langle n, i \rangle : O_{\langle n, i \rangle} \in \bar{G}, n < \omega, i < 2 \right\}$$

$$= \left\{ \langle n, i \rangle : O_{\langle n, i \rangle} \supseteq O_s \text{ for some } s \in G, n < \omega, i < 2 \right\}$$

$$= \left\{ \langle n, i \rangle : s(n) = i \text{ for some } s \in G, n < \omega, i < 2 \right\} = r.$$

In particular, if we let $M[\bar{G}] \subseteq V$ be the family of objects of the form $\tau_{\bar{G}}$ for some $\tau \in M^{\mathsf{B}}$, we have that $r, G, \bar{G} \in M[\bar{G}]$, and also that whenever $a, b \in M[\bar{G}]$, then also $\{a, b\}, \langle a, b \rangle \in M[\bar{G}]$.

Our future investigations will show that M^{B} gives a family of B-names for all elements of $M[\bar{G}]$, and that all the familiar operations on sets we can conceive are reflected in corresponding operations on M^{B}. This will render M^{B} a boolean valued model for ZFC, and $M[\bar{G}]$ a transitive model of ZFC. How will we be able to control the semantic of M^{B} and that of $M[\bar{G}]$? The guiding idea will be the following:

- The B-names define in M a family of "names" for the objects of $M[\bar{G}]$ which labels elements of $M[\bar{G}]$ via the map $\mathrm{val}_{\bar{G}}$.
- The first order properties which $M[\bar{G}]$ assigns to $\mathrm{val}_{\bar{G}}(\sigma)$ for a $\sigma \in M^{\mathsf{B}}$ are conditional on the choice of $\bar{G}$.
- These properties vary continuously with respect to $\mathrm{St}(\mathsf{B})$ as $\bar{G}$ ranges among the M-generic ultrafilters.
- These properties can be described inside M by means of a natural boolean valued semantics on the class M^{B} of B-names, a semantics which is first order definable in M.

For example let $\bar{H} \in \mathrm{St}(\mathsf{B})$ be an M-generic filter on B such that $O_{\langle 0, 1-r(0) \rangle} \in \bar{H}$ ($\bar{H}$ exists since there are densely many M-generic filters for B in V), $H = \{s \in 2^\omega : i(s) \in \bar{H}\}$, $t = \cup H$.

Since $\bar{H}$ is M-generic for B, we can also define

$$\mathrm{val}_{\bar{H}}(\tau) = \tau_{\bar{H}} = \{\sigma_{\bar{H}} : \tau(\sigma) \in \bar{H}\},$$

and we can check that $\mathrm{val}_{\bar{H}}(\check{a}) = a$ for all $a \in M$, but also that

$$\dot{r}_{\bar{H}} = t,$$

$$\dot{G}_{\bar{H}} = H,$$

$$\dot{H}_{\bar{H}} = \bar{H}.$$

This shows that certain properties of the object $\mathrm{val}_K(\tau)$ which is named by the B-name τ depend crucially on the decision an M-generic filter K makes. In our case, if $O_{\langle 0, 0 \rangle} \in K$, we get that $\dot{r}_K(0) = 0$, while if $O_{\langle 0, 1 \rangle} \in K$, $\dot{r}_K(0) = 1$. On the other hand, certain properties of τ cannot be changed by varying the M-generic filters for B. For example whichever K we choose, we will always get that $\dot{r}_K$ is a function in $(2^\omega)^V \setminus M$.

One can introduce in V the following forcing relation for $b \in \mathsf{B}$, $\phi(x_0, \ldots, x_n)$ a first order formula and $\tau_1, \ldots, \tau_n \in V^{\mathsf{B}}$:

$$b \Vdash \phi(\tau_1, \ldots, \tau_n)$$

if and only if

$$M[K] \models \phi(\text{val}_K(\tau_1), \ldots, \text{val}_K(\tau_n)) \text{ for all } M\text{-generic filters}$$

$$K \text{ for B such that } b \in K.$$

The intuition is that b decides (or "forces") certain facts (those described by ϕ) about the B-names $\tau_1, \ldots, \tau_n$ to be true in $M[K]$ of the objects $(\tau_1)_K, \ldots, (\tau_n)_K$, no matter how an M-generic filter $K \ni b$ evaluates $\tau_1, \ldots, \tau_n$. Formally

$$b \Vdash \phi(\tau_1, \ldots, \tau_n)$$

stands for

$$V \models \forall K \in \text{St}(\text{B}) \, [(K \text{ is } M\text{-generic for B} \wedge b \in K) \to \text{Sat}(M[K], \overline{\phi}, \langle (\tau_1)_K, \ldots, (\tau_n)_K \rangle)],$$

where $\text{Sat}(x, y, z)$ is the satisfaction predicate for structures of the form $(N, \in)$ introduced in Sect. A.4.

So far the above observations show among other things:

1. $1_\text{B} \Vdash \dot{r} : \check{\omega} \to \check{2}$ is a function.
2. For all $f \in (2^\omega)^M$, $1_\text{B} \Vdash \dot{G} \cap \check{D}_f \neq \emptyset$.
3. For all $i < 2, n \in \omega$, $O_{\langle n, i \rangle} \Vdash \dot{r}(\check{n}) = \check{i}$.
4. $b \Vdash \phi \wedge \psi$ iff $b \Vdash \phi$ and $b \Vdash \psi$.

This forcing relation tells us that when b is chosen by some G, all the properties which b assigns to a certain B-name will hold for the interpretation of that B-name by G. This is very useful and allows to compute in V what properties of a B-name τ are decided by a condition in B and in which ways. Moreover this forcing relation has the same flavor of the boolean valued semantics we met so far, and one of our main objectives (i.e., Cohen's forcing theorem) amounts to show that:

> The forcing relation on $\text{B}^+ \times \text{Form} \times (M^\text{B})^{<\omega}$ defined in V obeys to the laws given by Lemma 6.3.11, if (following the notation of the Lemma) we replace all over M/G by $M[G]$, and $G \in \text{St}(\text{B})$ by $G \in \text{St}(\text{B})$ *is M-generic for* B.

However we have a great problem to match for the above forcing relation:

> The semantic for M^B we defined above has not been defined inside M.

To define the forcing relation $b \Vdash \phi(\tau_1, \ldots, \tau_n)$, we need to be able to quantify over all G which are M-generic for B. This can be done meaningfully in V (where the above set is a dense subset of $\text{St}(\text{B})$); however in M the set of such G defines the empty set (since B is atomless), and $M[G]$ cannot be defined.

This problem causes serious difficulties if our aim is to endow M^B of the structure of a boolean valued model definable in M.

Can we expand the above forcing relation so to be able to give to M^B the structure of a B-valued model? Concretely, by means of the above forcing relation can we define inside M class functions $G_\phi^n : (M^B)^n \to B$ which assign a boolean value to formulae $\phi(x_0, \ldots, x_n)$ evaluated in the tuple $(\tau_0, \ldots, \tau_n)$ with assignment $x_i \mapsto \tau_i$?

In V we can define the set

$$A_\phi(\vec{\tau}) = \{b \in B^+ : V \models (b \Vdash \phi(\vec{\tau}))\}.$$

$A_\phi(\vec{\tau}) \in \mathcal{P}(B)$ is a subset of M, since $B \in M$ and M is transitive. However we have no special argument to expect that $A_\phi(\vec{\tau}) \in M$, since this set is defined using in V an instance of the comprehension axiom for the formula defining $\Vdash$; this formula requires to quantify over sets not in M. In particular if this set is in M, we must find some specific argument to be able to assert it.

Nonetheless it appears that the reasonable definition of a B-valued semantic for M^B is given by letting $[\![\phi(\vec{\tau})]\!] = \bigvee A_\phi(\vec{\tau})$ (as in the last item of the Forcing Lemma 6.3.11); for example this holds for:

1. $[\![\dot{r}(\check{n}) = \check{i}]\!] = \bigvee_B \{b : b \Vdash \dot{r}(\check{n}) = \check{i}\} = \bigvee_B \{b : b \leq O_{\langle n,i \rangle}\} = O_{\langle n,i \rangle}$.
2. $[\![\dot{r} : \check{\omega} \to \check{2}]\!] = \bigvee_B \{b : b \Vdash \dot{r} : \check{\omega} \to \check{2}\} = \bigvee_B \{b : b \leq 1_B\} = 1_B$.

In the above equalities we ended up having $A_\phi(\vec{\tau}) \in M$ for

$$\phi(\dot{r}, \check{n}, \check{i}) \equiv (\dot{r}(\check{n}) = \check{i})$$

and also for

$$\phi(\dot{r}, \check{\omega}, \check{2}) \equiv (\dot{r} : \check{\omega} \to \check{2} \text{ is a function}).$$

Is this a peculiarity of these formulae?

Let us work now under the assumption that $A_{x \in y}(\tau, \sigma)$, $A_{x=y}(\tau, \sigma)$, $A_{x \subseteq y}(\tau, \sigma)$ are in M for all $\sigma, \tau \in M^B$.

Fact 7.1.23 *Assume M is a transitive countable model of ZFC, and $B \in M$ is a complete boolean algebra such that M models B is complete. Assume further that*

$$A_R(\sigma, \tau) = \{b \in B : b \Vdash \sigma \mathrel{R} \tau\} \in M$$

for R among $\in, =, \subseteq$ and $\sigma, \tau \in M^B$. Then we can set $[\![\sigma \; R \; \tau]\!] = \bigvee A_R(\sigma, \tau)$, and we get that:

1. *For all $\tau, \sigma, \pi \in M^B$,*

$$[\![\tau = \tau]\!] = 1_B,$$
$$[\![\tau = \sigma]\!] = [\![\sigma = \tau]\!],$$
$$[\![\tau = \sigma]\!] \wedge [\![\sigma = \pi]\!] \leq [\![\tau = \pi]\!].$$

2. *For R among $\in, =$, and for all $\langle \tau_1, \tau_2 \rangle, \langle \sigma_1, \sigma_2 \rangle \in (M^B)^2$,*

$$[\![\tau_1 = \sigma_1]\!] \wedge [\![\tau_2 = \sigma_2]\!] \wedge [\![\tau_1 \; R \; \tau_2]\!] \leq [\![\sigma_1 \; R \; \sigma_2]\!].$$

In particular, letting $R^B(\tau, \sigma) = [\![\tau \; R \; \sigma]\!]$ for R among $\in, =$, $(M^B, =^B, \in^B)$ is in V a B-valued model.

Exercise 7.1.24 Prove the above inequalities. (HINT: First show that it suffices to prove that $b \Vdash \phi$ entails $b \Vdash \psi$ for ϕ, ψ formulae appearing in any of the above inequalities and for all $b \in B^+$ (where ϕ stands for the left-hand term of the inequality and ψ for the right-hand term). Then apply the definition of the forcing relation.)

We are led to the following driving questions:

1. Can we define in M class functions $A_R : (M^B)^2 \to \mathcal{P}(B)^M$ such that

$$A_R(\tau, \sigma) = \left\{ b \in B^+ : b \Vdash \sigma \; R \; \tau \right\}$$

   for R among $\in, =, \subseteq$?
2. Can we prove in general that $A_\phi(\vec{\tau}) \in M$ for all formulae $\phi(\vec{\tau})$ with parameters in M^B?
3. Assume both questions have a positive answer. Can we also prove that the boolean valued semantic for $\phi(\vec{\tau})$ given by $(M^B, \in^B, =^B, \subseteq^B)$ (where $[\![\sigma \; R^B \; \tau]\!] = \bigvee A_{xRy}(\sigma, \tau)$ for R among $=, \in, \subseteq$) assigns to each formula $\phi(\vec{\tau})$ the boolean value $\bigvee A_\phi(\vec{\tau})$?
4. Assume that the first question has a positive answer. Can we also prove that $(M^B, \in^B, =^B)$ is a *full* B-valued model in V?

We show in the next sections that all these questions have a positive answer, proving the following result:

M can define a structure of full B-valued model on M^B which assigns to any formula $\phi(x_1, \ldots, x_n)$ of the language of set theory a satisfaction class definable in M

$$\left\{ (\tau_1, \ldots, \tau_n, b) \in (M^B)^n \times B : M \models [\![\phi(\tau_1, \ldots, \tau_n)]\!] = b \right\}$$

with the feature that

$$M \models [\![\phi(\tau_1, \ldots, \tau_n)]\!] \geq_B b$$

if and only if

$$V \models b \Vdash \phi(\tau_1, \ldots, \tau_n).$$

In particular the forcing relation $b \Vdash \phi(\tau_1, \ldots, \tau_n)$ and the relations $R^B(\sigma, \tau) = [\![\sigma R \tau]\!]$ for $R \in \{=, \in, \subseteq\}$ can also be defined inside M making $(M^B, =^B, \in^B, \subseteq^B)$ a full B-valued model in V given by a quadruple of definable classes in M.

This will be done as follows: After having defined in M a boolean valued semantics on M^B, making M^B a full B-valued model for the first order language $\{\in, \subseteq\}$, we show that whenever G is M-generic for B, M^B/G is isomorphic to $M[G]$ via the map $[\tau]_G \mapsto \tau_G$ (where $[\sigma]_G = \{\tau \in M^B : [\![\tau = \sigma]\!] \in G\}$). By Łoś Theorem 6.3.7 and by the Forcing Lemmata 6.3.10 and 6.3.11, all the desired properties of M^B can be easily inferred, since the set of M-generic filters for St(B) is dense.

7.1.2 Internal Definition of Forcing: The Boolean Valued Semantics of V^B

To simplify matters and notations we will assume all over this section to be working in V, the standard model of ZFC which contains all sets, however all of our definitions and results can be declined and rephrased for any arbitrary first order model of ZFC since they will be based just on the assumption that V is a model of ZFC. We will need in the next sections the relativization of many of these definitions and results to a countable transitive set $M \in V$ which is itself a model of ZFC.

The aim of this section is to define a boolean semantic on the class V^B making it a full boolean valued model for the language $\{\in, \subseteq\}$. In the next section we will show that this semantics, when defined in a countable transitive model M of ZFC, induces the forcing relation on M^B defined in the previous section.

Definition 7.1.25 Let $\tau, \sigma \in V^B$. We define simultaneously by induction on the pairs $(\text{rk}(\tau), \text{rk}(\sigma))$ well ordered in type Ord by the square order[6] on Ord^2:

1.

$$[\![\tau \in \sigma]\!] = \bigvee_{\tau_0 \in \text{dom}(\sigma)} ([\![\tau = \tau_0]\!] \wedge \sigma(\tau_0)).$$

---

[6] The square order $<^2$ is given by $(\alpha, \beta) <^2 (\gamma, \delta)$ iff $\max\{\alpha, \beta\} < \max\{\gamma, \delta\}$ or $\max\{\gamma, \delta\} = \max\{\gamma, \delta\}$ and (α, β) is lexicographically below (γ, δ).

2.

$$\llbracket \tau \subseteq \sigma \rrbracket = \bigwedge_{\sigma_0 \in \mathrm{dom}(\tau)} (\tau(\sigma_0) \to \llbracket \sigma_0 \in \sigma \rrbracket) = \bigwedge_{\sigma_0 \in \mathrm{dom}(\tau)} (\neg \tau(\sigma_0) \vee \llbracket \sigma_0 \in \sigma \rrbracket).$$

3.

$$\llbracket \tau = \sigma \rrbracket = \llbracket \tau \subseteq \sigma \rrbracket \wedge \llbracket \sigma \subseteq \tau \rrbracket.$$

Remark 7.1.26 The definition of all three relations is by a simultaneous induction. More precisely let $F_j : V \times V^B \times V^B \to B$ for $j = 0, 1$ be defined by

$$\begin{cases} F_0(g, \tau, \sigma) = \bigvee_{\eta \in \mathrm{dom}(\sigma)} \sigma(\eta) \wedge g(\eta, \tau) \text{ if } & \text{for some } \alpha \\ & g : (V_\alpha^B)^2 \to B \text{ and } \mathrm{dom}(\sigma) \times \{\tau\} \subseteq \mathrm{dom}(g) \\ \\ F_0(g, \tau, \sigma) = 0_B & \text{otherwise;} \end{cases}$$

$$\begin{cases} F_1(g, \tau, \sigma) = \bigwedge_{\eta \in \mathrm{dom}(\tau)} (\tau(\eta) \to g(\eta, \sigma)) \text{ if } & \text{for some } \alpha \\ & g : (V_\alpha^B)^2 \to B \text{ and } \mathrm{dom}(\sigma) \times \{\tau\} \subseteq \mathrm{dom}(g) \\ \\ F_1(g, \tau, \sigma) = 0_B & \text{otherwise.} \end{cases}$$

Now let $G : (V^B)^2 \to B^3$ be defined by transfinite recursion by the following clauses:

$$G(\tau, \sigma) = (G_0(\tau, \sigma), G_1(\tau, \sigma), G_2(\tau, \sigma)),$$

where

$$G_0(\tau, \sigma) = F_0(G_2 \upharpoonright \mathrm{dom}(\sigma) \times \{\tau\}, \tau, \sigma),$$
$$G_1(\tau, \sigma) = F_1(G_0 \upharpoonright \mathrm{dom}(\tau) \times \{\sigma\}, \tau, \sigma),$$
$$G_2(\tau, \sigma) = G_1(\tau, \sigma) \wedge G_1(\sigma, \tau).$$

We leave to the reader to check that such a G is a definable class in V in the parameter $(B, \wedge, \vee, \neg, 0_B, 1_B)$, and that $G_0(\tau, \sigma) = \llbracket \tau \in \sigma \rrbracket$, $G_1(\tau, \sigma) = \llbracket \tau \subseteq \sigma \rrbracket$, $G_2(\tau, \sigma) = \llbracket \tau = \sigma \rrbracket$.

It can also be observed that the relations $\llbracket \tau \in \sigma \rrbracket$ and $\llbracket \tau \subseteq \sigma \rrbracket$ are Δ_1-definable in the parameter B and thus are absolute between M and V if M is a transitive model of ZFC to which B belongs. However this is slightly more subtle since B could be a complete boolean algebra in M, while it is not such in V, thus it is less transparent why the definition of $\llbracket \tau \in \sigma \rrbracket$ and $\llbracket \tau \subseteq \sigma \rrbracket$ which are using in an essential way the completeness of B can even be formulated in V where B might not be a complete boolean algebra. We may come back to this point later on when we will need to clarify it. The key observation to solve this issue being that $M \models b = \bigvee_B A$ iff $V \models b = \bigvee_B A$ for all $A \in M \cap \mathcal{P}(B)$.

Theorem 7.1.27 V^B *is a boolean valued model for set theory.*

Proof We have to check that V^B satisfies the four clauses of Definition 6.1.1, and the two additional items of Definition 7.1.1, i.e., we have to show that, for all $\tau, \sigma, \eta \in V^B$:

1. $[\![\tau = \tau]\!] = 1$.
2. $[\![\tau = \sigma]\!] = [\![\sigma = \tau]\!]$.
3. $[\![\tau = \sigma]\!] \wedge [\![\sigma = \eta]\!] \leq [\![\tau = \eta]\!]$.
4. $[\![\tau = \sigma]\!] \wedge [\![\sigma \ R \ \eta]\!] \leq [\![\tau \ R \ \eta]\!]$, for $R \in \{\in, \subseteq\}$.
5. $[\![\tau = \sigma]\!] \wedge [\![\eta \ R \ \tau]\!] \leq [\![\eta \ R \ \sigma]\!]$, for $R \in \{\in, \subseteq\}$.
6. $[\![\tau \subseteq \sigma]\!] \wedge [\![\sigma \subseteq \tau]\!] = [\![\tau = \sigma]\!]$.
7. $[\![\tau \in \sigma]\!] \wedge [\![\sigma \subseteq \eta]\!] \leq [\![\tau \in \eta]\!]$.

The proof of item 1 is by induction on $\mathrm{rk}(\tau)$, and we do it right away. We have

$$[\![\tau \subseteq \tau]\!] = \bigwedge_{\sigma \in \mathrm{dom}(\tau)} (\neg\tau(\sigma) \vee [\![\sigma \in \tau]\!]) =$$

$$\bigwedge_{\sigma \in \mathrm{dom}(\tau)} (\neg\tau(\sigma) \vee (\bigvee_{u \in \mathrm{dom}(\tau)} [\![\sigma = u]\!] \wedge \tau(u))) \geq \bigwedge_{\sigma \in \mathrm{dom}(\tau)} (\neg\tau(\sigma) \vee ([\![\sigma = \sigma]\!] \wedge \tau(\sigma))).$$

But $[\![\sigma = \sigma]\!] = 1$, because $\mathrm{rk}(\sigma)$ is below $\mathrm{rk}(\tau)$, and we can apply the inductive assumptions. Thus:

$$\bigwedge_{\sigma \in \mathrm{dom}(\tau)} (\neg\tau(\sigma) \vee ([\![\sigma = \sigma]\!] \wedge \tau(\sigma))) = \bigwedge_{\sigma \in \mathrm{dom}(\tau)} (\neg\tau(\sigma) \vee \tau(\sigma)) = 1,$$

i.e., $[\![\tau = \tau]\!] = 1$ as was to be shown.

Items 2 and 6 follow immediately from the definitions.

Next observe that if we can prove

$$[\![\tau \subseteq \sigma]\!] \wedge [\![\sigma \subseteq \eta]\!] \leq [\![\tau \subseteq \eta]\!] \tag{7.2}$$

for all triples (τ, σ, η), we also get items 3, as well as 4, 5 for the case of R being $\subseteq$, since:

$$[\![\tau = \sigma]\!] \wedge [\![\sigma \subseteq \eta]\!] \leq [\![\tau \subseteq \sigma]\!] \wedge [\![\sigma \subseteq \eta]\!] \leq [\![\tau \subseteq \eta]\!]$$

applying 7.2 to the triple (τ, σ, η) in the last inequality, which yields 4 for the case R being $\subseteq$. Similarly

$$[\![\tau = \sigma]\!] \wedge [\![\eta \subseteq \sigma]\!] \leq [\![\tau \subseteq \sigma]\!] \wedge [\![\eta \subseteq \sigma]\!] \leq [\![\eta \subseteq \tau]\!],$$

applying 7.2 to the triple (η, σ, τ) in the last inequality to infer 5 for the case R being $\subseteq$. 3 follows from 7.2 by a similar argument, left to the reader.

Next if we can prove 7 for all triples (τ, σ, η), we get 5 for the case of R being $\in$, since

$$[\![\tau = \sigma]\!] \wedge [\![\eta \in \tau]\!] \leq [\![\tau \subseteq \sigma]\!] \wedge [\![\eta \in \tau]\!] \leq [\![\eta \in \sigma]\!]$$

applying 7 to triple (η, τ, σ) in the last of the above inequalities.

Hence it suffices to prove 7, 7.2, 4 for the case of R being $\in$, i.e., the following three items:

(a) $[\![\tau \in \sigma]\!] \wedge [\![\sigma \subseteq \eta]\!] \leq [\![\tau \in \eta]\!]$.
(b) $[\![\tau = \sigma]\!] \wedge [\![\tau \in \eta]\!] \leq [\![\sigma \in \eta]\!]$.
(c) $[\![\tau \subseteq \sigma]\!] \wedge [\![\sigma \subseteq \eta]\!] \leq [\![\tau \subseteq \eta]\!]$.

We will prove (a), (b), (c) by means of a nested induction on the triples $(\mathrm{rk}(\tau), \mathrm{rk}(\sigma), \mathrm{rk}(\eta))$ ordered by the cube well order[7] on Ord^3; we will do the induction for (a), (b), (c) simultaneously; to prove each of these items for some triple of B-names, we will assume that all three properties (a), (b), (c) hold for all triples of B-names of lower rank in the cube ordering.

We will also use the following observation:

If (a), (b), (c) hold for all triples up to a given rank in the cube order, we get that 3, 4, 5, 6, 7 hold for all these triples.

This is the case since the arguments we gave above leading from any of (a), (b), (c) to some of 3, 4, 5, 6, 7 can be repeated verbatim for the triples at hand since no inductive assumption is needed to carry these arguments:

(a): $[\![\tau \in \sigma]\!] \wedge [\![\sigma \subseteq \eta]\!]$ is equal to

$$\bigvee_{\sigma_0 \in \mathrm{dom}(\sigma)} ([\![\tau = \sigma_0]\!] \wedge \sigma(\sigma_0)) \wedge \bigwedge_{\sigma_1 \in \mathrm{dom}(\sigma)} (\neg \sigma(\sigma_1) \vee [\![\sigma_1 \in \eta]\!]).$$

The latter is equal to

$$\bigvee_{\sigma_0 \in \mathrm{dom}(\sigma)} \bigwedge_{\sigma_1 \in \mathrm{dom}(\sigma)} [([\![\tau = \sigma_0]\!] \wedge \sigma(\sigma_0) \wedge \neg \sigma(\sigma_1)) \vee ([\![\tau = \sigma_0]\!] \wedge \sigma(\sigma_0)$$

$$\wedge [\![\sigma_1 \in \eta]\!])].$$

---

[7] The cube order $<^3$ is given by $(\alpha, \beta, \gamma) <^2 (\eta, \delta, \nu)$ iff $\max\{\alpha, \beta, \gamma\} < \max\{\eta, \delta, \nu\}$ or $\max\{\alpha, \beta, \gamma\} = \max\{\eta, \delta, \nu\}$ and (α, β, γ) is lexicographically below (η, δ, ν).

Now $\sigma(\sigma_0) \wedge \neg\sigma(\sigma_0)) = 0$ for any $\sigma_0 \in \mathrm{dom}(\sigma)$ and σ_0, σ_1 both range among $\mathrm{dom}(\sigma)$. Hence for all $\sigma_0 \in \mathrm{dom}(\sigma)$

$$\bigwedge_{\sigma_1 \in \mathrm{dom}(\sigma)} [(\llbracket \tau = \sigma_0 \rrbracket \wedge \sigma(\sigma_0) \wedge \neg\sigma(\sigma_1)) \vee (\llbracket \tau = \sigma_0 \rrbracket \wedge \sigma(\sigma_0) \wedge \llbracket \sigma_1 \in \eta \rrbracket)]$$

$$\leq \llbracket \tau = \sigma_0 \rrbracket \wedge \sigma(\sigma_0) \wedge \llbracket \sigma_0 \in \eta \rrbracket$$

$$\leq \llbracket \tau = \sigma_0 \rrbracket \wedge \llbracket \sigma_0 \in \eta \rrbracket .$$

This gives that

$$\llbracket \tau \in \sigma \rrbracket \wedge \llbracket \sigma \subseteq \eta \rrbracket \leq$$

$$\leq \bigvee_{\sigma_0 \in \mathrm{dom}(\sigma)} \llbracket \tau = \sigma_0 \rrbracket \wedge \llbracket \sigma_0 \in \eta \rrbracket$$

$$\leq \bigvee_{\sigma_0 \in \mathrm{dom}(\sigma)} \llbracket \tau \in \eta \rrbracket$$

$$= \llbracket \tau \in \eta \rrbracket .$$

For the last inequality we have used the inductive hypothesis (b) on the triple (τ, σ_0, η) which is below the triple (τ, σ, η) in the cube order on Ord^3.

(b): Let $t \in \mathrm{dom}(\eta)$; we have

$$\llbracket \sigma = \tau \rrbracket \wedge \llbracket \tau = t \rrbracket \wedge \eta(t) \leq \llbracket \sigma = t \rrbracket \wedge \eta(t)$$

applying the inductive assumption on item 3 to the triple (σ, τ, t) which is below the triple (σ, τ, η) in the cube order. Thus:

$$\llbracket \tau = \sigma \rrbracket \wedge \llbracket \sigma \in \eta \rrbracket =$$

$$= \llbracket \sigma = \tau \rrbracket \wedge \llbracket \sigma \in \eta \rrbracket$$

$$= \llbracket \sigma = \tau \rrbracket \wedge (\bigvee_{t \in \mathrm{dom}(\eta)} \llbracket \tau = t \rrbracket \wedge \eta(t))$$

$$= \bigvee_{t \in \mathrm{dom}(\eta)} (\llbracket \sigma = \tau \rrbracket \wedge \llbracket \tau = t \rrbracket \wedge \eta(t))$$

$$\leq \bigvee_{t \in \mathrm{dom}(\eta)} (\llbracket \sigma = t \rrbracket \wedge \eta(t))$$

$$= \llbracket \sigma \in \eta \rrbracket .$$

(c): Let $t \in \mathrm{dom}(\tau)$. We can apply the inductive assumption (a) on the triple (t, σ, η) which is below the triple (τ, σ, η) in the cube order on Ord^3 to get

$$[\![t \in \sigma]\!] \wedge [\![\sigma \subseteq \eta]\!] \leq [\![t \in \eta]\!] .$$

This gives that

$$[\![\tau \subseteq \eta]\!] = \bigwedge_{t \in \mathrm{dom}(\tau)} \tau(t) \to [\![t \in \eta]\!] \geq \bigwedge_{t \in \mathrm{dom}(\tau)} \tau(t) \to ([\![t \in \sigma]\!] \wedge [\![\sigma \subseteq \eta]\!]).$$

The latter is equal to

$$(\bigwedge_{t \in \mathrm{dom}(\tau)} \tau(t) \to [\![t \in \sigma]\!]) \wedge (\bigwedge_{t \in \mathrm{dom}(\tau)} \tau(t) \to [\![\sigma \subseteq \eta]\!]).$$

Now observe that

$$(\bigwedge_{t \in \mathrm{dom}(\tau)} \tau(t) \to [\![\sigma \subseteq \eta]\!]) = [\![\sigma \subseteq \eta]\!] \vee (\bigwedge_{t \in \mathrm{dom}(\tau)} \neg \tau(t)) \geq [\![\sigma \subseteq \eta]\!] ,$$

while

$$\bigwedge_{t \in \mathrm{dom}(\tau)} \tau(t) \to [\![t \in \sigma]\!] = [\![\tau \subseteq \sigma]\!] .$$

We conclude that

$$[\![\tau \subseteq \eta]\!] \geq [\![\tau \subseteq \sigma]\!] \wedge [\![\sigma \subseteq \eta]\!] ,$$

as was to be shown.

The proof is complete. □

Remark 7.1.28 The above proof can be formalized in the following manner: Letting $G : (V^{\mathsf{B}})^2 \to \mathsf{B}^3$ the class function defining the relations $[\![\tau = \sigma]\!] = G_0(\tau, \sigma), [\![\tau \subseteq \sigma]\!] = G_1(\tau, \sigma), [\![\tau = \sigma]\!] = G_2(\tau, \sigma)$, one shows by recursion on the appropriate (pair or triple of) rank(s) that:

1. $G_1(\tau, \tau) = G_2(\tau, \tau) = 1_\mathsf{B}$.
2. $G_2(\tau, \sigma) = G_2(\sigma, \tau)$.
3. $G_1(\tau, \sigma) \wedge G_1(\sigma, \tau) = G_2(\sigma, \tau)$.
4. $G_2(\tau, \sigma) \wedge G_2(\sigma, \eta) \leq G_2(\eta, \tau)$.
5. $G_2(\tau, \sigma) \wedge G_j(\sigma, \eta) \leq G_j(\tau, \eta)$ for $j = 0, 1$.
6. $G_2(\tau, \sigma) \wedge G_j(\tau, \eta) \leq G_j(\sigma, \eta)$ for $j = 0, 1$.
7. $G_0(\tau, \sigma) \wedge G_1(\sigma, \eta) \leq G_1(\tau, \eta)$.

Once we have shown that in V the class V^B with the classes G_0, G_1, G_2 for $\in^B, \subseteq^B, =^B$ satisfies the clauses for a boolean valued model given in Definition 6.1.1, we can give an interpretation to all formulae of the first order language $\{\in, \subseteq, =\}$ assigning by recursion to each formula ϕ its boolean satisfaction class G_ϕ as follows:

Definition 7.1.29 For each formula $\phi(x_0, \ldots, x_n)$ formula in displayed free variables in the language $\{\in, \subseteq, =\}$, we let

$$G_\phi^n : (V^B)^n \to B \qquad\qquad (\tau_0, \ldots, \tau_n) \mapsto [\![\phi(\tau_0, \ldots, \tau_n)]\!]_B$$

be the class defined by the following constraints on θ, ψ as θ, ψ vary among the subformulae of ϕ:

$$[\![\psi(\tau_0, \ldots \tau_n)]\!]_B \wedge_B [\![\theta(\tau_0, \ldots, \tau_m)]\!]_B = [\![\psi(\tau_0, \ldots \tau_n) \wedge \theta(\tau_0, \ldots, \tau_m)]\!]_B .$$

$$[\![\psi(\tau_0, \ldots \tau_n)]\!]_B \vee_B [\![\theta(\tau_0, \ldots, \tau_m)]\!]_B = [\![\psi(\tau_0, \ldots \tau_n) \vee \theta(\tau_0, \ldots, \tau_m)]\!]_B .$$

$$\neg_B [\![\psi(\tau_0, \ldots \tau_n)]\!]_B = [\![\neg\psi(\tau_0, \ldots \tau_n)]\!]_B .$$

$$[\![\exists x_j \psi(\tau_0, \ldots, \tau_{j-1}, x, \tau_j, \ldots, \tau_n)]\!]_B$$
$$= \bigvee_{\sigma \in V^B} [\![\psi(\tau_0, \ldots, \tau_{j-1}, \sigma, \tau_j, \ldots, \tau_n)]\!]_B .$$

Remark 7.1.30 We think it is important for this definition to be pedantic. A detailed account of the above clauses is the following for $\phi(x_0, \ldots, x_n)$ formula in the language $\{\in, \subseteq, =\}$ in displayed free variables:

$\phi \equiv x_i \ R \ x_j$ **for R among** $\in, \subseteq, =$ **and** $i \leq j \leq n$:

$$G_\phi^n = \{(\tau_0, \ldots, \tau_n, b) : V \models [\![\tau_i \ R \ \tau_j]\!]_B = b\} .$$

$\phi \equiv \psi(x_0, \ldots, x_n) \wedge \theta(x_0, \ldots, x_n)$: We set G_ϕ^n to be the class of $(\tau_0, \ldots, \tau_n, b)$ in $V^{n+1} \times B$ such that

$$V \models \exists c, d \in B \, (b = c \wedge_B d) \wedge G_\psi^n(\tau_0, \ldots, \tau_n, c) \wedge G_\theta^n(\tau_0, \ldots, \tau_n, d).$$

That is:

$$b = [\![\psi(\tau_0, \ldots \tau_n)]\!]_B \wedge_B [\![\theta(\tau_0, \ldots, \tau_n)]\!]_B .$$

$\phi \equiv \neg \psi (x_0, \ldots, x_n)$:

$$G_\phi^n = \left\{ (\tau_0, \ldots, \tau_n, b) : V \models \exists c \in \mathbf{B} \, (b = \neg c) \wedge G_\psi^n (\tau_0, \ldots, \tau_n, c) \right\}.$$

That is:

$$b = \neg \, [\![\psi (\tau_0, \ldots \tau_n)]\!]_\mathbf{B} \, .$$

$\phi \equiv \exists x_j \psi (x_0, \ldots, x_n)$: Letting

$$\theta (x_0, \ldots, x_{j-1}, x_{j+1}, \ldots, x_n, z)$$

be the formula

$$\forall u [\exists y \, G_\psi^{n+1} (x_0, \ldots, x_{j-1}, u, x_{j+1}, \ldots, x_n, y) \rightarrow z \geq_\mathbf{B} y],$$

set G_ϕ^n to be the class of $(\tau_0, \ldots, \tau_{j-1}, \tau_{j+1} \ldots, \tau_n, c)$ such that

$$V \models \exists \sigma \, G_\psi^{n+1} (\tau_0, \ldots, \tau_{j-1}, \sigma, \tau_{j+1}, \ldots, \tau_n, c)$$
$$\wedge \theta (\tau_0, \ldots \tau_{j-1}, \tau_{j+1}, \ldots, \tau_n, c).$$

That is, $(\tau_0, \ldots, \tau_{j-1}, \tau_{j+1} \ldots, \tau_n, c) \in G_\phi^n$ if and only if

$$c = \bigvee_{\sigma \in V^\mathbf{B}} [\![\psi (\tau_0, \ldots \tau_{j-1}, \sigma, \tau_{j+1}, \ldots, \tau_n)]\!]_\mathbf{B} \, .$$

With these definitions we have that $\left\langle V^\mathbf{B}, \in^\mathbf{B}, \subseteq^\mathbf{B}, =^\mathbf{B} \right\rangle$ is a $\mathbf{B}$-valued model. Note also that each class G_ϕ^n is a definable class in V, but (it can be shown that) the collection of classes

$$\left\{ G_\phi^n : \phi \text{ a formula of } \mathcal{L}, \, n \in \omega \right\}$$

cannot be represented as a definable class in V. We now show that $\left\langle V^\mathbf{B}, \in^\mathbf{B}, \subseteq^\mathbf{B}, =^\mathbf{B} \right\rangle$ is full, which formally amounts to show that:

For all formulae $\phi \equiv \exists x_j \psi (x_0, \ldots, x_n)$, $n \in \omega$, and all $(\tau_0, \ldots, \tau_{j-1}, \tau_{j+1}, \ldots \tau_n, b) \in G_\phi^n$, there exists $\sigma \in V^\mathbf{B}$ such that

$$(\tau_0, \ldots, \tau_{j-1}, \sigma, \tau_{j+1}, \ldots \tau_n, b) \in G_\psi^{n+1},$$

and

$$b \geq d$$

for all d such that for some $\tau \in V^B$

$$(\tau_0, \ldots, \tau_{j-1}, \tau, \tau_{j+1}, \ldots \tau_n, d) \in G_\psi^{n+1},$$

and informally to assert that

$$\bigvee_{\tau \in V^B} \llbracket \psi(\tau_0, \ldots, \tau_{j-1}, \tau, \tau_{j+1}, \ldots \tau_n) \rrbracket = \llbracket \psi(\tau_0, \ldots, \tau_{j-1}, \sigma, \tau_{j+1}, \ldots \tau_n) \rrbracket$$

for some $\sigma \in V^B$.

Toward this aim we prove that V^B satisfies the mixing property:

Lemma 7.1.31 (Mixing Lemma) *Let* B *be a boolean algebra. Let* A *be an antichain of* B *and for any* $a \in A$ *let* τ_a *be an element of* V^B. *Then there exists some* $\tau \in V^B$ *such that* $a \le \llbracket \tau = \tau_a \rrbracket$ *for all* $a \in A$.

In the proof of the above Lemma, we will need Exercise 3.3.5 on boolean algebras.

Proof Let $D = \bigcup_{a \in A} \mathrm{dom}(\tau_a)$ and, for every $t \in D$, let

$$\tau(t) = \bigvee \{a \wedge \tau_a(t) : a \in A \wedge t \in \mathrm{dom}(\tau_a)\}.$$

Since A is an antichain and by the definition of $\tau(t)$, we have that

$$a \wedge \tau(t) = a \wedge \tau_a(t)$$

for any $a \in A$ and any $t \in \mathrm{dom}(\tau_a)$. So, by Exercise 3.3.5, for any $a \in A$,

$$\text{For all } t \in \mathrm{dom}(\tau_a) \ (a \le \tau_a(t) \leftrightarrow \tau(t)). \tag{7.3}$$

On the other hand

$$\text{For all } t \in D \setminus \mathrm{dom}(\tau_a) \ (a \wedge \tau(t) = 0) \tag{7.4}$$

holds since for such elements t

$$\tau(t) = \bigvee \{\tau_b(t) \wedge b : b \ne a, \ b \in A, \ t \in \mathrm{dom}(\tau_b)\} \le \left(\bigvee A\right) \setminus a.$$

□

Now we use 7.3 and 7.4 to obtain that:

Claim 7.1.31.1 *The following holds for* $a \in A$:

$$a \le \llbracket \tau_a \subseteq \tau \rrbracket,$$

$$a \leq [\![\tau \subseteq \tau_a]\!] \, .$$

Proof We use Eq. 7.3 to infer that $a \leq [\![\tau_a \subseteq \tau]\!]$ as follows: First of all

$$[\![t \in \tau]\!] \geq \tau(t)$$

for all $t \in D = \mathrm{dom}(\tau)$, so we have

$$a \leq \tau_a(t) \rightarrow \tau(t) \leq \tau_a(t) \rightarrow [\![t \in \tau]\!]$$

for any $a \in A$ and any $t \in \mathrm{dom}(\tau_a) \subseteq D$. So we have

$$a \leq \bigwedge_{t \in \mathrm{dom}(\tau_a)} \tau_a(t) \rightarrow \tau(t) \leq \bigwedge_{t \in \mathrm{dom}(\tau_a)} \tau_a(t) \rightarrow [\![t \in \tau]\!] = [\![\tau_a \subseteq \tau]\!] \, .$$

This proves the first inequality of the claim.

To prove the second inequality of the claim, it is enough to show that

$$a \leq \tau(t) \rightarrow [\![t \in \tau_a]\!]$$

for all $t \in D$. We prove it using 7.4 as follows:

- If $t \in D \setminus \mathrm{dom}(\tau_a)$, then 7.4 gives that

$$a \wedge \tau(t) = 0.$$

If we combine it with Exercise 3.3.5, we get that

$$a \leq \tau(t) \rightarrow b$$

for any $b \in \mathsf{B}$ and $t \in D \setminus \mathrm{dom}(\tau_a)$. In particular the following holds:

$$\text{for all } t \in D \setminus \mathrm{dom}(\tau_a) \; a \leq \tau(t) \rightarrow [\![t \in \tau_a]\!] \, . \tag{7.5}$$

- If $t \in \mathrm{dom}(\tau_a)$, we can follow the pattern we have seen in the proof of the first inequality to get (using 7.3):

$$\text{for all } t \in \mathrm{dom}(\tau_a) \; (a \leq \tau(t) \rightarrow [\![t \in \tau_a]\!]). \tag{7.6}$$

Thus by 7.5, 7.6 we get

$$a \leq \bigwedge_{t \in \mathrm{dom}(\tau)} \tau(t) \rightarrow [\![t \in \tau_a]\!] = [\![\tau \subseteq \tau_a]\!] \, .$$

The second inequality of the claim is proved. $\qquad\square$

The proof of the Mixing Lemma is completed.   □

The following would be an instance of Proposition 6.3.14 if V^B were a set. We include a proof since V^B is not a set.

Theorem 7.1.32 (Maximum Principle) *For any boolean algebra* B, V^B *is full in the strong form asserting that for all formulae* $\varphi(x, \bar{y})$ *and* $\bar{\tau} \in (V^B)^{<\omega}$

$$[\![\exists x \varphi(x, \bar{\tau})]\!] = [\![\varphi(\sigma, \bar{\tau})]\!]$$

for some $\sigma \in V^B$.

Proof

$$[\![\exists x \varphi(x, \bar{\tau})]\!] \geq [\![\varphi(\sigma, \bar{\tau})]\!]$$

holds always. So we want to show that

$$[\![\varphi(\sigma, \bar{\tau})]\!] \geq [\![\exists x \varphi(x, \bar{\tau})]\!]$$

for some $\sigma \in V^B$. Let

$$u_0 = [\![\exists x \varphi(x, \bar{\tau})]\!] > 0_B.$$

Let

$$D = \{u \in B^+ : \text{ there is some } \sigma_u \in V^B \text{ such that } u \leq [\![\varphi(\sigma_u, \bar{\tau})]\!]\}.$$

D is dense and open below u_0 in B^+. Let A be a maximal antichain of D (A exists applying Exercise 3.11.4 to the boolean algebra B $\restriction u_0$); clearly,

$$\bigvee \{u : u \in A\} = u_0.$$

Now we can appeal to the mixing lemma to find $\sigma \in V^B$ such that $[\![\sigma = \sigma_u]\!] \geq u$ for any $u \in A$. Thus for each $u \in A$ we have

$$u \leq [\![\sigma = \sigma_u]\!] \wedge [\![\varphi(\sigma_u, \bar{\tau})]\!] \leq [\![\varphi(\sigma, \bar{\tau})]\!] .$$

Therefore

$$[\![\exists x \varphi(x, \bar{\tau})]\!] = u_0 = \bigvee A \leq [\![\varphi(\sigma, \bar{\tau})]\!] .$$

The proof is complete.   □

The remainder of this section is not of key importance for the development of our core results.

Fact 7.1.33 V^{B} *satisfies the Axiom of Extensionality.*

Proof Let $\tau, \sigma \in V^{\mathsf{B}}$. We want to prove that:

$$\llbracket \forall u (u \in \tau \leftrightarrow u \in \sigma) \to \tau = \sigma \rrbracket = 1.$$

By Lemma 3.3.3, it is enough to show that:

$$\llbracket \forall u (u \in \tau \leftrightarrow u \in \sigma) \rrbracket \leq \llbracket \tau = \sigma \rrbracket.$$

We observe that if $a \leq a'$, then $(a' \to b) \leq (a \to b)$. Thus for any $u \in \mathrm{dom}(\tau)$ we have

$$(\llbracket u \in \tau \rrbracket \to \llbracket u \in \sigma \rrbracket) \leq (\tau(u) \to \llbracket u \in \sigma \rrbracket),$$

and therefore

$$\bigwedge_{u \in V^{\mathsf{B}}} (\llbracket u \in \tau \rrbracket \to \llbracket u \in \sigma \rrbracket) \leq \bigwedge_{u \in \mathrm{dom}(\tau)} (\llbracket u \in \tau \rrbracket \to \llbracket u \in \sigma \rrbracket)$$

$$\leq \bigwedge_{u \in \mathrm{dom}(\tau)} (\tau(u) \to \llbracket u \in \sigma \rrbracket).$$

The left-hand side of the above equation is equal to $\llbracket \forall u (u \in \tau \to u \in \sigma) \rrbracket$, while the right-hand side is the definition of $\llbracket \tau \subseteq \sigma \rrbracket$. Consequently:

$$\llbracket \forall u (u \in \tau \leftrightarrow u \in \sigma) \rrbracket \leq \llbracket \tau = \sigma \rrbracket.$$

The proof is completed.   □

Summing up, so far we have proved that V^{B} is a full boolean valued model for set theory which satisfies the Axiom of Extensionality. We will later see that it satisfies all the axioms of **ZFC**.

We now connect the **B**-names for elements of 2^ω with the boolean valued model $C(St(\mathsf{B}), 2^\omega)$ we examined in Sect. 6.4.3.

Definition 7.1.34 Assume $\llbracket \tau : \check{\lambda} \to \check{2} \rrbracket_{\mathsf{B}} = 1_{\mathsf{B}}$ and endow the function space 2^λ with the product topology. Define $f_\tau : St(\mathsf{B}) \to 2^\lambda$ as

$$f_\tau(G)(\alpha) = i \iff \llbracket \tau(\check{\alpha}) = \check{i} \rrbracket \in G.$$

Now assume $f : St(\mathsf{B}) \to 2^\lambda$ is a continuous function, and then define

$$\tau_f = \{ \langle (\check{\alpha}, i), \{ G : f(G)(\alpha) = i \} \rangle : \alpha < \lambda, i < 2 \} \in V^{\mathsf{B}}.$$

Observe that

$$\{G : f(G)(\alpha) = i\} = f^{-1}[N_{\alpha,i}],$$

where $N_{\alpha,i} = \{g \in 2^\lambda : g(\alpha) = i\}$. Since f is continuous, then $f^{-1}[N_{\alpha,i}]$ is clopen, and so it is an element of the Boolean algebra.

Proposition 7.1.35 *Let* B *be a cba. Assume* $\left[\!\!\left[\tau : \check{\lambda} \to \check{2}\right]\!\!\right]_{\mathsf{B}} = 1_{\mathsf{B}}$ *and* $f : \mathrm{St}(\mathsf{B}) \to$ 2^λ *is continuous. Then:*

1. $\tau_f \in V^{\mathsf{B}}$.
2. $f_\tau : \mathrm{St}(\mathsf{B}) \to 2^\lambda$ *is continuous.*
3. $[\![\tau_{f_\tau} = \tau]\!]_{\mathsf{B}} = 1_{\mathsf{B}}$.
4. $f_{\tau_f} = f$.

Proof

1. By definition.
2. We just need to check that the preimage of a basic open set is a basic open set. Fix α, i,

$$f_\tau^{-1}[N_{\alpha,i}] = \{G : \left[\!\!\left[\tau(\check{\alpha}) = \check{i}\right]\!\!\right] \in G\} \in \mathrm{Cl}((\,)\,St\,(\mathsf{B})),$$

   since by definition

$$[\![\tau_f(\alpha) = i]\!] \in G \iff G \in \{H : f(H)(\alpha) = i\} \iff f(G)(\alpha) = i.$$

3. By definition for any G

$$[\![\tau_{f_\tau}(\alpha) = i]\!] \in G \iff f_\tau(G)(\alpha) = i \iff [\![\tau(\alpha) = i]\!] \in G.$$

   Then for any α, i

$$\left[\!\!\left[\tau_{f_\tau}(\check{\alpha}) = \check{i}\right]\!\!\right] = \left[\!\!\left[\tau(\check{\alpha}) = \check{i}\right]\!\!\right].$$

   Therefore $[\![\tau_{f_\tau} = \tau]\!]_{\mathsf{B}} = 1_{\mathsf{B}}$.
4. As in the proof of point 2, we can observe that

$$[\![\tau_f(\alpha) = i]\!] \in G \iff G \in \{H : f(H)(\alpha) = i\} \iff f(G)(\alpha) = i.$$

   Moreover we have that

$$f_{\tau_f}(G)(\alpha) = i \iff [\![\tau_f(\alpha) = i]\!] \in G;$$

   hence we are done.

$\square$

7.2 Cohen's Forcing Theorem

The goal of this section is to give a positive answer to the questions of Sect. 7.1.1 regarding the definability inside M of the forcing relation. We show that for a countable transitive model M of ZFC and a $B \in M$ which is in M a cba, the forcing relation defined externally in Sect. 7.1.1 is induced by the boolean valued semantic defined internally on M^B (by relativizing to M all the results of Sect. 7.1.2 for V^B) and allows to control the first order theory of the models $M[G]$ we introduced in Sect. 7.1.1. We will also give examples on how this identification can greatly simplify several computations.

Lemma 7.2.1 *Assume $M \in V$ is a transitive set such that $M \vDash$ ZFC and $B \in M$ is such that M models B is a complete boolean algebra. Then V models that*

$$\left\langle M^B = (V^B)^M, (\in^B)^M, (=^B)^M, (\subseteq^B)^M \right\rangle$$

is a full B-valued model.

Notice that this occurs regardless of the fact that B is a complete boolean algebra in V, which is never the case if B is infinite and M is countable (see Exercise 7.1.15).

Proof Since M is a model of ZFC, Theorem 7.1.27 applied in M shows that M models that $\in^B, =^B, \subseteq^B$ are binary relations on the class M^B satisfying the clauses of Definition 6.1.1. In particular V models that

$$\left\langle M^B, (\in^B)^M, (=^B)^M, (\subseteq^B)^M \right\rangle$$

is a B-valued model. Moreover, by the Mixing Lemma 7.1.31 and the Maximum Principle 7.1.32 applied in M (which is a model of ZFC), we get that for all formulae $\phi(x_0, \ldots, x_n)$ and all $\tau_1, \ldots, \tau_n \in M$, there exists a $\tau_0 \in M$ such that

$$M \vDash \llbracket \phi(\tau_0, \tau_1, \ldots, \tau_n) \rrbracket_B = \bigvee_{\sigma \in M^B} \llbracket \phi(\sigma, \tau_1, \ldots, \tau_n) \rrbracket_B,$$

i.e.,

$$V \vDash \llbracket \phi(\tau_0, \tau_1, \ldots, \tau_n) \rrbracket_B^M = \bigvee_{\sigma \in M^B} \llbracket \phi(\sigma, \tau_1, \ldots, \tau_n) \rrbracket_B^M.$$

In particular V models that

$$\left\langle M^B, (\in^B)^M, (=^B)^M, (\subseteq^B)^M \right\rangle$$

is a full B-valued model. $\qquad \square$

Notation 7.2.2 Let $M \vDash \mathsf{ZFC}$ be transitive and $\mathsf{B} \in M$ be such that M models B *is a complete boolean algebra*. Let $G \subseteq \mathsf{B}$ be any ultrafilter in V. Denote by R_G the binary relation $(R^\mathsf{B})^M /_G$ on $M^\mathsf{B} /_G$ for R among $=, \in, \subseteq$.

Assuming M, B, G are as in the notation above, since $\langle M^\mathsf{B}, (R^\mathsf{B})^M : R \in \{=, \subseteq, \in\} \rangle$ is a full B-valued model in V, and by Łoś Theorem 6.3.7, we get that $(M^\mathsf{B} /_G, \in_G, \subseteq_G)$ is a Tarski model for the language $\mathcal{L} = \{\in, \subseteq\}$ such that

$$(M^\mathsf{B} /_G, \in_G, \subseteq_G) \vDash \phi$$

if and only if $\llbracket \phi \rrbracket_\mathsf{B}^M \in G$.

A pair of comments:

- In the definition of $M^\mathsf{B} /_G$, it is possible that $G \notin M$. However, also in this case the definition makes sense.
- There is no reason why the classes $[x]_G$ should be definable in M if $G \notin M$, and thus $M^\mathsf{B} /_G$ in general is not a definable class in M. We shall see that this is exactly what occurs if G is M-generic for an atomless boolean algebra.
- If $G \in M$, then the classes $[x]_G$ are definable in M, and in M one can define three classes which taken as a triple in V define an isomorphic copy of the structure $(M^\mathsf{B} /_G, \in_G, \subseteq_G)$ as the extension of a formula in the parameter B.
- We have a clear meaning of what is the boolean valued model

$$\left\langle M^\mathsf{B}, (\in^\mathsf{B})^M, (=^\mathsf{B})^M, (\subseteq^\mathsf{B})^M \right\rangle$$

in V, since all the relevant objects are now sets in V. On the other hand inside M, we can speak of (i.e., formalize) the satisfaction predicate for each single formula of the language (as a class definable in M), but we cannot speak simultaneously inside M of the family of classes of M given by the satisfaction predicates for formulae of the language.
- From now on for the sake of simplicity, we denote by $\llbracket \phi \rrbracket$ the boolean value $\llbracket \phi \rrbracket_\mathsf{B}^M$ as M ranges over countable transitive models of ZFC and $\mathsf{B} \in M$ among the (complete in M) boolean algebras of M.

We now start to plug in some of the material developed in Chap. 5. First of all, by Lemma 5.1.2, if M is a countable transitive model of ZFC and $\mathsf{B} \in M$ is a boolean algebra, there exists an ultrafilter G M-generic for B.

We want to study which is the relationship between $M[G]$ and $M^\mathsf{B} /_G$. These observations were made in Sect. 7.1.1:

Assume M is a countable transitive model of ZFC and $\mathsf{B} \in M$ is an atomless boolean algebra which M models to be complete. Then:

- The family of M-generic filters in $\mathrm{St}(\mathsf{B})$ forms a dense subset of $\mathrm{St}(\mathsf{B})$, i.e., for any $b \in \mathsf{B}^+$, there exists an ultrafilter G M-generic, with $b \in G$ (i.e., $G \in N_b$), and each such $G \notin M$.

- For any G M-generic of B we can define in V by recursion on M^{B}:

$$\tau_G = \{\sigma_G : \tau(\sigma) \in G\},$$

for any given $\tau \in M^{\mathsf{B}}$.
We also define

$$M[G] = \{\tau_G : \tau \in M^{\mathsf{B}}\}.$$

- We have already seen that $M[G]$ is transitive, $M \subseteq M[G]$, $G \in M[G]$.

We will now proceed to identify the models $(M[G], \in)$ and $(M^{\mathsf{B}}/G, \in_G)$ under the assumption that G is M-generic for some complete boolean algebra $\mathsf{B} \in M$. This is the content of Theorem 7.2.4 below. In case $G \in \mathrm{St}(\mathsf{B})$ is not M-generic, $M[G]$ is still a transitive set, and $(M^{\mathsf{B}}/G, \in_G)$ is a well defined Tarski model, but it can be shown that the two structures cannot be isomorphic.

We need this preparatory lemma:

Lemma 7.2.3 *Let $M \vDash$ ZFC, $M \in V$, and transitive. Let B be a cba in M. Assume $G \in V$ is an M-generic filter for B. The following holds:*

(a) Assume $b \in G$, $X \in M$, and $X \subseteq \mathsf{B}$ is predense below b. Then $G \cap X \neq \emptyset$.
(b) Assume $A \in M$ and $A \subseteq G$. Then $\bigwedge A \in G$.

Proof We proceed as follows:

(a): Let

$$E = \{c \in \mathsf{B} : M \models c \leq \neg b \vee (c \leq b \wedge c \in\downarrow X)\}.$$

We leave to the reader to check that $E \in M$, M models that E is dense, and that $c \in G \cap E$ if $c \leq b$ and $c \in\downarrow X$.

(b) Suppose $A \in M$, $A \subseteq G$, and $M \vDash \bigwedge A = c$. Since M satisfies the Axiom of Choice, we can write $A = \{a_\xi : \xi < \gamma\} \in M$ for some $\gamma \in M$. Let $b_\xi = \bigwedge_{\alpha < \xi} a_\alpha$. The sequence $\langle b_\xi : \xi \leq \gamma \rangle \in M$ is decreasing, and $c = b_\gamma$. Assume toward a contradiction that there is $\xi \leq \gamma$ least ordinal such that $b_\xi \notin G$. Then $\neg b_\xi \in G$. Set $c_\alpha = \neg b_\xi \wedge b_\alpha$. Then $c_\alpha \in G$ for all $\alpha < \xi$, $\{c_\alpha : \alpha < \xi\} \in M$, and $\bigwedge_{\alpha < \xi} c_\alpha = 0_{\mathsf{B}}$. Thus $\bigvee_{\alpha < \xi} \neg c_\alpha = 1_{\mathsf{B}}$ and $\{\neg c_\alpha : \alpha < \xi\} \in M$ as well. Since G is M-generic, $\neg c_\alpha \in G$ for some $\alpha < \xi$. Then $c_\alpha \wedge \neg c_\alpha = 0_{\mathsf{B}} \in G$, a contradiction which proves the lemma.

$\square$

Theorem 7.2.4 *Let M be a transitive model of ZFC, B be a complete boolean algebra in M, and G be an M-generic filter for B. Then*

$$\pi_G^M : M^{\mathsf{B}}/G \to M[G]$$

$$[\tau]_G \mapsto \tau_G$$

defines an isomorphism between the structures $(M^{\mathbb{B}}/_G, \in_G)$ *and* $(M[G], \in)$. *In particular* π_G^M *is the Mostowski collapse of the well-founded extensional relation* $\in_G$ *on* $M^{\mathbb{B}}/_G$.

Proof It suffices to prove:

1. $[\![\tau \in \sigma]\!] \in G \Leftrightarrow \tau_G \in \sigma_G$.
2. $[\![\tau = \sigma]\!] \in G \Leftrightarrow \tau_G = \sigma_G$.

We prove both items by induction on $(\mathrm{rk}^{\mathbb{B}}(\tau), \mathrm{rk}^{\mathbb{B}}(\sigma))$ with respect to the square order on Ord^2.

1. ($\Rightarrow$) Suppose $[\![\tau \in \sigma]\!] \in G$. By definition

$$[\![\tau \in \sigma]\!] = \bigvee_{u \in \mathrm{dom}(\sigma)} \sigma(u) \wedge [\![u = \tau]\!].$$

Let $b_u = \sigma(u) \wedge [\![\tau = u]\!]$. Notice that $\{b_u : u \in \mathrm{dom}(\sigma)\} \in M$ is predense under $[\![\tau \in \sigma]\!] \in G$. So[8] we can appeal to (a) of Lemma 7.2.3 to find $u \in \mathrm{dom}(\sigma)$ such that $b_u = [\![\tau = u]\!] \wedge \sigma(u) \in G$. Thus $[\![\tau = u]\!] \in G$, so we can apply the inductive assumption 2 on the pair (τ, u) which has lower rank than (τ, σ) in the square order on Ord^2. We conclude that $\tau_G = u_G$. Now observe that $u_G \in \sigma_G = \{v_G : \sigma(v) \in G\}$ since $u \in \mathrm{dom}(\sigma)$ and $\sigma(u) \in G$.

($\Leftarrow$) Suppose $\tau_G \in \sigma_G$. Then there is $u \in \mathrm{dom}(\sigma)$ such that $\sigma(u) \in G$ and $\tau_G = u_G$. Therefore, applying again 2 on the pair (τ, u) which has lower rank than the pair (τ, σ), we get that $[\![\tau = u]\!] \in G$. So

$$[\![\tau = u]\!] \wedge \sigma(u) \in G$$

as well. Now we can observe that

$$[\![\tau = u]\!] \wedge \sigma(u) \leq \bigvee_{v \in \mathrm{dom}(\sigma)} [\![\tau = v]\!] \wedge \sigma(v) = [\![\tau \in \sigma]\!].$$

Therefore $[\![\tau \in \sigma]\!] \in G$.

2. ($\Rightarrow$) Suppose $[\![\tau = \sigma]\!] \in G$. Observe that $[\![\tau \subseteq \sigma]\!] \in G$ gives that

$$\neg\tau(u) \vee [\![u \in \sigma]\!] \geq [\![\tau \subseteq \sigma]\!]$$

is also in G for all $u \in \mathrm{dom}(\tau)$. This gives that $[\![u \in \sigma]\!] \in G$ for all $u \in \mathrm{dom}(\tau)$ such that $\tau(u) \in G$. Since the pairs (u, σ) are of lower rank than the pair (τ, σ) for all such u, we can apply the first item to these pairs to get that

---

[8] Everywhere in this proof we appeal to Lemma 7.2.3, we are crucially using the assumption that G is M-generic.

$u_G \in \sigma_G$ for all $u \in \text{dom}(\tau)$ such that $\tau(u) \in G$. This gives that $\tau_G \subseteq \sigma_G$. The other inclusion is proved in exactly the same manner.

($\Leftarrow$) Suppose $[\![\tau \neq \sigma]\!] \in G$. W.l.o.g. we can suppose that $[\![\tau \not\subseteq \sigma]\!] \in G$. But

$$[\![\tau \not\subseteq \sigma]\!] = \bigvee_{u \in \text{dom}(\tau)} \tau(u) \wedge \neg [\![u \in \sigma]\!].$$

Since G is M-generic, we can appeal to (a) of Lemma 7.2.3 to find $u \in \text{dom}(\tau)$ such that

$$\tau(u) \wedge [\![u \notin \sigma]\!] \in G.$$

Applying the first item on the pair (u, σ) which has lower rank than (τ, σ), we get that $u_G \notin \sigma_G$, while $u_G \in \tau_G$ since $\tau(u) \in G$. Hence $\tau_G \not\subseteq \sigma_G$, which also gives that $\tau_G \neq \sigma_G$.

The proof is complete. □

Summing up, we can prove:

Theorem 7.2.5 (Cohen's Forcing Theorem) *Let M be a countable transitive model of* ZFC *and* B $\in M$ *be a boolean algebra which M models to be complete. Then for all formulae $\phi(x_1, \ldots, x_n)$ in the free variables $x_1, \ldots, x_n$ and all $\tau_1, \ldots, \tau_n \in M^{\mathsf{B}}$:*

1. *For all $b \in$ B $[\![\varphi(\tau_1, \ldots, \tau_n)]\!] \geq b$ if and only if we have that*

$$M[G] \vDash \varphi((\tau_1)_G, \ldots, (\tau_n)_G)$$

 for all G M-generic filter for B *with $b \in G$.*
2. *For all G M-generic filter for* B

$$M[G] \vDash \varphi((\tau_1)_G, \ldots, (\tau_n)_G)$$

 if and only if $[\![\varphi(\tau_1, \ldots, \tau_n)]\!] \in G$.

Proof We sketch just some parts of the proof leaving the others as an instructive exercise for the reader:

1. ($\Rightarrow$) Suppose $b \leq [\![\phi]\!]$. Let G be M-generic for B with $b \in G$. Then $[\![\phi(\tau_1, \ldots, \tau_n)]\!] \in G$ as well, and thus

$$M^{\mathsf{B}}/G \vDash \phi([\tau_1]_G, \ldots, [\tau_n]_G)$$

by Theorem 6.3.7. Since the map $[\tau]_G \mapsto \tau_G$ is an isomorphism, we also get
that

$$M[G] \models \phi((\tau_1)_G, \ldots, (\tau_n)_G).$$

($\Leftarrow$) is left to the reader.

2. It is an immediate consequence of the isomorphism of the structures M^B/G and
   $M[G]$ and of Theorem 6.3.7 applied in V to the full B-valued model M^B.

The proof is complete. $\square$

Definition 7.2.6 (Cohen's Forcing Relation) Let M be a countable transitive
model of ZFC and $B \in M$ be a complete boolean algebra. For each formula
$\phi(x_0, \ldots, x_n)$, $\tau_0, \ldots, \tau_n \in M^B$, and $b \in B^+$ the forcing relation is defined in
V by

$$b \Vdash \phi(\tau_0, \ldots, \tau_n) \ (b \text{ forces } \phi(\tau_0, \ldots, \tau_n))$$

if and only if

$M[K] \models \phi((\tau_0)_K, \ldots, (\tau_n)_K)$ for all M-generic filters K for B such that $b \in K$.

Lemma 7.2.7 *Let M be a countable transitive model of ZFC and $B \in M$ be a
complete boolean algebra. For each formula $\phi(x_0, \ldots, x_n)$, $\tau_0, \ldots, \tau_n \in M^B$, and
$b \in B^+$, the following are equivalent:*

1. $b \Vdash \phi(\tau_0, \ldots, \tau_n)$.
2. The set of K M-generic for B such that $b \in K$ and

$$M[K] \models \phi((\tau_0)_K, \ldots, (\tau_n)_K)$$

is dense in N_b.
3. $M \models b \leq_B [\![\phi(\tau_0, \ldots, \tau_n)]\!]$.

Proof A useful exercise for the reader. $\square$

We are almost ready to prove that every axiom of ZFC is valid in M^B and that CH
is independent from the ZFC-axioms (more precisely from the theory ZFC+*there
exists a countable transitive model of* ZFC). To do this we will often appeal to
Cohen's forcing theorem. So let us explain how we are going to use it.

7.2.1 How to Use Cohen's Forcing Theorem

Recurring examples of how we will use of the Cohen forcing theorem include the
following:

Example 7.2.8 Given ordinals γ and β, consider a forcing statement of the following form:

$$\left[\!\!\left[\dot{f} : \check{\gamma} \to \check{\beta} \text{ is a function} \right]\!\!\right] > 0_{\mathsf{B}}.$$

We define for each $\alpha < \gamma$ a set in M

$$A_\alpha = \left\{ b_\eta \in \mathsf{B}^+ : M \models b_\eta = \left[\!\!\left[\dot{f} : \check{\gamma} \to \check{\beta} \text{ is a function} \right]\!\!\right] \wedge \left[\!\!\left[\dot{f}(\check{\alpha}) = \check{\eta} \right]\!\!\right] \right\} \in M$$

(where η ranges among the ordinals below β for which $b_\eta > 0_{\mathsf{B}}$).

We want to argue that M models that A_α is an antichain, proving the stronger assertion that for $\eta \neq \nu$ both in β $b_\eta \wedge b_\nu = 0_{\mathsf{B}}$.

Toward this aim we proceed as follows: We assume by contradiction that we can find $\eta \neq \nu < \beta$ such that $c = b_\eta \wedge b_\nu > 0_{\mathsf{B}}$. We pick G M-generic with $c \in G$, and we get that $\left[\!\!\left[\dot{f} : \check{\gamma} \to \check{\beta} \text{ is a function} \right]\!\!\right]$, $\left[\!\!\left[\dot{f}(\check{\alpha}) = \check{\eta} \right]\!\!\right]$, $\left[\!\!\left[\dot{f}(\check{\alpha}) = \check{\nu} \right]\!\!\right]$ are all in G. Then by the forcing theorem

$$M[G] \models \dot{f}_G : \gamma \to \beta \text{ is a function},$$

$$M[G] \models \dot{f}_G(\alpha) = \eta,$$

$$M[G] \models \dot{f}_G(\alpha) = \nu.$$

This gives that

$$M[G] \models \nu = \eta,$$

and thus that $\nu = \eta$, which contradicts our assumption that $\nu \neq \eta$. A direct argument carried entirely in M (without ever appealing to the forcing theorem) yielding that A_α is an antichain can also be found but is much more convoluted.

Lemma 7.2.9 *Assume $M \models$ ZFC and is countable, and $\mathsf{B} \in M$ is a cba in M. Then for all $\tau, \sigma \in M^{\mathsf{B}}$, there exists $\eta : \mathrm{dom}(\sigma) \to \mathsf{B}$ in M^{B} such that*

$$[\![\tau \subseteq \sigma]\!] \leq [\![\tau = \eta]\!].$$

Proof Given $\tau, \sigma \in M^{\mathsf{B}}$, set

$$\eta = \left\{ \langle u, \sigma(u) \wedge [\![u \in \sigma \leftrightarrow u \in \tau]\!] \rangle : u \in \mathrm{dom}(\sigma) \right\} \in M^{\mathsf{B}}.$$

First of all we prove that

$$[\![\eta \subseteq \sigma]\!] = 1_{\mathsf{B}} :$$

Fix G M-generic for **B**, pick $a \in \eta_G$, and then $a = u_G$ for some $u \in \mathrm{dom}(\eta) = \mathrm{dom}(\sigma)$ with $[\![u \in \sigma \leftrightarrow u \in \tau]\!] \wedge \sigma(u) \in G$. Hence $\sigma(u) \in G$ as well, giving that $u_G \in \sigma_G$. Since this holds for all $a \in \eta_G$, we conclude that $M[G] \models \eta_G \subseteq \sigma_G$ for all G M-generic for **B**. We conclude by the forcing theorem.

Now assume G is M-generic for **B** with $[\![\tau \subseteq \sigma]\!] \in G$. We show that $\eta_G = \tau_G$. By the forcing theorem, this suffices to prove the lemma:

$\eta_G \subseteq \tau_G$ Assume $a \in \eta_G$. Then $a = u_G$ for some $u \in \mathrm{dom}(\eta) = \mathrm{dom}(\sigma)$ and $\eta(u) = [\![u \in \sigma \leftrightarrow u \in \tau]\!] \wedge \sigma(u) \in G$. Since $\sigma(u) \in G$ and

$$[\![u \in \sigma]\!] = \bigvee \{v \in \mathrm{dom}(\sigma) : [\![u = v]\!] \wedge \sigma(v)\} \geq [\![u = u]\!] \wedge \sigma(u) = \sigma(u) \in G,$$

we get that $[\![u \in \sigma]\!] \in G$. Since $[\![u \in \sigma \leftrightarrow u \in \tau]\!] \in G$ as well, we conclude that $[\![u \in \tau]\!] \in G$ as well, giving that $u_G \in \tau_G$ by the forcing theorem.

$\eta_G \supseteq \tau_G$ assume $a \in \tau_G$. Then $a = v_G$ for some $v \in \mathrm{dom}(\tau)$ with $\tau(v) \in G$. Since $[\![\tau \subseteq \sigma]\!] \in G$, we get that $M[G] \models \tau_G \subseteq \sigma_G$, hence we also get that $a = u_G$ for some $u \in \mathrm{dom}(\sigma)$ with $\sigma(u) \in G$. We get that $\sigma(u) \wedge \tau(v) \wedge [\![u = v]\!] \wedge [\![\tau \subseteq \sigma]\!] \in G$. Now:

- $\tau(v) \leq \bigvee_{t \in \mathrm{dom}(\tau)} [\![v = t]\!] \wedge \tau(t) = [\![v \in \tau]\!]$. Hence $[\![v \in \tau]\!] \in G$.
- $[\![u = v]\!] \in G$. Therefore $[\![u = v]\!] \wedge [\![v \in \tau]\!] \in G$ as well, and

$$[\![u \in \tau]\!] \geq [\![u = v]\!] \wedge [\![v \in \tau]\!] \in G.$$

  Hence $[\![u \in \tau]\!] \in G$.
- $\sigma(u) \leq \bigvee_{t \in \mathrm{dom}(\sigma)} [\![u = t]\!] \wedge \sigma(t) = [\![u \in \sigma]\!]$. Hence $[\![u \in \sigma]\!] \in G$.

We conclude that

$$[\![u \in \tau]\!] \wedge [\![u \in \sigma]\!] \in G.$$

But

$$[\![u \in \tau]\!] \wedge [\![u \in \sigma]\!] \leq [\![u \in \tau \leftrightarrow u \in \sigma]\!].$$

Hence

$$[\![u \in \tau \leftrightarrow u \in \sigma]\!] \in G$$

as well. Then $\sigma(u) \wedge [\![u \in \tau \leftrightarrow u \in \sigma]\!] \in G$, yielding that

$$a = u_G \in \eta_G = \{t_G : t \in \mathrm{dom}(\sigma) \text{ and } \sigma(t) \wedge [\![t \in \tau \leftrightarrow t \in \sigma]\!] \in G\},$$

as was to be shown.

$\square$

7.3 Independence of CH

In this section we prove the independence of the Continuum Hypothesis from the axioms of ZFC using the forcing method over a countable transitive model M of ZFC. We assume throughout this section that $M[G]$ models ZFC whenever G is M-generic for some $B \in M$ which M models to be a complete boolean algebra. This will be proved in full detail in the next section (see Sect. 7.4). In order to appreciate the full power of the forcing theorem, we believe it is more instructive to understand how this theorem allows us to compute the truth value of specific statements in forcing extensions of M (i.e., models of the form $M[G]$ with G M-generic for a cba $B \in M$). This is what we do in this section using the forcing theorem to compute the truth values of CH in two distinct forcing extensions of M.

We first show that if G is M-generic for $\mathrm{RO}(2^{\omega_2 \times \omega})^M$, then CH fails in $M[G]$. Next we show that there is $B \in M$ such that M^B models CH. Combined with the results of Sect. 7.4, these two proofs will give the independence of CH with respect to the theory ZFC over the theory ZFC+*there is a countable transitive model of* ZFC. It is possible to convert these proofs in a proof of the independence of CH from ZFC right away from ZFC using arguments rooted in the reflection properties of V (see [19, Sections IV.7, VII.1]).

7.3.1 A Model of ¬CH

The following lemma describes the key application of CCC-ness in the forcing construction of Cohen.

Lemma 7.3.1 *Assume M is a countable transitive model of* ZFC *and M models that B is a CCC complete boolean algebra. Then*

$$M[G] \models \mathrm{cf}(\kappa) = \alpha \Leftrightarrow M \models \mathrm{cf}(\kappa) = \alpha$$

for all M-generic filters G for B and $\alpha \leq \kappa \in \mathrm{Ord}^M$.

Proof Note that since $M \subseteq M[G]$ if $f : \lambda \to \kappa$ is cofinal and in M, f witnesses that $\mathrm{cf}(\kappa)^{M[G]} \leq \lambda$.

To conclude that $\mathrm{cf}(\kappa)^{M[G]} = \mathrm{cf}(\kappa)^M$, it is enough to restrict our attention to the case of κ being a regular cardinal in M (so that $\mathrm{cf}(\kappa)^M = \kappa$) and to show the following: □

Fact 7.3.2 *Assume κ is a regular uncountable cardinal in M and G is M-generic for B. Then $M[G]$ models that every function $\sigma_G \in M[G]$ from λ to κ is bounded (i.e., has range contained in some $\beta < \kappa$) for each $\lambda < \kappa$.*

If the fact holds, κ is regular in M if and only if it is regular in $M[G]$, and therefore M and $M[G]$ compute the same way all cofinalities.

Proof Let $\sigma \in M^{\mathsf{B}}$, and $b \in G$ be such that M models

$$\llbracket \sigma \text{ is a function from } \check{\lambda} \text{ to } \check{\kappa} \rrbracket = b.$$

For every $\eta < \lambda$ consider the set:

$$A_\eta = \{\beta < \kappa : M \models \llbracket \sigma(\check{\eta}) = \check{\beta} \rrbracket > 0_{\mathsf{B}}\}.$$

We notice that the sequence $\{A_\eta : \eta < \omega\} \in M$, since it is the extension in M of a formula with parameters in M by the following argument: Let

$$X = \left\{ \langle \eta, \beta, c \rangle : M \models \llbracket \sigma(\check{\eta}) = \check{\beta} \rrbracket = c > 0_{\mathsf{B}} \right\}.$$

Then $X \in M$, since it is a subset of $\lambda \times \kappa \times \mathsf{B}$ defined by an application of the comprehension axiom in M. Therefore for each $\eta < \lambda$, $A_\eta \in M$ as well, since

$$A_\eta = \{\beta : \exists c \, \langle \eta, \beta, c \rangle \in X\},$$

and $(A_\eta : \eta < \lambda) \in M$ since

$$(A_\eta : \eta < \lambda) = \{\langle \eta, u \rangle : M \models \beta \in u \leftrightarrow \exists b \, \langle \eta, \beta, b \rangle \in X\}.$$

Thus $\{A_\eta : \eta < \lambda\} = \mathrm{ran}((A_\eta : \eta < \lambda)) \in M$ as well. $\square$

Claim 7.3.2.1 *M models that every A_η is at most countable for all $\eta < \lambda$. Moreover $\sigma_H(\eta) \in A_\eta$ for all $\eta < \lambda$, and any H M-generic for B with*

$$b = \llbracket \sigma : \check{\lambda} \to \check{\kappa} \text{ is a function} \rrbracket \in H.$$

Proof Let

$$W_\eta = \{b_\beta^\eta = \llbracket \sigma(\check{\eta}) = \check{\beta} \rrbracket \wedge b : \beta \in A_n\}.$$

Then: $\square$

Subclaim 7.3.2.1 *The sequence $\{W_\eta : \eta < \lambda\} \in M$, and M models that each W_η is a countable antichain in M.*

Proof The first part of the subclaim is a useful exercise for the reader. For the second part it is enough to show that M models that each W_η is an antichain. Since M models B is CCC, we conclude that M models that each W_η is a countable antichain.

Assume W_η is not an antichain and find $\gamma \neq \beta \in A_\eta$ with $c = b_\gamma^\eta \wedge b_\beta^\eta > 0_B$. Let H be M-generic for B with $c \in H$. Since

$$c \leq b = \left[\!\left[\sigma : \check{\lambda} \to \check{\kappa} \text{ is a function} \right]\!\right],$$

$M[H]$ models that $\sigma_H : \lambda \to \kappa$ is a function by the forcing theorem. On the other hand since

$$c \leq b_\beta^\eta \leq \left[\!\left[\sigma(\check{\eta}) = \check{\beta} \right]\!\right], b_\gamma^\eta \leq \left[\!\left[\sigma(\check{\eta}) = \check{\gamma} \right]\!\right]$$

again by the forcing theorem, we get that $\sigma_H(\eta) = \beta$ and $\sigma_H(\eta) = \gamma$. Since $M[H]$ models that σ_H is a function we get that $\beta = \gamma$, a contradiction.   □

To complete the proof of the claim, observe the following:

- For each $\eta < \lambda$ the map $\phi_\eta : A_\eta \to W_\eta$ given by $\beta \mapsto b_\beta^\eta$ is in M, since

$$\phi_\eta = \{\langle \beta, b \rangle : \langle \eta, \beta, b \rangle \in X\},$$

  and is injective: Assume $c = b_\gamma^\eta = b_\beta^\eta$, then pick H M-generic for B with $c \in M$ to get that in $M[H]$, $\sigma_H : \lambda \to \kappa$ is a function, and $\gamma = \sigma_H(\eta) = \beta$. In particular $M \models \gamma = \beta$ as well.
  Hence M models that A_η is countable, since it is mapped injectively in a countable set by a map in M.
- For each $\eta < \lambda$ and H M-generic for B with

$$b = \left[\!\left[\sigma : \check{\lambda} \to \check{\kappa} \text{ is a function} \right]\!\right] \in H$$

  we have that $\sigma_H(\eta) \in A_\eta$, since $\sigma_H(\eta) = \beta$ iff $\left[\!\left[\sigma(\check{\eta}) = \check{\beta} \right]\!\right] \wedge b \in H \cap W_\eta$, giving that $\beta \in A_\eta$.

The proof of the claim is completed.   □

To conclude the proof of the fact, observe that M models that $\{A_\eta : \eta < \lambda\} \in M$ is for M a family of size λ of countable subsets of κ, which is a regular cardinal for M. Hence M models that the union of the A_η has size at most $\lambda \times \aleph_0 \leq \lambda < \kappa$.
We conclude (again by the regularity of κ in M) that for some $\beta < \kappa$,

$$M \models A = \bigcup\{A_\eta : \eta < \kappa\} \subseteq \beta < \kappa.$$

The fact is proved, since we get that in $M[G]$

$$\sigma_G[\lambda] \subseteq A \subseteq \beta < \kappa.$$

□

The proof of the lemma is completed.   □

Now let us choose in M the poset $\mathsf{RO}(2^{\omega_2 \times \omega})^M$ (which M models to be CCC by Proposition 5.3.4 applied in M). We can use the facts proved so far to check the following:

Theorem 7.3.3 *Assume M is a countable transitive model of* ZFC *and let*

$$\mathsf{B} = \mathsf{RO}(2^{\omega_2 \times \omega})^M.$$

Then $M[G] \models \neg\mathsf{CH}$ for all M-generic filters G for B.

Proof Set:

$$\tau = \{\langle \mathrm{op}(\mathrm{op}(\check{\alpha}, \check{n}), \check{i}), N_{\langle \alpha, n, i \rangle} \rangle : \alpha < \omega_2^M, n \in \omega, i < 2\} \in M^{\mathsf{B}},$$

where

$$N_{\langle \alpha, n, i \rangle} = \left\{ f \in 2^{\omega_2 \times \omega} \cap M : f(\alpha, n) = i \right\}.$$

Let G be M-generic for B. Then

$$g = \tau_G = \{\langle \langle \alpha, n \rangle, i \rangle : N_{\langle \alpha, n, i \rangle} \in G\}.$$

For all $\alpha < \omega_2^M$ and $n < \omega$, define $g_\alpha(n) = g(\alpha, n)$.   □

Claim 7.3.3.1 *$M[G]$ models that $g : \omega_2^M \times \omega \to 2$ is a total function. Moreover, $g_\alpha \neq g_\beta$ are distinct elements of $2^\omega \cap M[G]$ for all $\alpha < \beta < \omega_2^M$.*

Assume the claim is proved. Since M models that $\mathsf{RO}(2^{\omega_2 \times \omega})^M$ has the CCC (applying Corollary 5.3.6 inside M), by Lemma 7.3.1 we get that $M[G]$ models that ω_2^M is the second uncountable cardinal. By the claim

$$M[G] \models |2^\omega| \geq \omega_2^M,$$

thus CH fails in $M[G]$.

We are left with proof of the claim.

Proof Let for any $s \in Fn(\omega_2 \times \omega, 2)^M$

$$N_s = \left\{ f \in 2^{\omega_2 \times \omega} \cap M : s \subseteq f \right\}.$$

We can apply Exercise 5.2.8 in M to get that the sets:

- $D_{n,\alpha} = \{N_s : s \in Fn(\omega_2 \times \omega, 2)^M, \ (\alpha, n) \in \mathrm{dom}(s)\}$
- $E_{\alpha,\beta} = \{N_s : s \in Fn(\omega_2 \times \omega, 2), \ \exists n \ s(\alpha, n) \neq s(\beta, n)\}$

are dense in $\mathsf{RO}(2^{\omega_2 \times \omega})^M$ for all $\alpha \neq \beta < \omega_2^M$ and $n < \omega$.

Our definitions now give that:

- $M[G] \models (\alpha, n) \in \mathrm{dom}(g)$ for all $\alpha < \omega_2^M$ and $n < \omega$ since $D_{n,\alpha} \cap G \neq \emptyset$ for all such n, α.
- $M[G] \models (\alpha, n, i), (\alpha, n, j) \in g$ iff $i = j$: On the one hand $(\alpha, n, i) \in g$ iff $N_{\langle \alpha, n, i \rangle} \in G$, and on the other hand $N_{\langle \alpha, n, i \rangle}$ and $N_{\langle \alpha, n, j \rangle}$ are compatible conditions in $\mathrm{RO}(2^{\omega_2 \times \omega})^M$ iff $i = j$ for all $\alpha < \omega_2^M, n < \omega$.
- $M[G] \models g(\alpha, n) \neq g(\beta, n)$ for some n, since $E_{\alpha, \beta} \cap G \neq \emptyset$ for all $\alpha < \beta < \omega_2^M$.

The claim follows immediately from the above observations. □

The theorem is proved. □

7.3.2 A Model of CH

In this section we prove that $CH + ZFC$ is consistent relative to the theory $ZFC+$ *there is a countable transitive model of* ZFC.

Definition 7.3.4 Let κ, λ be infinite cardinals.

A boolean algebra B is κ-*distributive* if for all collections $\{D_\alpha : \alpha < \kappa\}$ of dense open sets in B^+ we have that

$$D = \bigcap_{\alpha < \kappa} D_\alpha \text{ is an open dense subset of } \mathsf{B}^+.$$

B is $<\lambda$-*distributive* if it is κ-distributive for all $\kappa < \lambda$.

Definition 7.3.5 Let λ be an infinite cardinal. A preorder $(P, <)$ is $< \lambda$-*closed* if for every $\gamma < \lambda$, every decreasing sequence $(p_\alpha)_{\alpha < \gamma}$ contained in P has a lower bound in P.

$< \omega_1$-closed posets are said to be *countably closed*, and $< \omega_1$-distributive boolean algebras are said to be *countably distributive*.

Lemma 7.3.6 *Assume* $(P, \leq_P)$ *is a separative* $< \lambda$-*closed poset. Then* $\mathrm{RO}(P)$ *is* $< \lambda$-*distributive.*

The assumption that P is separative is redundant, but the proof without this assumption is slightly more intricate, and we will use the lemma just for separative posets P, and thus we prove the lemma using this assumption.

Proof Let $i : P \to \mathrm{RO}(P) = \mathsf{B}$ be the dense embedding of P into its boolean completion provided by Theorem 4.2.4. Since P is separative, i is injective, and $i(p) \leq_{\mathsf{B}} i(q)$ if and only if $p \leq_P q$, by Corollary 4.2.7, assume

$$\{D_\alpha : \alpha < \gamma\}$$

is a family of dense open subsets of $\mathsf{RO}(P)$ for some $\gamma < \lambda$. It is immediate to check that

$$D = \bigcap \{D_\alpha : \alpha < \gamma\}$$

is open. We need to show that D is dense, i.e., given $b \in \mathsf{B}^+$, we need to find $q \leq_\mathsf{B} b$ in D.

Build a decreasing chain of conditions $\{p_\alpha : \alpha \leq \gamma\} \subseteq P$ by recursion as follows:

- Choose p_0 such that $i(p_0) \leq b$ and $p_0 \in D_0$ (which is possible since $i[P]$ is a dense subset of B^+).
- Given $p_\alpha \in P$, let $s \in D_{\alpha+1}$ be such that $s \leq_\mathsf{B} i(p_\alpha)$ and find $p_{\alpha+1} \in P$ such that $i(p_{\alpha+1}) \leq_\mathsf{B} s$ (which is possible since $i[P]$ is a dense subset of B^+). Then $p_{\alpha+1} \leq_P p_\alpha$ (since $i(p_{\alpha+1}) \leq_\mathsf{B} i(p_\alpha)$, and P is separative), and $i(p_{\alpha+1}) \in D_{\alpha+1}$.
- Given $\langle p_\beta : \beta < \alpha \rangle \subseteq P$ with $\beta < \gamma$ limit, first of all we notice that, by our construction, $\langle p_\beta : \beta < \alpha \rangle$ is a descending sequence in P. Since P is $< \lambda$-closed, we have that $\langle p_\beta : \beta < \alpha \rangle$ has a lower bound $r \in P$ refining each p_β. Now refine $i(r)$ to some $s \in D_\alpha$ and find $p_\alpha \in P$ such that $i(p_\alpha) \leq_\mathsf{B} s$ (which is possible since $i[P]$ is a dense subset of B^+). Then $i(p_\alpha) \in D_\alpha$, and p_α is a lower bound for the chain $\{p_\xi : \xi < \alpha\}$, since $i(p_\alpha) \leq_P s \leq_P i(p_\beta)$ for all $\beta < \alpha$, and P is separative.
- Let u be a lower bound for the descending sequence $\langle p_\beta : \beta < \gamma \rangle \subseteq P$.

Then

$$q = i(u) \in D = \bigcap \{D_\alpha : \alpha < \gamma\}$$

since $0_\mathsf{B} < i(u) \leq i(p_\alpha) \in D_\alpha$ for all $\alpha < \gamma$.

Since $b \geq_\mathsf{B} u$ is arbitrary, the proof is completed. $\square$

Definition 7.3.7 Given an uncountable cardinal κ, let

$$P_\kappa = \{f : \alpha \to \kappa : f \text{ is an injection and } \alpha < \omega_1\}$$

ordered by $f \leq_{P_\kappa} g$ iff $f \supseteq g$.

Fact 7.3.8 $(P_\kappa, \leq_{P_\kappa})$ is $< \omega_1$-closed and separative.

Proof First of all notice that f, g are incompatible in P_κ if and only if they disagree on some j in $\mathrm{dom}(g) \cap \mathrm{dom}(f)$, else their union is a common refinement.

We prove both properties of P as follows:

P_κ **is separative:** Assume $f \not\le g$, and then $g \not\sqsubseteq f$. In particular, either $\mathrm{dom}(g) \subseteq \mathrm{dom}(f)$, in which case f and g are already incompatible, or there is $i \in \mathrm{dom}(g) \setminus \mathrm{dom}(f)$. In this case we let $h : i + 1 \to \kappa$ be defined by the requirements:

- $h \supseteq f$.
- $h \restriction (i + 1 \setminus \mathrm{dom}(f)) \to (\kappa \setminus (\mathrm{ran}(f) \cup \mathrm{ran}(g)))$ is injective.

Since κ is an uncountable cardinal and $\mathrm{ran}(f) \cup \mathrm{ran}(g)$ is a countable subset of κ, $\kappa \setminus (\mathrm{ran}(f) \cup \mathrm{ran}(g))$ has size κ, and thus at most countable set $i + 1 \setminus \mathrm{dom}(f)$ can be injected inside it.

We conclude that $h \supseteq f$, and $h \in P_\kappa$ since it is an injective function with domain a countable ordinal, moreover h is incompatible with g, since $h(i) \notin \mathrm{ran}(g)$, and thus $h(i) \ne g(i)$.

P_κ **is countably closed:** Assume we have a decreasing sequence

$$\{f_\alpha : \alpha < \gamma\}$$

of elements of P_κ indexed by some countable ordinal γ. Let $f = \bigcup_{\alpha < \gamma} f_\alpha$; we show that $f \in P$ is a lower bound for all the f_α: It is enough to show that f is also an element of P_κ, and this is the case since its domain is a countable ordinal (a countable union of countable ordinals is a countable ordinal), and f is injective, since it is the coherent union of injective functions.

$\square$

Let M be a countable transitive model of ZFC and κ be a regular cardinal in M. Consider the partial order $P = (P_\kappa)^M$ in M. Then M models that P is countably closed and separative.

Let $B = \mathrm{RO}(P)^M$ and $i : P \to B$ in M be a canonical injection of P in its boolean completion. Then M models that B is countably distributive, applying Lemma 7.3.6 inside M to P and B. We will show the following:

Theorem 7.3.9 *Let M be a countable transitive model of ZFC and $\kappa \ge 2^{\aleph_0}$ be a regular cardinal in M. Consider the partial order $P = (P_\kappa)^M$ in M and let $B = \mathrm{RO}(P)^M$.*

Then M models that $[\![CH]\!]_B = 1_B$.

The theorem will be an immediate consequence of the following proposition:

Proposition 7.3.10 *Assume M, κ, P, B are as in the assumptions of Theorem 7.3.9, and G is M-generic for P.*

Then:

1. $(\omega_1)^M = (\omega_1)^{M[G]}$.
2. $M[G] \cap 2^\omega = M \cap 2^\omega$.
3. $M[G]$ models that there is a bijection of $(\omega_1)^M$ with κ.

Assume the proposition has been proved. Here is the proof of Theorem 7.3.9:

Proof Let $h : 2^\omega \cap M \to \kappa$ in M be a bijection of $(2^\omega)^M = 2^\omega \cap M = (2^\omega)^{M[G]}$ with κ, and then $g \circ h : (2^\omega)^{M[G]} \to (\omega_1)^M$ is a bijection and $(\omega_1)^M = (\omega_1)^{M[G]}$, i.e., $M[G]$ models h is a bijection of the powerset of ω with the first uncountable cardinal, as was to be shown. $\square$

We now prove Proposition 7.3.10:

Proof We first prove item 2: $\square$

Proof Let $\dot{r} \in M^B$ be a B-name such that $\left[\!\left[\dot{r} : \check{\omega} \to \check{2}\right]\!\right] = b > 0_B$. Define:

$$D_n = \{f \in B : M \models \exists i < 2 \ f \le \left[\!\left[\dot{r}(\check{n}) = \check{i}\right]\!\right]\}.$$

By an application of the forcing theorem, we can prove that each $D_n \in M$ is an open dense subset of B^+ below b as follows: For each $q \in B^+$ refining b, pick $G \in M$ such that $q \in G$. Then $b \in G$ gives that

$$M[G] \models \dot{r}_G : \omega \to 2,$$

and thus for some $i < 2$

$$M[G] \models \dot{r}_G(n) = i,$$

yielding that

$$s = \left[\!\left[\dot{r}(\check{n}) = \check{i}\right]\!\right] \wedge q \in G,$$

and thus $0_B < s \in D_n$ refines q.

This gives that M models that for all $q \le b$ there exists $0_B < s \le_B q$ in D_n. Thus M models that each D_n is dense. Notice also that the sequence

$$\{D_n : n \in \omega\} \in M.$$

Toward this aim, observe that

$$X = \left\{\langle n, q, i\rangle \in \omega \times B \times 2 : 0_B < q \le_B b \wedge \left[\!\left[\dot{f}(\check{n}) = \check{i}\right]\!\right]\right\} \in M,$$

since it is obtained by applying comprehension in M to define a subset of $\omega \times B \times 2$. Now

$$\langle D_n : n \in \omega\rangle = (\langle n, c\rangle : n \in \omega, \forall x(x \in c \leftrightarrow \exists i < 2 \langle n, x, i\rangle \in X) \in M$$

is obtained as a subset of $(\omega \times \mathcal{P}(B))^M$, applying comprehension in M once again.

Since M models that B is countably distributive, we get that

$$D_{\dot{r}} = \bigcap \{ D_n : n \in \omega \}$$

is also in M and is open dense below b. We claim the following:

$$D_{\dot{r}} = \{ q \leq_{\mathsf{B}} b : \exists s \in 2^\omega \cap M \text{ such that } g \Vdash \dot{r} = \check{s} \}. \tag{7.7}$$

To this aim choose $c \leq_{\mathsf{B}} b$ arbitrarily. Find $q \leq_{\mathsf{B}} c$ in $D_{\dot{r}}$, which is possible since $D_{\dot{r}}$ is open dense. Then $q \in D_n$ for all $n \in \omega$. In M, we can let for each $q \in D_{\dot{r}}$

$$f_q = \left\{ \langle n, i \rangle \in \omega \times 2 : M \models \left[\!\left[\dot{r}(\check{n}) = \check{i} \right]\!\right] \geq q \right\}.$$

Then $f_q \in M$ applying the comprehension axiom in M to isolate f_q as a subset of $\omega \times 2$, defined by a property in the parameters $\mathsf{B}, \dot{r}, q$.

We claim that $q \leq_{\mathsf{B}} \left[\!\left[\dot{r} = \check{f_q} \right]\!\right]$. To this aim let G be M-generic with $q \in G$. Then $b \in G$ yields that

$$M[G] \models \dot{r}_G : \omega \to 2.$$

Let $i_n = \dot{r}_G(n)$ for each $n \in \omega$. Then $\left[\!\left[\dot{r}(\check{n}) = \check{i_n} \right]\!\right] \in G$ for all $n \in \omega$. Now $q \in D_n \cap G$ for all n, and for each n, $q \in D_n$ entails that $q \leq_{\mathsf{B}} \left[\!\left[\dot{r}(\check{n}) = \check{i} \right]\!\right]$ if and only if $f_q(n) = i$, by definition of f_q. Therefore since $\dot{r}_G$, f_q are functions, we must have that $f_q(n) = i_n$ for all n, i.e., that

$$M[G] \models \dot{r}_G = f_q.$$

Since this occurs for all M-generic filters G for B to which q belongs, we get that $q \leq_{\mathsf{B}} \left[\!\left[\dot{r} = \check{f_q} \right]\!\right]$. This concludes the proof that 7.7 holds.

In particular we get that:

- $D_{\dot{r}}$ is open dense below b for all $\dot{r} \in M^{\mathsf{B}}$ such that $\left[\!\left[\dot{r} : \omega \to 2 \right]\!\right] = b$.
- For any condition q in $D_{\dot{r}}$ $q \leq_{\mathsf{B}} \left[\!\left[\dot{r} = \check{f_q} \right]\!\right]$.

This gives that for any $r = \dot{r}_G \in 2^\omega \cap M[G]$, we have that

$$\left[\!\left[\dot{r} : \check{\omega} \to \check{2} \text{ is a function} \right]\!\right] = b \in G;$$

this gives that $G \cap D_{\dot{r}}$ is non-empty, yielding that $r = \dot{r}_G = f_q \in 2^\omega \cap M$ for some $q \in G \cap D_{\dot{r}}$. Item 2 of the proposition is proved. □

Exercise 7.3.11 Prove item 1 of the proposition. (HINT: Follow the pattern of the proof of item 2 of the proposition. Now start from $\dot{r}$ a B-name for a function from ω into ω_1, and argue once again that $\dot{r}_G \in M$ whenever G is M-generic for P.)

We now prove item 3 of the proposition:

Proof Let $i : P \to B = \mathsf{RO}(P)^M$ in M be the canonical dense embedding of P into its boolean completion according to M.

Let $\dot{g} \in M^B$ be such that

$$\dot{g} = \{\langle \mathrm{op}(\check{j}, f\check{(j)}), i(f) \rangle : f \in P, \ j \in \mathrm{dom}(f)\}.$$

We claim that $g = \dot{g}_G$ is a bijection of $(\omega_1)^M$ into κ. To this aim observe that

$$f \subseteq g \leftrightarrow i(f) \in G$$

for all $f \in P$, since

$$\langle j, \alpha \rangle \in g$$

if and only if there is some $i(f) \in G$ such that

$$\left\langle \mathrm{op}(\check{j}, \check{\alpha}), i(f) \right\rangle \in \dot{g}$$

if and only if

$$f(j) = \alpha \text{ for some (any) } i(f) \in G \text{ with } j \in \mathrm{dom}(f).$$

This gives immediately that g is an injective function, since it is the coherent union of injective functions. Moreover for all $\alpha < \kappa$ and all $\xi < (\omega_1)^M$ the following sets are easily seen to be dense and in M:

$$D_\alpha = \{i(f) : f \in P, \ \alpha \in \mathrm{ran}(f)\},$$

$$E_\xi = \{i(f) : f \in P, \ \xi \in \mathrm{dom}(f)\}.$$

Observe that whenever G is an M-generic filter for B, $G \cap D_\alpha \neq \emptyset$ iff $\alpha \in \mathrm{ran}(\dot{g}_G)$ and $G \cap E_\xi \neq \emptyset$ iff $\xi \in \mathrm{dom}(\dot{g}_G)$.

This gives that $M[G]$ models that $\dot{g}_G$ is a bijection of $(\omega_1)^M$ with κ. The proof of item 3 of the proposition is completed.   □

The proof of the proposition is completed.   □

Modulo the results of the next section, the consistency of $\mathsf{CH} + \mathsf{ZFC}$ has been established.

7.4 M^B **Models ZFC**

In this section we prove that the axioms of ZFC are valid in any model M^B, whenever M is a countable transitive model of ZFC and B is a boolean algebra that belongs to M and which M models to be complete.

Theorem 7.4.1 *Assume M is a transitive countable model of* ZFC *in V and* B $\in M$ *is such that M models* B *is a complete boolean algebra. Then*

$$M \models [\![\phi]\!]_B = 1_B$$

for every axiom ϕ of ZFC.

Remark 7.4.2 We can actually prove a stronger result stating that whenever M is *any* model of ZFC and B $\in M$ is a boolean algebra which M models to be complete, then

$$M \models [\![\phi]\!]_B = 1_B$$

for every axiom ϕ of ZFC. That is, we can remove the assumption that M is countable and transitive in the above theorem. However the proof of this latter result is slightly more involved since we cannot appeal to the forcing theorem to obtain it. As we will see below, the forcing theorem plays a crucial role in most of the arguments to follow.

Proof We show that $M[G]$ satisfies all ZFC-axioms whenever G is M-generic for B. The proof can be completed appealing to Theorem 7.2.5. From now on we assume that G is an M-generic filter for B.   □

Extensionality $M[G] \subseteq V$, and hence $(M[G], \in)$ models the Extensionality Axiom since the latter is a Π_1-sentence holding in V.

Foundation $M[G]$ is a transitive set contained in V, so $M[G]$ models the Axiom of Foundation, since the latter is expressed by the universal sentence

$$\forall f \ [(f \text{ is a function } \wedge \ \text{dom}(f) = \omega) \rightarrow \exists n \in \omega (f(n) \notin f(n+1))]$$

in signature $\in_{\Delta_0}$ and holds in V (hence we can apply Lemma A.3.2 to $M[G] \subseteq V$).

Infinity $\omega = \check{\omega}_G \in M[G]$.

Pairing   Let $\sigma_G, \tau_G \in M[G]$. Given $\sigma, \tau \in M^B$, let

$$\text{up}(\sigma, \tau) = \{\langle \sigma, 1 \rangle, \langle \tau, 1 \rangle\}.$$

Then $\text{up}(\sigma, \tau) = \{\sigma_G, \tau_G\}$ (since $\rho(\sigma) = \rho(\tau) = 1_B \in G$) is a witness for the pairing axiom (see also Exercise 7.1.21).

Union   Given $\sigma_G \in M[G]$, we let

$$\tau = \{\langle \rho, 1_{\mathsf{B}}\rangle : \exists u \in \text{dom}(\sigma)(\rho \in \text{dom}(u))\}.$$

Then τ_G is a witness for the union axiom for σ_G, since:

$$\bigcup \sigma_G = \{a : \exists u_G \in \sigma_G (a \in u_G)\}$$

$$= \{a : \exists u \in \text{dom}(\sigma)(\sigma(u) \in G \wedge a \in u_G)\}$$

$$\subseteq \{a : \exists u \in \text{dom}(\sigma)(a \in u_G)\}$$

$$= \{\rho_G : \exists u \in \text{dom}(\sigma)(\rho \in \text{dom}(u) \wedge u(\rho) \in G)\}$$

$$\subseteq \{\rho_G : \exists u \in \text{dom}(\sigma)(\rho \in \text{dom}(u))\} = \tau_G.$$

Powerset   Let $\sigma_G \in M[G]$. Set

$$\mathcal{P}^{\mathsf{B}}(\sigma) = \big\{\langle \tau, [\![\tau \subseteq \sigma]\!]\rangle : \text{dom}(\tau) = \text{dom}(\sigma)\big\} \in M,$$

applying the comprehension axiom in M to the set

$$\Big\{\langle \tau, b\rangle \in \mathcal{P}(\text{dom}(\sigma) \times \mathsf{B})^M \times \mathsf{B}\Big\} \in M$$

and the formula

$$\phi(z, \sigma, \mathsf{B}) \equiv \exists x, y \left[(z = \langle x, y\rangle) \wedge (x : \text{dom}(\sigma) \to \mathsf{B} \text{ is a B-name}) \wedge (y = [\![x \subseteq \sigma]\!]_{\mathsf{B}})\right].$$

We claim that

$$M[G] \models (\mathcal{P}^{\mathsf{B}}(\sigma))_G = \{a \in M[G] : a \subseteq \sigma_G\}$$

for all G M-generic for B.
Both inclusions follow by Lemma 7.2.9:

$\subseteq$: $\eta_G \in (\mathcal{P}^{\mathsf{B}}(\sigma))_G$ for some $\eta \in \text{dom}(\mathcal{P}^{\mathsf{B}}(\sigma))$ iff $[\![\eta \subseteq \sigma]\!] = \mathcal{P}^{\mathsf{B}}(\sigma)(\eta) \in G$,
giving that $\eta_G \subseteq \sigma_G$ by the forcing theorem.

$\supseteq$: Assume $a \subseteq \sigma_G$ for some $a \in M[G]$. Then $a = \tau_G$ with $[\![\tau \subseteq \sigma]\!] \in G$,
by the forcing theorem. By Lemma 7.2.9 we get that there is $\eta \in M^{\mathsf{B}}$ with
$\text{dom}(\eta) = \text{dom}(\sigma)$ such that $[\![\tau = \eta]\!] \geq [\![\tau \subseteq \sigma]\!]$. Since $[\![\tau \subseteq \sigma]\!] \in G$, we get
that $[\![\tau = \eta]\!] \in G$ as well; hence $\tau_G = \eta_G$. However $\eta_G \in (\mathcal{P}^{\mathsf{B}}(\sigma))_G$, since
$\text{dom}(\eta) = \text{dom}(\sigma)$ and

$$[\![\eta \subseteq \sigma]\!] \geq [\![\tau \subseteq \sigma]\!] \wedge [\![\tau = \eta]\!] \in G.$$

Comprehension   Let $\varphi(x, z)$ be a formula, $a \in M[G]$, $\sigma_G = a$, and $\bar{\tau}_G = \bar{d}$, where $\bar{\tau} = (\tau_1, \ldots, \tau_n)$ and $\bar{\tau}_G = ((\tau_1)_G, \ldots, (\tau_n)_G)$. Let

$$e = \{c \in a : M[G] \models \varphi(c, \bar{d})\}.$$

We must show that the definable class e in $M[G]$ is an element of $M[G]$. Let

$$\eta = \{\langle \rho, [\![\varphi(\rho, \bar{\tau})]\!] \wedge [\![\rho \in \sigma]\!]) \rangle : \rho \in \mathrm{dom}(\sigma)\} \in M^B.$$

We claim that $\eta_G = e$.

$\eta_G \subseteq e$:   We have that

$$\eta_G = \left\{\rho_G : \rho \in \mathrm{dom}(\sigma) \text{ and } [\![\varphi(\rho, \bar{\tau})]\!] \wedge [\![\rho \in \sigma]\!] \in G\right\}.$$

Hence $\rho_G \in e$ for all $\rho_G \in \eta_G$, since

$$M[G] \models \varphi(\rho_G, \bar{\tau}_G) \wedge \rho_G \in \sigma_G$$

by the forcing theorem.

$\eta_G \supseteq e$: $u \in e$ iff $u \in \sigma_G$ and

$$M[G] \models \phi(u, \bar{\tau}_G).$$

First observe that $u \in \sigma_G$ iff $u = v_G$ for some $v \in \mathrm{dom}(\sigma)$ with $\sigma(v) \in G$. Moreover,

$$\sigma(v) \leq \bigvee_{u \in \mathrm{dom}(\sigma)} [\![u = v]\!] \wedge \sigma(u) = [\![v \in \sigma]\!],$$

and hence $[\![v \in \sigma]\!] \in G$.

The forcing theorem also gives that $[\![\phi(\rho, \bar{\tau})]\!] \in G$ for some ρ with $\rho_G = u$ as

$$M[G] \models \phi(u, \bar{\tau}_G).$$

Since $\rho_G = u = v_G$, we get that $[\![\rho = v]\!] \in G$ and

$$[\![\phi(v, \bar{\tau})]\!] \geq [\![\phi(\rho, \bar{\tau})]\!] \wedge [\![\rho = v]\!] \in G.$$

We conclude that

$$\eta(v) = [\![\phi(v, \bar{\tau})]\!] \wedge [\![v \in \sigma]\!] \in G.$$

Hence $u = \nu_G \in \eta_G$, concluding the proof.

Collection   Let $\psi(x, y, \bar{\tau}_G)$ be a formula and $\sigma_G \in M[G]$ be a set such that

$$M[G] \models \forall x \in \sigma_G \, \exists y \, \phi(x, y, \bar{\tau}_G),$$

where $\bar{\tau}_G = ((\tau_1)_G, \ldots, (\tau_n)_G)$. By the forcing theorem we have that

$$[\![\forall x \in \sigma \, \exists y \, \phi(x, y, \bar{\tau})]\!] \in G.$$

In M consider the definable class (in parameters $\mathsf{B}, \sigma, \tau_1, \ldots, \tau_n$) given by the binary relation

$$R^* \subseteq \operatorname{dom}(\sigma) \times M^{\mathsf{B}}$$

of ordered pairs $\langle \eta, \nu \rangle$ such that

$$b_{\eta, \nu} = [\![\phi(\eta, \nu, \tau_1, \ldots, \tau_n)]\!] \wedge \sigma(\eta) > 0_{\mathsf{B}}.$$

Find in M a set $Z \subseteq M^{\mathsf{B}}$ with $Z \in M$ such that for any $\eta \in \operatorname{dom}(\sigma)$ and $\nu \in M^{\mathsf{B}}$ there is $\nu^* \in Z$ such that $b_{\eta, \nu} = b_{\eta, \nu^*}$.

We can find the required set Z applying the collection principle and the comprehension principle in M to the binary relation S contained in

$$(\operatorname{dom}(\sigma) \times \mathsf{B}) \times M^{\mathsf{B}}$$

and given by ordered pairs $\langle (\eta, b), \nu \rangle$ such that $b = b_{\eta, \nu}$.

This is possible by the following argument: The domain A of S is definable in M in parameters $\mathsf{B}, \sigma, \tau_1, \ldots, \tau_n$ and is a subset of $\operatorname{dom}(\sigma) \times \mathsf{B}$; hence A in M by comprehension; the binary relation S is a definable class in M from parameters $\mathsf{B}, \sigma, \tau_1, \ldots, \tau_n$ and its domain is A; therefore the Collection Principle applied in M to S gives a set Z as required.

Let

$$\mu = \left\{ \langle \operatorname{op}(\eta, \nu), b_{\eta, \nu} \rangle : \langle \eta, \nu \rangle \in R^*, \nu \in Z, \eta \in \operatorname{dom}(\sigma) \right\}.$$

Since $\mu \in M$ and $\mu : M^{\mathsf{B}} \to \mathsf{B}$ is a function, we get that $\mu \in M^{\mathsf{B}}$.

Claim 7.4.2.1 *The following holds.*

$$M[G] \models \forall x \in \sigma_G \, \exists y \in \operatorname{ran}(\mu_G) \, \psi(x, y, \bar{\tau}_G).$$

This suffices to realize the required instance of the Collection Principle in $M[G]$ as witnessed by $\text{ran}(\mu_G) \in M[G]$, modulo the following:

Exercise 7.4.3 Assume N is transitive, $a \in N$, and N models Extensionality, Pairing, Union, Comprehension. Then $\text{ran}(a) \in N$.

So we are left with the proof of the claim:

Proof Pick any $a \in \sigma_G$. Then for some $\eta \in \text{dom}(\sigma)$ with $\sigma(\eta) \in G$ we have that $a = \eta_G$.

Note also that, since $a \in \sigma_G$,

$$M[G] \models \exists y \, \psi(a, y, \bar{\tau}_G).$$

Therefore for some $\nu \in M^B$

$$M[G] \models \psi(\eta_G, \nu_G, \bar{\tau}_G).$$

This gives that

$$b_{\nu,\eta} = [\![\phi(\eta, \nu, \bar{\tau})]\!] \wedge \sigma(\nu) \in G.$$

By choice of Z there is $\nu^* \in Z$ such that $b_{\eta,\nu^*} = b_{\nu,\eta} \in G$.

This gives that

$$\langle \eta_G, \nu_G^* \rangle = \text{op}(\eta, \nu^*)_G \in \mu_G$$

as $b_{\eta,\nu^*} = \mu(\text{op}(\nu, \eta^*)) \in G$.

Therefore

$$M[G] \models \nu_G^* \in \text{ran}(\mu_G) \wedge \psi(a, \nu_G^*, \bar{\tau}_G)$$

as $b_{\eta,\nu^*} = \sigma(\eta) \wedge [\![\psi(\eta, \nu^*, \tau_1, \ldots, \tau_n)]\!] \in G$.

Since a was chosen arbitrarily in σ_G, we are done.   □

Choice   We prove that $M[G]$ satisfies that every set can be well ordered. More precisely, we show that for all $X \in M[G]$, there exists an injection $f : X \to \text{Ord}$ belonging to $M[G]$.

Let $\tau_G = X \in M[G]$. Since $M \models \text{ZFC}$, there exists $f \in M$ such that $M \models f : \tau \to \beta$ is a bijection. Let

$$\tau^* = \{\langle \text{op}(\sigma, \check{\alpha}), 1_B \rangle : \sigma \in \text{dom}(\tau), \, f(\sigma) = \alpha\} \in M^B.$$

Then:

$$\tau_G^* = \{\mathrm{op}(\sigma, \check{\alpha})_G : \sigma \in \mathrm{dom}(\tau),\ f(\sigma) = \alpha\}$$
$$= \{(\sigma_G, f(\sigma)) : \sigma \in \mathrm{dom}(\tau)\} = R.$$

Notice that

$$R \subseteq \{\sigma_G : \sigma \in \mathrm{dom}(\tau)\} \times \beta$$

may not be a functional relation (there could be distinct $\sigma, \sigma' \in \mathrm{dom}(\tau)$ such that $\sigma_G = \sigma'_G$, with $f(\sigma) \neq f(\sigma')$). Notice also that $\tau_G \subseteq \mathrm{dom}(R)$: Indeed

$$\tau_G = \{\sigma_G : \tau(\sigma) \in G\} \subseteq \{\sigma_G : \langle\sigma_G, f(\sigma)\rangle \in R : \sigma \in \mathrm{dom}(\tau)\} = \mathrm{dom}(R).$$

By what we have shown so far $M[G] \models$ ZF. In particular we can use the comprehension axiom in $M[G]$ to refine R to a functional relation g with the same domain letting:

$$g = \{(\sigma_G, \xi) \in R : \forall\gamma \in \beta(\sigma_G, \gamma) \in R \Rightarrow \xi \leq \gamma\}.$$

Clearly, $R \in M[G]$ implies that $g \in M[G]$ by comprehension applied in $M[G]$. We leave to the reader to check that $g : \mathrm{dom}(R) \to \mathrm{Ord}$ is an injective function. Hence, g witnesses that $X = \tau_G$ can be well ordered in $M[G]$.

The proof that all axioms of ZFC hold in $M[G]$ is complete.   □

Corollary 7.4.4 *Assume M is a transitive countable model of* ZFC *and G is M-generic for a* B $\in M$ *which M models to be a complete boolean algebra. Then $M[G]$ is the smallest transitive model N of* ZFC *with $N \supseteq M$ and $G \in N$. Moreover, $M[G] \cap \mathrm{Ord} = M \cap \mathrm{Ord}$.*

Proof $G = \dot{G}_G \in M[G]$, where $\dot{G} \in M^{\mathsf{B}}$ is the B-name $\left\{\left(\check{b}, b\right) : b \in \mathsf{B}^+\right\}$, and $M \subseteq M[G]$ since $\check{a}_G = a$ for all $a \in M$. In particular $M[G] \subseteq N$ for all N transitive model of ZFC containing M and with $G \in N$. Since $M[G] \models$ ZFC and is transitive, we are done.

For the second part of the corollary (i.e., the assertion that $M[G] \cap \mathrm{Ord} = M \cap \mathrm{Ord}$), we proceed as follows: Since $M[G] \models$ ZFC, we get that

$$\mathrm{Ord} \cap M[G] = \left\{\mathrm{rk}(\tau_G) : \tau \in M^{\mathsf{B}}\right\}.$$

An easy induction shows that $\mathrm{rk}(\tau_G) \leq \mathrm{rk}(\tau)$ for all $\tau \in M[G]$; moreover $\mathrm{rk}(\check{\alpha}_G) = \mathrm{rk}(\alpha) = \alpha$ for all limit $\alpha \in M \cap \mathrm{Ord}$. We get that

$$M \cap \mathrm{Ord} = \left\{ \check{\alpha}_G : \alpha \in M \cap \mathrm{Ord} \right\} \subseteq \mathrm{Ord} \cap M[G] = \left\{ \mathrm{rk}(\tau_G) : \tau \in M^B \right\} \subseteq M \cap \mathrm{Ord}.$$

$\square$

Appendix A
Absoluteness for Set Theoretic Concepts

This appendix deals with a delicate set theoretic issue: that of establishing which set theoretic concepts are simple and which are not. We develop here a first order formalization of set theory which reflects in the Lévy complexity of a formula the set theoretic complexity of the concept it formalizes.

We organize this chapter in three parts:

- We abstractly analyze first order $\mathcal{L}$-theories T under definable expansions of the language $\mathcal{L}$. We outline that expansions of the language $\mathcal{L}$ which make a Δ_1-definable concept of T as axiomatized in $\mathcal{L}$ logically equivalent to an atomic formula in the expanded language do not change the Lévy complexity of any other concept of T as formalized in the expanded language.
- We outline that for the specific case of set theory the set theoretic concepts which can be formalized by $\in$-formulae where all quantifiers are bounded to range on a set (the so-called Δ_0-formulae) express simple set theoretic concepts. Plugging in the results of the first part, we obtain that the Δ_1-properties of set theory as formalized in a language $\in_{\Delta_0}$ which has predicate symbols for all Δ_0-formulae formalize simple set theoretic concepts. On the model theoretic side, we note that transitive models of suitable fragments of ZFC compute correctly the relevant basic set theoretic concepts and agree on their meaning. Then we proceed to outline the main set theoretic concepts that are simple, i.e., the Δ_1-properties with respect to set theory as formalized in $\in_{\Delta_0}$, among these we find the notion of well-foundedness, the concepts defined by means of transfinite recursion on a Δ_1-property (e.g., transitivity, Mostowski collapse, the Tarski satisfaction predicate,...). We also outline which set theoretic concepts are not simple (e.g., not Δ_1 for $\in_{\Delta_1}$), among which the concepts of cardinality, cofinality, and powerset.
- In the last part of the chapter we briefly introduce the structures H_κ. For a given cardinal κ, the structure H_κ consists of the elements of V which can be generated by sets of size less than κ, i.e., elements whose transitive closure has size smaller than κ. $H_{\aleph_0}$ is the collection of hereditarily finite sets, and $H_{\aleph_1}$ is the collection

© The Author(s), under exclusive license to Springer Nature Switzerland AG 2024
M. Viale, *The Forcing Method in Set Theory*, La Matematica per il 3+2 168,
https://doi.org/10.1007/978-3-031-71660-7

of hereditarily countable sets. In general the universe can be stratified as the increasing union of the H_κ as κ ranges over the infinite cardinals. We will see that for many purposes it is more convenient to work in H_κ rather than in V_κ. Furthermore $H_\kappa = V_\kappa$ when κ is strongly inaccessible. It can also be shown that $H_{\aleph_0}$ is well suited to develop in a set theoretic context recursion theory, while $H_{\aleph_1}$ is well suited to develop in a set theoretic context second order number theory.

A.1   Model Theoretic Absoluteness

We introduce the following terminology:

Notation A.1.1

- Given a first order language $\mathcal{L}$, an $\mathcal{L}$-formula is quantifier free if no quantifiers occur in it. We denote by $\Sigma_0(\mathcal{L})$ and $\Pi_0(\mathcal{L})$ the quantifier free $\mathcal{L}$-formulae; inductively an $\mathcal{L}$-formula is $\Sigma_{n+1}(\mathcal{L})$ if it is logically equivalent to a formula of the form $\exists \vec{x}\, \psi$ with $\psi\ \Pi_n(\mathcal{L})$, and it is $\Pi_{n+1}(\mathcal{L})$ if its negation is $\Sigma_{n+1}(\mathcal{L})$. We just write Σ_n and Π_n rather than $\Sigma_n(\mathcal{L})$ and $\Pi_n(\mathcal{L})$ when $\mathcal{L}$ is clear from the context.
- $\mathcal{L}_\forall$ denotes the universal $\mathcal{L}$-formulae.
  Likewise we interpret $\mathcal{L}_\exists, \mathcal{L}_{\forall\exists}, \ldots$.
- $\mathcal{L}_{\forall\vee\exists}$ denotes the boolean combinations of universal $\mathcal{L}$-formulae, i.e., formulae (which are logically equivalent to formulae) of type $\psi \wedge \phi$ or $\psi \vee \phi$ with $\phi\ \Sigma_1$ and $\psi\ \Pi_1$.
  Likewise we interpret $\mathcal{L}_{\forall\exists\vee\exists\forall}, \ldots$.
- Given a first order theory T in language $\mathcal{L}$, $T_\forall^{\mathcal{L}}$ denotes the sentences in $\mathcal{L}_\forall$ which are logical consequences of T; likewise we interpret $T_\exists^{\mathcal{L}}, T_{\forall\exists}^{\mathcal{L}}, T_{\forall\vee\exists}^{\mathcal{L}}, \ldots$. If the language of T is clear, we omit the superscript $\mathcal{L}$ and just write $T_\forall, \ldots$.
- We often write $\mathfrak{M} \models \phi(\vec{a})$ rather than $\mathfrak{M} \models \phi(\vec{x})[\vec{x}/\vec{a}]$ when $\mathfrak{M}$ is an $\mathcal{L}$-structure, $\vec{a} \in \mathfrak{M}^{<\omega}$, and ϕ is an $\mathcal{L}$-formula.
- We often denote an $\mathcal{L}$-structure $\mathfrak{M} = (M, R^{\mathfrak{M}} : R \in \mathcal{L})$ by $(M, \mathcal{L}^{\mathfrak{M}})$ (or even by $(M, \mathcal{L}^M)$, identifying a structure with its domain when no confusion on the interpretation of the symbols of $\mathcal{L}$ in M can arise).
- We often identify an $\mathcal{L}$-structure $\mathfrak{M} = (M, \mathcal{L}^M)$ with its domain M and an ordered tuple $\vec{a} \in M^{<\omega}$ with its set of elements.
- We let the atomic diagram $\Delta_0(\mathfrak{M})$ of an $\mathcal{L}$-model $\mathfrak{M} = (M, \mathcal{L}^M)$ be the family of quantifier free sentences $\phi(\vec{a})$ in language $\mathcal{L} \cup M$ (where M is considered as a set of fresh constants added to the language $\mathcal{L}$ and the interpretation of each new constant $a \in M$ is given by a itself) such that $\mathfrak{M} \models \phi(\vec{a})$.
- $\sqsubseteq$ denotes the substructure relation between structures.

- $\mathfrak{M} \prec_n \mathfrak{N}$ indicates that $\mathfrak{M}$ is a Σ_n-elementary substructure of $\mathfrak{N}$, i.e., $\mathfrak{M} \sqsubseteq \mathfrak{N}$, and for all Σ_n-formulae $\psi(\vec{x})$ and $\vec{a} \in M^{<\omega}$

$$\mathfrak{M} \models \psi(\vec{a}) \text{ if and only if } \mathfrak{N} \models \psi(\vec{a}).$$

We omit the n to denote full elementarity.
- We often write $T + S$ or $T + \psi$ to denote the $\mathcal{L}$-theories $T \cup S$ or $T \cup \{\psi\}$.
- When writing $\psi(x_0, \ldots, x_n)$, we assert that all free variables occurring in ψ are among $x_0, \ldots, x_n$, but not all of them need appear in ψ.
  In particular, for a fixed $\mathcal{L}$-formula ψ, we write $\psi(x_0, \ldots, x_n)$ *in displayed free variables* to intend that ψ defines an $n + 1$-dimensional subset of an $\mathcal{L}$-structure, even if the free variables occurring in the string ψ are just some of those in the list $x_0, \ldots, x_n$.
- Given an $\mathcal{L}$-formula $\phi(x_0, \ldots, x_n)$ in displayed free variables and an $\mathcal{L}$-structure $\mathfrak{M}$ with domain M, we let

$$\phi(\mathfrak{M}) = \left\{ \vec{a} \in M^{n+1} : \mathfrak{M} \models \phi(\vec{a}) \right\}.$$

Notation A.1.2 Given a language $\mathcal{L}$, let $\phi(x_0, \ldots, x_n)$ be an $\mathcal{L}$-formula in displayed free variables.
   We let:

- R_ϕ^{n+1} be a new $n + 1$-ary relation symbol.
- f_ϕ^n be a new n-ary function symbol.[1]

We also let

$$\mathsf{AX}_{\phi,n+1}^0 := \forall \vec{x}[\phi(\vec{x}) \leftrightarrow R_\phi^{n+1}(\vec{x})],$$

$$\mathsf{AX}_{\phi,n+1}^1 := \forall x_1, \ldots, x_n, y\,(\phi(y, x_1, \ldots, x_n) \leftrightarrow f_\phi^n(x_1, \ldots, x_n) = y).$$

for $\phi(x_0, \ldots, x_n)$ having the displayed free variables.
   Let $\mathsf{Form}_\mathcal{L}$ denote the set of $\mathcal{L}$-formulae. For $A \subseteq \mathsf{Form}_\mathcal{L} \times \omega \times 2$:

- $\mathcal{L}_A$ is the language obtained by adding to $\mathcal{L}$ relation symbols R_ϕ^n for the $(\phi, n, 0) \in A$ and function (or constant) symbols f_ϕ^n for the $(\phi, n, 1) \in A$.
- $T_{\mathcal{L},A}$ is the $\mathcal{L}_A$-theory having as axioms the sentences $\mathsf{AX}_{\phi,n}^i$ for $(\phi, n, i) \in A$.

---

[1] As usual we confuse 0-ary function symbols with constants.

Note the following:

- For any $\mathcal{L}$-theory T, let $A = \mathsf{Form}_{\mathcal{L}} \times \omega \times \{0\}$ and $\mathcal{L}^* = \mathcal{L}_A$; then $T^* = T + T_{\mathcal{L},A}$ is an $\mathcal{L}^*$-theory admitting quantifier elimination (the Morleyization of T, see Proposition A.1.3 below). Furthermore any $\mathcal{L}$-structure admits exactly one extension to an $\mathcal{L}^*$-structure which is a model of $T_{\mathcal{L},A}$.
- For any $\mathcal{L}$-theory T and any $\mathcal{L}$-formula $\phi(x_0, \ldots, x_n)$ in displayed free variables such that T models $\forall \vec{x} \, \exists! y \, \phi(y, \vec{x})$, any $\mathcal{L}$-structure $\mathfrak{M}$ with domain M admits exactly one expansion to an $\mathcal{L}_{\{\phi\} \times \{n+1\} \times \{1\}}$-structure which models $\mathsf{AX}^1_{\phi, n+1}$.

We are interested in analyzing what happens when the Morleyization process on some $\mathcal{L}$-theory T is performed on arbitrary subsets A of $\mathsf{Form}_{\mathcal{L}} \times \omega \times 2$ where each $(\phi, n, 1) \in A$ is such that T models $\forall \vec{x} \, \exists! y \, \phi(\vec{x}, y)$.

Proposition A.1.3 *Given a language $\mathcal{L}$, consider the language $\mathcal{L}^*$ which adds an n-ary predicate symbol R^{n+1}_ϕ for any $\mathcal{L}$-formula $\phi(x_0, \ldots, x_n)$ with displayed free variables.*

Let $T_{\mathcal{L}}$ be the following $\mathcal{L}^*$-theory:

- *For all quantifier free $\mathcal{L}$-formulae $\phi(x_0, \ldots, x_n)$ in displayed free variables and $n \in \omega$,*

$$\forall x_0, \ldots, x_n \, (\phi(\vec{x}) \leftrightarrow R^{n+1}_\phi(\vec{x})).$$

- *For all $\mathcal{L}$-formulae $\phi \wedge \psi(x_0, \ldots, x_m)$, $\phi(x_0, \ldots, x_m)$, $\psi(x_0, \ldots, x_m)$ in displayed free variables and $m \in \omega$ in displayed free variables,*

$$\forall x_0, \ldots, x_m \, [R^{m+1}_{\phi \wedge \psi}(x_0, \ldots, x_m) \leftrightarrow (R^{m+1}_\psi(x_0, \ldots, x_m) \wedge R^{m+1}_\phi(x_0, \ldots, x_m))].$$

- *For all $\mathcal{L}$-formulae $\phi(x_0, \ldots, x_m)$ in displayed free variables and $m \in \omega$,*

$$\forall \vec{x} \, [R^{m+1}_{\neg \phi}(\vec{x}) \leftrightarrow \neg R^{m+1}_\phi(\vec{x})].$$

- *For all $\mathcal{L}$-formulae $\phi(x_0, \ldots, x_j, y, x_{j+1}, \ldots, x_m)$ in displayed free variables and all $j \leq m \in \omega$,*

$$\forall x_0, \ldots, x_m \, [\exists y \, R^{m+2}_\phi(x_0, \ldots, x_j, y, x_{j+1}, \ldots, x_m)$$

$$\leftrightarrow R^{m+1}_{\exists y \phi}(x_0, \ldots, x_j, x_{j+1}, \ldots, x_m)].$$

Then any $\mathcal{L}$-structure $\mathfrak{N}$ admits a unique extension to an $\mathcal{L}^*$-structure $\mathfrak{N}^*$ which models $T_{\mathcal{L}}$. Moreover every $\mathcal{L}^*$-formula is $T_{\mathcal{L}}$-equivalent to an atomic $\mathcal{L}^*$-formula. In particular for any $\mathcal{L}$-model $\mathfrak{N}$, the algebras of its $\mathcal{L}$-definable subsets and of the $\mathcal{L}^*$-definable subsets of $\mathfrak{N}^*$ are the same.

Proof By an easy induction one can prove that any $\mathcal{L}$-formula $\phi(x_0, \ldots, x_n)$ is $T_{\mathcal{L}}$-equivalent to the atomic $\mathcal{L}^*$-formula $R^{n+1}_\phi(x_0, \ldots, x_n)$.

Another simple inductive argument brings that any $\mathcal{L}^*$-formula $\phi(\vec{x})$ is $T_{\mathcal{L}}$-equivalent to the $\mathcal{L}$-formula obtained by replacing (in the unique reasonable way to preserve meaning) all strings of type $R_\psi^{m+1}(t_0(\vec{x}_0), \ldots t_m(\vec{x}_m))$ occurring in ϕ by the $\mathcal{L}$-formula $\psi(t_0(\vec{x}_0), \ldots t_m(\vec{x}_m))$. Note also that there are no new function symbols in $\mathcal{L}^*$; hence all terms appearing in some $\mathcal{L}^*$-formula are $\mathcal{L}$-terms.

Combining these observations together, we get that any $\mathcal{L}^*$-formula is equivalent to an atomic $\mathcal{L}^*$-formula.

$T_{\mathcal{L}}$ forces the $\mathfrak{N}^*$-interpretation of any relation symbol $R_\phi(\vec{x})$ in $\mathcal{L}^* \setminus \mathcal{L}$ to be the $\mathfrak{N}$-interpretation of the $\mathcal{L}$-formula $\phi(\vec{x})$ to which it is $T_{\mathcal{L}}$-equivalent. □

The passage from $\mathcal{L}$-structures to $\mathcal{L}^*$-structures which model $T_{\mathcal{L}}$ can have effects on the embeddability relation; for example, assume $\mathfrak{M} \sqsubseteq \mathfrak{N}$ is a non-elementary embedding of $\mathcal{L}$-structures; then $\mathfrak{M}^* \not\sqsubseteq \mathfrak{N}^*$: If the nonatomic $\mathcal{L}$-formula $\phi(\vec{a})$ in parameter $\vec{a} \in \mathfrak{M}^{<\omega}$ holds in $\mathfrak{M}$ and does not hold in $\mathfrak{N}$, the atomic $\mathcal{L}^*$-formula $R_\phi(\vec{a})$ holds in $\mathfrak{M}^*$ and does not hold in $\mathfrak{N}^*$.

Let us also introduce the notation we will use to handle the substructure relation over expanded languages. The following conventions supplement Notation A.1.2.

Notation A.1.4 Let $\mathcal{L}$ be a language, $(M, \mathcal{L}^M)$ an $\mathcal{L}$-structure, and B a subset of $\text{Form}_{\mathcal{L}} \times \omega \times 2$. Then, $(M, \mathcal{L}_B^M)$ is the unique extension of $(M, \mathcal{L})$ defined in accordance with Notation A.1.2 which satisfies $T_{\mathcal{L},B}$. In particular $(M, \mathcal{L}_B^M)$ is a shorthand for $(M, S^M : S \in \mathcal{L}_B)$.

If $(N, \mathcal{L}^N)$ is a substructure of $(M, \mathcal{L}^M)$, we also write $(N, \mathcal{L}_B^M)$ as a shorthand for $(N, S^M \restriction N : S \in \mathcal{L}_B)$.

It is a delicate matter to establish whether for $\mathfrak{M} \sqsupseteq \mathfrak{N}$, $(N, \mathcal{L}_B^M)$ and $(N, \mathcal{L}_B^N)$ are the same structure, as this depends on whether $\mathfrak{M}, \mathfrak{N}$ compute the same way the concepts formalized by B.

Definition A.1.5 Given $\mathcal{L}$-structures $\mathfrak{M} \sqsubseteq \mathfrak{N}$, an $\mathcal{L}$-formula $\phi(\vec{x})$ is absolute for $\mathfrak{M}, \mathfrak{N}$ if for all $\vec{b} \in \mathfrak{M}^{\vec{x}}$

$$\mathfrak{M} \models \phi(\vec{b})$$

if and only if

$$\mathfrak{N} \models \phi(\vec{b}).$$

From now on we feel free to omit oftentimes the superscript n in R_ϕ^n, f_ϕ^n when the arity of those symbols is clear from the context.

The following exercise will only be needed in the proof of the reflection Theorem B.2.3:

Exercise A.1.6 Assume

$$\mathfrak{M} \sqsubseteq \mathfrak{N} \sqsubseteq \mathfrak{P}$$

are such that

$$\mathfrak{M} \prec_n \mathfrak{P}$$

and

$$\mathfrak{N} \prec_n \mathfrak{P}.$$

Then

$$\mathfrak{M} \prec_n \mathfrak{N}$$

holds as well. (HINT: This is basic model theory. Proceed to prove the thesis by induction on $m \leq n$. The base case $m = 0$ is granted.)

A.1.1   Preservation of the Substructure Relation

The following addresses the matter of absoluteness for substructures:

Lemma A.1.7 *Assume $\mathfrak{M} \sqsupseteq \mathfrak{N}$ are $\mathcal{L}$-structures which are both models of an $\mathcal{L}$-theory T. Assume $B \subseteq \mathsf{Form}_{\mathcal{L}} \times \omega \times 2$ is such that:*

- *Any $(\phi, n, 0) \in A$ is such that ϕ is $\Delta_1(T)$.*
- *Any $(\psi, n, 1) \in A$ is such that ψ is $\Sigma_1(T)$ and such that T proves*

$$\forall x_0, \ldots, x_{n-1} \, \exists! y \, \psi(y, x_0, \ldots, x_{n-1}).$$

Then $(N, \mathcal{L}_B^M) = (N, \mathcal{L}_B^N) \sqsubseteq (M, \mathcal{L}_B^M).$

Proof The unique delicate part of the proof is to check that $(N, \mathcal{L}_B^M) = (N, \mathcal{L}_B^N)$.

Σ_1-formulae are upward absolute between substructures, while Π_1-formulae are downward absolute; hence $\Delta_1(T)$-properties are absolute between $(N, \mathcal{L}_B^N)$ and $(M, \mathcal{L}_B^M)$. This shows that $(R_\psi^n)^M = (R_\psi^n)^N \cap M^{<\omega}$ for any $(\psi, n, 0) \in B$.

Now note that if $(\psi, n, 1) \in B$, with ψ being $\exists \vec{z} \, \phi(y, \vec{z}, \vec{x})$ and ϕ quantifier free, $\phi(\vec{z}, \vec{x}, y)$ is absolute between $\mathfrak{M} \sqsupseteq \mathfrak{N}$, being it a quantifier free $\mathcal{L}$-formula. Hence $\phi(\mathfrak{M}) \cap \mathfrak{N}^{|\vec{x}|+|\vec{z}|+1} = \phi(\mathfrak{N})$, being ϕ quantifier free.

Furthermore recall that

$$T \models \forall \vec{x} \, \exists! y \, \exists \vec{z} \, \phi(y, \vec{z}, \vec{x}). \tag{A.1}$$

Since both structures satisfy T, each can define

$$g^{\mathfrak{M}} : M^{|\vec{x}|} \to M$$

$$\vec{a} \mapsto c \text{ if and only if } \mathfrak{M} \models \phi(c, \vec{b}, \vec{a}) \text{ for some } \vec{b}$$

and

$$g^{\mathfrak{N}} : N^{|\vec{x}|} \to N$$

$$\vec{a} \mapsto c \text{ if and only if } \mathfrak{N} \models \phi(c, \vec{b}, \vec{a}) \text{ for some } \vec{b}.$$

The following holds:

Claim A.1.7.1 $g^{\mathfrak{M}} \restriction N^{|\vec{x}|} = g^{\mathfrak{N}}$.

Proof Assume towards a contradiction that for $c \neq d$ in M, and $\vec{a} \in N^{|\vec{x}|}$, $g^{\mathfrak{M}}(\vec{a}) = c$, while $g^{\mathfrak{N}}(\vec{a}) = d$.

Then there would be $\vec{b}_0 \in \mathfrak{M}^{|\vec{z}|}$, $\vec{b}_1 \in \mathfrak{N}^{|\vec{z}|}$ such that

$$\mathfrak{M} \models \phi(c, \vec{b}_0, \vec{a}),$$

and

$$\mathfrak{N} \models \phi(d, \vec{b}_1, \vec{a}).$$

Now since ϕ is quantifier free, also

$$\mathfrak{M} \models \phi(d, \vec{b}_1, \vec{a}).$$

This contradicts

$$\mathfrak{M} \models \forall \vec{x} \, \exists! y \, \exists \vec{z} \, \phi(y, \vec{z}, \vec{x})$$

as witnessed by $\vec{x} = \vec{a}$.

The proof is completed.   □

A.1.2   Absoluteness for Provably Δ_1-Properties

Definition A.1.8 Let S be an $\mathcal{L}$-theory. An $\mathcal{L}$-formula $\phi(\vec{x})$ is $\Delta_1(S)$ if there are quantifier free $\mathcal{L}$-formulae $\psi_\phi(\vec{x}, \vec{y})$ and $\theta_\phi(\vec{x}, \vec{z})$ such that S proves

$$\forall \vec{x} \, [\phi(\vec{x}) \leftrightarrow \forall \vec{y} \, \psi_\phi(\vec{x}, \vec{y}) \leftrightarrow \exists \vec{z} \, \theta_\phi(\vec{x}, \vec{z})].$$

The following result condenses the relevant facts about Δ_1-properties and allows to formalize the informal assertion that "Δ_1 on Δ_1 remains Δ_1."

Theorem A.1.9 *Assume T is an $\mathcal{L}$-theory and $A \subseteq \mathrm{Form}_{\mathcal{L}} \times \omega \times 2$ is such that:*

- *ψ is $\Delta_1(T)$ for any $(\psi, n, 0) \in A$.*
- *θ is a Σ_1-formula and $T \models \forall \vec{x} \exists! y\, \theta(\vec{x}, y)$ for any $\mathcal{L}$-formula $\theta(\vec{x}, y)$ with $(\theta, n, 1) \in A$.*

Let ϕ be an $\mathcal{L}$-formula which is $\Delta_1(T + T_{\mathcal{L}, A})$.
Then ϕ is $\Delta_1(T)$.

We split the proof of the theorem in two parts to handle separately the new relation symbols and the new function symbols.

Lemma A.1.10 *Assume T is an $\mathcal{L}$-theory and $A \subseteq \mathrm{Form}_{\mathcal{L}} \times \{0\}$.*
Let ϕ be an $\mathcal{L}$-formula which is $\Delta_1(T + T_{\mathcal{L}, A})$. Then ϕ is $\Delta_1(T)$.

Proof Let $B = \{(\psi_j, 0) : j = 1, \ldots, n\}$ list all the formulae in A whose corresponding relation symbol occurs in ϕ. By interpolation ϕ is $\Delta_1(T + T_{\mathcal{L}, B})$. We need to show that it is also $\Delta_1(T)$: Assume

$$T + T_{\mathcal{L}, B} \vdash \phi(\vec{x}) \leftrightarrow \exists \vec{z}\, \eta_\phi(\vec{z}, \vec{x}) \leftrightarrow \forall \vec{u}\, \nu_\phi(\vec{u}, \vec{x})$$

with η_ϕ, ν_ϕ quantifier free in $T_{\mathcal{L}, B}$. By induction on the quantifier free $\mathcal{L}_B$-formulae, we show that there are a Σ_1-formula η for $\mathcal{L}$ and a Π_1-formula ν for $\mathcal{L}$ such that

$$T + T_{\mathcal{L}, B} \vdash \forall \vec{z}, \vec{x}\, (\eta_\phi(\vec{z}, \vec{x}) \leftrightarrow \eta(\vec{z}, \vec{x}))$$

and

$$T + T_{\mathcal{L}, B} \vdash \forall \vec{u}, \vec{x}\, (\nu_\phi(\vec{u}, \vec{x}) \leftrightarrow \nu(\vec{u}, \vec{x})).$$

To do so we may assume that $\eta_\phi(\vec{z}, \vec{x})$ is (logically equivalent to) a disjunction of conjunctions of atomic formulae or negated atomic formulae of $\mathcal{L}_B$; then if R_{ψ_i} appears in any such conjunction as an atomic formula, we substitute it with the Σ_1-formula for $\mathcal{L}$ which is equivalent to R_{ψ_i}; if R_{ψ_i} appears in any such conjunction as a negated atomic formula, we substitute it with the Π_1-formula for $\mathcal{L}$ which is equivalent to R_{ψ_i}. In this way we replaced $\eta_\phi(\vec{u}, \vec{x})$ with a $T + T_{\mathcal{L}, B}$-equivalent Σ_1-formula η for $\mathcal{L}$. Accordingly we operate on ν_ϕ to find ν a $T + T_{\mathcal{L}, B}$-equivalent Π_1-formula for $\mathcal{L}$. Hence we get that

$$T + T_{\mathcal{L}, B} \vdash \forall \vec{x}\, (\phi(\vec{x}) \leftrightarrow \forall \vec{u}\, \nu(\vec{u}, \vec{x}) \leftrightarrow \exists \vec{z}\, \eta(\vec{z}, \vec{x})).$$

Since none of the formulae on the right-hand side of $\vdash$ contains symbols not in $\mathcal{L}$, we get (by interpolation) that

$$T \vdash \forall \vec{x} \, (\phi(\vec{x}) \leftrightarrow \forall \vec{u} \, \nu(\vec{u}, \vec{x}) \leftrightarrow \exists \vec{z} \, \eta(\vec{z}, \vec{x})),$$

as was to be shown. $\square$

Lemma A.1.11 *Let T be an $\mathcal{L}$-theory and ϕ be a Σ_1-formula for $\mathcal{L}$ such that $T \vdash \forall \vec{x} \exists ! y \phi(\vec{x}, y)$.*

Assume $\theta(\vec{u})$ is $\Delta_1(T + T_{\mathcal{L},A})$ for $\theta(\vec{u})$ an $\mathcal{L}$-formula and $A = \{(\phi, 1)\}$. Then we have that θ is $\Delta_1(T)$.

Proof The key point is to analyze the complexity of the formula $y = t(x_1, \ldots, x_n)$ for t an $\mathcal{L}_A$-term. We can prove the following:

Claim A.1.11.1 *For any $\mathcal{L}_A$-term $t(x_1, \ldots, x_n)$ in displayed variables, there are a Π_1-formula $\theta_t(x_1, \ldots, x_n)$ and a Σ_1-formula $\psi_t(x_1, \ldots, x_n)$ for $\mathcal{L}$ such that*

$$T + T_{\mathcal{L},A} \models \forall x_1, \ldots, x_n, y \, [\psi_t(y, x_1, \ldots, x_n) \leftrightarrow t(x_1, \ldots, x_n)$$
$$= y \leftrightarrow \theta_t(y, x_1, \ldots, x_n)]. \tag{A.2}$$

Proof We proceed by induction on the depth[2] of the $\mathcal{L}_A$-term t. If t is a term of depth 0, then t is a constant or a variable and there is almost nothing to prove (i.e., we can let ψ_t and θ_t be the atomic $\mathcal{L}$-formula $y = t$).

Now assume the claim holds for all terms of depth n. Let

$$t = f(t_1(x_1, \ldots, x_n), \ldots, t_k(x_1, \ldots, x_n))$$

be a term of depth $n + 1$ with f a function symbol of $\mathcal{L}_A$. By inductive assumptions there are $\mathcal{L}$-formulae $\theta_{t_j}(y_j, x_1, \ldots, x_n)$ and $\psi_{t_j}(y_j, x_1, \ldots, x_n)$ for $j = 1, \ldots, k$ which are, respectively, Π_1 for $\mathcal{L}$ and Σ_1 for $\mathcal{L}$ and such that

$$T + T_{\mathcal{L},A} \models \forall x_1, \ldots, x_n, y \, \left[\psi_{t_j}(y, x_1, \ldots, x_n) \leftrightarrow t_j(x_1, \ldots, x_n)\right.$$
$$\left. = y \leftrightarrow \theta_{t_j}(y, x_1, \ldots, x_n)\right]. \tag{A.3}$$

Note that A.3 also entails that

$$T + T_{\mathcal{L},A} \models \forall x_1, \ldots, x_n \exists ! y \, \psi_{t_j}(y, x_1, \ldots, x_n) \tag{A.4}$$

and

$$T + T_{\mathcal{L},A} \models \forall x_1, \ldots, x_n \exists ! y \, \theta_{t_j}(y, x_1, \ldots, x_n). \tag{A.5}$$

---

[2] That is, the height of the syntactic tree describing the term t using the function symbols of $\mathcal{L}_A$.

This gives that $y = f(t_1(x_1, \ldots, x_n), \ldots, t_k(x_1, \ldots, x_n))$ is $T + T_{\mathcal{L},A}$-equivalent to the Σ_1-formula for $\mathcal{L}_A$

$$\psi_t^*(x_1, \ldots, x_n) := \exists y_1, \ldots, y_k \left[\bigwedge_{j=1}^{k} \psi_j(y_j, x_1, \ldots, x_n) \wedge y = f(y_1, \ldots, y_k) \right] \tag{A.6}$$

and to the Π_1-formula for $\mathcal{L}_A$

$$\theta_t^*(x_1, \ldots, x_n) := \forall y_1, \ldots, y_k \left[\bigwedge_{j=1}^{k} \psi_j(y_j, x_1, \ldots, x_n) \rightarrow y = f(y_1, \ldots, y_k) \right]. \tag{A.7}$$

If f is a function symbol of $\mathcal{L}$, we let ψ_t be ψ_t^* and θ_t be θ_t^*. These are $\mathcal{L}$-formulae, since $y = f(y_1, \ldots, y_k)$ is already an atomic $\mathcal{L}$-formula, and all the other symbols occurring in A.6 and A.7 are also in $\mathcal{L}$, and we easily get that A.2 holds for $\psi_t := \psi_t^*, \theta_t := \theta_t^*$; else f is f_ϕ (therefore $k = n$), and we are considering the atomic $\mathcal{L}_A$-formula

$$y = f_\phi(y_1, \ldots, y_n).$$

Now observe that (using $T \vdash \forall \vec{x} \exists! y \phi(\vec{x}, y)$ and that ϕ is a Σ_1-formula for $\mathcal{L}$)

$$y = f_\phi(y_1, \ldots, y_n)$$

is $T + T_{\mathcal{L},A}$-equivalent to the Σ_1-formula for $\mathcal{L}$

$$\exists z \, (\phi(z, y_1, \ldots, y_n) \wedge z = y) \tag{A.8}$$

and is $T + T_{\mathcal{L},A}$-equivalent to the Π_1-formula for $\mathcal{L}$

$$\forall z \, (\phi(z, y_1, \ldots, y_n) \rightarrow z = y). \tag{A.9}$$

Also in this case we are done: Replacing in A.6 the $\mathcal{L}_A$-formula $f_\phi(x_1, \ldots, x_n) = y$ with the $\mathcal{L}$-formula A.8 gives a Σ_1-formulation (by a formula in $\mathcal{L}$) relative to $T + T_{\mathcal{L},A}$ of $f(x_1, \ldots, x_n) = y$, while replacing in A.7 $f_\phi(x_1, \ldots, x_n) = y$ with the $\mathcal{L}$-formula A.9 gives a Π_1-formulation (by a formula in $\mathcal{L}$) relative to $T + T_{\mathcal{L},A}$ of $f(x_1, \ldots, x_n) = y$. We let ψ_t and θ_t be the $\mathcal{L}$-formulae obtained from ψ_t^* and θ_t^* by this substitution. A minimal variant of the argument given above shows that ψ_t and θ_t are $T + T_{\mathcal{L},A}$-equivalent to $y = t(x_1, \ldots, x_n)$ and are, respectively, Σ_1 or Π_1.

The claim is proved. $\square$

Now any quantifier free $\mathcal{L}_A$-formula $\theta(\vec{x})$ can be written as $\psi(t_1(\vec{x}), \ldots, t_n(\vec{x}))$ with $\psi(y_1, \ldots, y_n)$ a quantifier free $\mathcal{L}$-formula with only relation symbols occurring in it.

Clearly $\theta(\vec{x})$ is logically equivalent to the Σ_1-formula

$$\exists y_1, \ldots, y_n \, [\psi(y_1, \ldots, y_n) \wedge (\bigwedge_{i=1}^{n} t_i(\vec{x}) = y_i)]$$

and to the Π_1-formula

$$\forall y_1, \ldots, y_n \, [(\bigwedge_{i=1}^{n} t_i(\vec{x}) = y_i) \rightarrow \psi(y_1, \ldots, y_n)].$$

By the above claim (combined with the above representation of quantifier free $\mathcal{L}_A$-formulae by means of $\Delta_1(T + T_{\mathcal{L},A})$-properties where their Σ_1 and Π_1 representations are given by formulae with all terms appearing only as substitutions of x in the atomic formula $x = y$), we can reduce any $\mathcal{L}$-formula θ which is $\Delta_1(T + T_{\mathcal{L},A})$ to a $T + T_{\mathcal{L},A}$-equivalent Σ_1-formula for $\mathcal{L}$ and to a $T + T_{\mathcal{L},A}$-equivalent Π_1-formula for $\mathcal{L}$. Therefore θ is $\Delta_1(T)$ by interpolation, as was to be shown. $\square$

We are in the position to prove the theorem:

Proof Given the $\mathcal{L}$-formula ϕ which is $\Delta_1(T + T_{\mathcal{L},A})$, we let $\psi_j(\vec{x}_j)$ for $j = 1, \ldots, n$ and $\theta_i(\vec{x}_i, y)$ for $i = 1, \ldots, m$ be the $\mathcal{L}$-formulae such that $R_{\psi_j}^{m_j}$ and $f_{\theta_i}^{m_i}$ occur in ϕ.

Note that:

- The ψ_js are all $\Delta_1(T)$.
- The θ_is are Σ_1-formulae for $\mathcal{L}$ such that T proves that $\forall \vec{x}_i \, \exists! y \, \theta_i(\vec{x}_i, y)$.

Letting $B_0 = \{(\psi_j, m_j, 0) : j = 1, \ldots, n\}$, it suffices to show that ϕ is $\Delta_1(T + T_{\mathcal{L},B_0})$ and then invoke Lemma A.1.10 to conclude that ϕ is $\Delta_1(T)$.

We can proceed by induction to show that every $\Delta_1(T + T_{\mathcal{L},B_{n+1}})$-property can be reduced to a $\Delta_1(T + T_{\mathcal{L},B_n})$-property, where $B_n = B_0 \cup \{(\theta_i, m_i, 1) : i = 1, \ldots, n\}$ for $0 < n \leq m$.

To handle the inductive step, we just need to invoke Lemma A.1.11 relative to θ_{n+1}, with $\mathcal{L}$ replaced by $\mathcal{L}_{B_n}$ and T replaced by $T + T_{\mathcal{L},B_n}$. $\square$

A.2 Set Theoretic Absoluteness for Properties Defined by Bounded Quantification

A.2.1 What Is the Right Language for Set Theory?

The standard axioms of ZFC in the $\in$-language are clearly sufficient to provide a first order axiomatization of set theory. However a closer inspection reveals that many simple set-theoretic concepts are not formalized by simple $\in$-formulae.

Consider, for example, the notion of ordered pair. While we informally write $x = \langle y, z \rangle$ to mean that *x is the ordered pair with first component y and second component z*, in set theoretic terms this statement hides a nontrivial coding of the concept of ordered pair, for example, by means of Kuratowski's definition: $x = \{\{y\}, \{y, z\}\}$. A proper definition of the concept of ordered pair in the $\in$-language can then be given by the following $\in$-formula:

$$\exists t \exists u \, [\forall w \, (w \in x \leftrightarrow w = t \vee w = u) \wedge \forall v \, (v \in t \leftrightarrow v = y)$$
$$\wedge \forall v \, (v \in u \leftrightarrow v = y \vee v = z)].$$

It is clear that the meaning of this $\in$-formula is hardly recognizable with a rapid glance (unlike $x = \langle y, z \rangle$). Moreover, from a purely logical perspective, its Lévy complexity is already Σ_2. This clashes with our understanding that the concept of ordered pair is simple. Indeed, we do not regard the notion of ordered pair as a complex concept, contrary to other more complicated and theoretically loaded ones like that of uncountability, or many of the properties of the continuum (such as its correct place in the hierarchy of uncountable cardinals). In a similar vein other very basic notions such as being a function, a binary relation, or the domain or the range of a function are formalized by rather complicated $\in$-formulae, from the point of view of both their readability and their Lévy complexity.

The standard solution adopted in set theory textbooks[3] is to regard as basic all those $\in$-formulae in which the quantifiers are bounded to range over the elements of some set, that is, the Δ_0-formulae. In order to make these observations precise, we need to be extremely cautious on our notational conventions.

We introduce the basic languages and fragments of set theory we will always include in any language of interest to us.

Definition A.2.1 The family Δ_0 of *bounded* $\in$-formulae is the smallest family of $\in$-formulae which:

* Contains the atomic formulae
* Is closed under boolean combinations
* Is such that for all ϕ in the family so are $\forall x \in \phi := \forall x(x \in y \rightarrow \phi)$ and $\exists x \in \phi := \exists x(x \in y \wedge \phi)$

---

[3] See, for example, [19, Chapter IV, Def. 3.5] or [16, Def. 12.9].

Fact A.2.2 *The following concepts are expressible by an $\in$-formula which is Δ_0; hence by an atomic $\in_{\Delta_0}$-formula:*

- $x \subseteq y$.
- $x = \{y, z\}$.
- $x = \langle y, z \rangle$.
- $x = \bigcup y$.
- $x = \bigcap y$.
- $x = y \setminus z$.
- $x = y \times z$.
- x *is a binary relation.*
- x *is a function.*
- x *is a surjection.*
- x *is an injection.*
- x *is a bijection.*
- $x = \mathrm{dom}(y)$.
- $x = \mathrm{ran}(y)$.
- $x \in \mathrm{dom}(y)$.
- $x \in \mathrm{ran}(y)$.
- x *is an n-ary relation (i.e., every $y \in x$ is a function with domain n).*
- x *is transitive.*
- w *is a function and $w(x) = y$.*
- x *is the empty set.*
- x *is an ordinal.*
- x *is a successor ordinal.*
- x *is a limit ordinal.*
- x *is the natural number n (for any n in $\mathbb{N}$).*
- x *is ω (ω is the set of natural numbers $\mathbb{N}$ or the first infinite ordinal).*
- $x = y^\frown z$ *(for strings a, b—i.e., those sets that are also called finite tuples, or functions with domain a natural number, or finite sequences, or . . .—$a^\frown b$ denotes the string obtained by concatenation of the string a to the string b).*
- . . .

Furthermore also the following Gödel operations (which we borrow from [16, Def. 13.6]) have a Δ_0-definable graph in models $(V, \in)$ of ZF^-*:*

$$G_1(X, Y) = \{X, Y\} \qquad\qquad G_2(X, Y) = X \times Y$$
$$G_3(X, Y) = \{(x, y) : x \in X, y \in Y, x \in y\} \quad G_4(X, Y) = X \setminus Y$$
$$G_5(X, Y) = X \cap Y \qquad\qquad G_6(X) = \bigcup X = \{z : \exists y \in X\, z \in y\}$$
$$G_7(X) = \{x : (x, y) \in X\} \qquad G_8(X) = \{(x, y) : (y, x) \in X\}$$
$$G_9(X) = \{(x, y, z) : (x, z, y) \in X\} \qquad G_{10}(X) = \{(x, y, z) : (y, z, x) \in X\}.$$

Proof This is a very useful exercise for the reader. $\qquad\qquad\square$

Notation A.2.3 We let $\in_{\Delta_0}$ be $\in_D$ for $D \subseteq \mathsf{Form}_\in \times \omega \times 2$ extending the set $\Delta_0 \times \omega \times \{0\}$ with the pairs $(\phi, n, 1)$ as $\phi(x_0, \ldots, x_{n-1}, x)$ ranges over the following Δ_0-formulae:

- The Δ_0-formulae $\phi_\omega(x)$, $\phi_\emptyset(x)$ defining $\emptyset$ and ω in any model of ZF^- (also we denote by ω and $\emptyset$ the constants $f_{\phi_\emptyset}$ and f_{ϕ_ω})
- The Δ_0-formulae $\phi_i(\vec{x}, y)$ as G_i ranges over the operations $G_1, \ldots, G_{10}$ as defined in Fact A.2.2 and $\phi_i(\vec{x}, y)$ is the Δ_0-formula defining the graph of G_i in any $\in$-model of[4] ZF^-.

We let T_{Δ_0} be given by the axioms:

$$\forall \vec{x} \, (R_{\forall z \in y \phi}(y, z, \vec{x}) \leftrightarrow \forall z(z \in y \to R_\phi(y, z, \vec{x}))), \tag{A.10}$$

$$\forall \vec{x} \, [R_{\phi \wedge \psi}(\vec{x}) \leftrightarrow (R_\phi(\vec{x}) \wedge R_\psi(\vec{x}))], \tag{A.11}$$

$$\forall \vec{x} \, [R_{\neg \phi}(\vec{x}) \leftrightarrow \neg R_\phi(\vec{x})], \tag{A.12}$$

$$\forall x \, (x \notin \emptyset), \tag{A.13}$$

ω is a non-empty ordinal whose elements are all successor ordinals or $\emptyset$, (A.14)

$$\forall \vec{x} \, \exists! y \, (y = G_i(\vec{x})), \tag{A.15}$$

$$\forall \vec{x} \, \forall y \, [y = G_i(\vec{x}) \leftrightarrow R_{\phi_i}(\vec{x}, y)] \tag{A.16}$$

for the Gödel operations $G_1, \ldots, G_{10}$.

We axiomatize suitable fragments of the $\in$-theory $\mathsf{ZFC} + T_{\Delta_0}$ as follows:

- $\mathsf{Z}^-_{\Delta_0}$ stands for the $\in_{\Delta_0}$-theory given by:

  (a) The Extensionality Axiom

  $$\forall x, y \, [(y \subseteq x \wedge x \subseteq y) \leftrightarrow x = y].$$

  (b) The FoundationAxiom

  $$\forall x \, [x = \emptyset \vee \exists y \in x \, \forall z \in x \, (z \notin y)].$$

  (c) T_{Δ_0}.

- Z_{Δ_0} enriches $\mathsf{Z}^-_{\Delta_0}$ adding the powerset axiom

  $$\forall x \, \exists y \, [\forall z \, (z \subseteq x \leftrightarrow z \in y)].$$

---

[4] In models of ZF^- the Gödel operations $G_1, \ldots, G_{10}$ as listed and defined in [16, Def. 13.6] and their compositions have as graph the extension of a Δ_0-formula (by [16, Lemma 13.7]).

- $ZC^-_{\Delta_0}$ enriches $Z^-_{\Delta_0}$ adding the Axiom of Choice AC

$$\forall x\, \exists f\, \left[(f \text{ is a bijection}) \wedge \mathrm{dom}(f) = x \wedge (\mathrm{ran}(f) \text{ is an ordinal})\right].$$

- $ZF^-_{\Delta_0}$ enriches $Z^-_{\Delta_0}$ adding the Collection Principle for all $\in_{\Delta_0}$-formulae.
- $ZFC^-_{\Delta_0}, ZF_{\Delta_0}, ZFC_{\Delta_0}$ are defined as expected.

Remark A.2.4 We took the pain of giving an explicit axiomatization of $Z^-_{\Delta_0}$ using Extensionality, Foundation, and axioms A.10,...,A.16 because this axiomatization is given by Π_2-sentences of $\in_{\Delta_0}$; hence it is preserved by Σ_1-substructures. Note that AC is a Π_2-axiom of $\in_{\Delta_0}$, while the powerset axiom and the replacement schema for a quantifier free $\in_{\Delta_0}$-formula are both Π_3.

An inductive argument shows that $ZF^- + T_{\in,D}$ (where D is the subset of $\mathrm{Form}_\in \times \omega \times 2$ used in Not. A.2.3 to define $\in_{\Delta_0}$) is logically equivalent to ZF^- enriched with axioms A.10,...,A.16.

Exercise A.2.5 Prove the above remark.

We now introduce the terminology to handle set theory formalized in languages expanding $\in_{\Delta_0}$.

Notation A.2.6 Let $\mathcal{L} \supseteq \in_{\Delta_0}$. For an $\mathcal{L}$-formula $\phi(\vec{x}, \vec{y}, \vec{z})$:

- The *Collection Principle* for ϕ ($\mathsf{Rep}(\phi)$) states

$$\forall \vec{z}\, \forall X$$

$$[$$

$$(\forall x \in X \exists y\, \phi(x, y, \vec{z}))$$

$$\rightarrow$$

$$\exists R\, (R \text{ is a binary relation} \wedge \mathrm{dom}(R) = X \wedge \forall \langle x, y \rangle \in R\, \phi(x, y, \vec{z}))$$

$$].$$

$\mathsf{Rep}_{\mathcal{L}}$ holds if $\mathsf{Rep}(\phi)$ holds for all $\mathcal{L}$-formulae ϕ.
- $ZF^-_{\mathcal{L}}$ is $Z^-_{\Delta_0} + \mathsf{Rep}_{\mathcal{L}}$.
- Accordingly we define $ZFC_{\mathcal{L}}, ZFC^-_{\mathcal{L}}, ZF_{\mathcal{L}}, ZFC_{\mathcal{L}},\ldots$
- We write ZFC_{Δ_0} rather than $ZFC_{\mathcal{L}}$ when $\mathcal{L} = \in_{\Delta_0}$, etc.
- If $A \subseteq \mathrm{Form}_\in \times 2$ is such that $\in_{\Delta_0} \subseteq \in_A$, we write ZFC^-_A rather than $ZFC^- + T_{\in,A},\ldots$

Clearly (the suitable fragment of) $ZFC + T_{\in,A}$ is logically equivalent to (the suitable fragment of) ZFC_A.

The following is a trivial consequence of the results of Sect. A.1:

Fact A.2.7 *Let $\phi(\vec{x}, y)$ and $\psi(\vec{x})$ be $\in$-formulae which are $\Delta_1(T)$ for some $\in_{\Delta_0}$-theory $T \supseteq \mathsf{ZF}^-_{\Delta_0}$ and such that*

$$T \models \forall \vec{x} \exists! y \phi(\vec{x}, y).$$

Then any $\Delta_1(T + \mathsf{AX}^1_\phi + \mathsf{AX}^0_\psi)$ is also $\Delta_1(T)$.

The following is a very useful consequence of the Collection Principle:

Lemma A.2.8 *Given $\mathcal{L} \supseteq \in_{\Delta_0}$ and $T \supseteq \mathsf{ZF}^-_{\mathcal{L}}$ an $\mathcal{L}$-theory, any $\mathcal{L}$-formula ψ which is $\Sigma_n(T)$ is such that $\forall x \in y\, \psi$ is still $\Sigma_n(T)$.*

Proof Assume ψ is $\exists z_0, \ldots, z_n \theta(x, x_0, \ldots, x_m, z_0, \ldots, z_n)$ with $\theta(x, \vec{x}, \vec{z})$ a Π_{n-1}-formula.

By the Collection Principle for the $\mathcal{L}$-formula

$$\forall x \in y\, \exists u \left[u \text{ is an } m+1\text{-tuple} \wedge \theta(x, u(0), \ldots, u(m), z_0, \ldots, z_n) \right],$$

we get that

$$T + \forall x \in y\, \psi$$

models

$$\exists w [\tag{A.17}$$

$$(w \text{ is an } m+2\text{-ary relation})$$

$$\wedge$$

$$(\forall x \in y\, \exists v \in w\, v(0) = x)$$

$$\wedge$$

$$(\forall v \in w\, (v(0) \in y \wedge \theta(v(0), v(1), \ldots, v(m+1), z(1), \ldots, z(n))))$$

$$].$$

Now note that

$$\forall v \in w\, [v(0) \in y \wedge \theta(v(0), v(1), \ldots, v(m+1), z(1), \ldots, z(n))]$$

is Π_{n-1}, while (w *is an* $m+2$-*ary relation*) and ($\forall x \in y\, \exists v \in w\, v(0) = x$) are atomic $\in_{\Delta_0}$-formulae.

Now it is not hard to check that $\forall x \in y\, \psi$ is T-equivalent to the Σ_n-formula A.17. $\qquad \square$

Notation A.2.9 Given $\mathcal{L} \supseteq \in_{\Delta_0}$, $T \supseteq \mathsf{ZF}_{\mathcal{L}}^-$ an $\mathcal{L}$-theory, and ψ an $\mathcal{L}$-formula such that

$$T \vdash \forall x, \vec{z} \, \exists ! y \, \phi(x, y, \vec{z}).$$

$\theta_\phi(u, v, \vec{z})$ is the $\mathcal{L}$-formula describing in an $\mathcal{L}$-model $(V, \mathcal{L}^V)$ of T the graph of the unique $\mathcal{L}$-definable function $G : \mathrm{Ord}^V \times V^{\vec{z}} \to V$ such that $G(\alpha, \vec{b}) = a$ if and only if

$$(V, \mathcal{L}) \models \phi(G \upharpoonright \alpha, a, \vec{b}).$$

We have the following:

Lemma A.2.10 *Assume $\mathcal{L} \supseteq \in_{\Delta_0}$ and $T \supseteq \mathsf{ZF}_{\mathcal{L}}^-$ is an $\mathcal{L}$-theory and ϕ an $\mathcal{L}$-formula which is $\Delta_1(T)$ and such that*

$$T \vdash \forall z_0 \ldots, z_n, x \exists ! y \phi(x, \vec{z}, y).$$

Let $\theta_\phi(x, \vec{z}, y)$ be the $\mathcal{L}$-formula given by Not. A.2.9. Then $\theta_\phi(x, y, \vec{z})$ is $\Delta_1(T)$.

Proof Let $\psi(w, z_0, \ldots, z_n)$ say that:

- w is a function.
- $\mathrm{dom}(w) = \beta \times \{a\}$ with β an ordinal and a the tuple $\langle z_0, \ldots, z_n \rangle$.
- $\phi(w \upharpoonright \alpha \times \{a\}, \alpha, a(0), \ldots, a(n), w(\alpha ^\frown a))$ holds for all $\alpha \in \beta$ (recall that $\alpha ^\frown a$ denotes the string obtained by concatenation of the 1-tuple $\langle a \rangle$ to the string a).

Note that in a model $(V, \mathcal{L}^V)$ of T, $G(x, \vec{z}) = y$ is expressed both by

$$\exists w \, [\psi(w, \vec{z}) \wedge (x ^\frown \vec{z}) \in \mathrm{dom}(w) \wedge w(x ^\frown \vec{z}) = y]$$

and by

$$\forall w \, [(\psi(w, \vec{z}) \wedge (x ^\frown \vec{z}) \in \mathrm{dom}(w)) \to w(x ^\frown \vec{z}) = y].$$

Now by Lemma A.2.8,

$$\forall u \in \mathrm{dom}(w) \, \phi(w \upharpoonright u(0) \times \{u(1)\}, u(0), a(0), \ldots, a(n), w(u))$$

is also $\Delta_1(T)$: Note that the above formula can be expressed both as

$$\forall y \, \forall u \in y \, (y = \mathrm{dom}(w) \to \phi(w \upharpoonright u(0) \times \{u(1)\}, u(0), a(0), \ldots, a(n), w(u))$$

and as

$$\exists y \, (\mathrm{dom}(w) = y \wedge \forall u \in y \, \phi(w \restriction u(0) \times \{u(1)\}, u(0), a(0), \ldots, a(n), w(u)).$$

Therefore $\theta_\phi(x, y, \vec{z})$ is $\Delta_1(T + \mathsf{AX}^1_\phi)$.

Applying the results of Sect. A.1, we get that $\theta_\phi(x, y, \vec{z})$ is $\Delta_1(T)$.   □

Remark A.2.11 Note that Lemma A.2.10 establishes that a certain definitional schema of transfinite recursion (as defined in Notation A.2.9) over a Δ_1-property yields a Δ_1-property. With minor arrangements left to the reader, one can check that any reasonable definitional schema of transfinite recursion applied to a Δ_n (Σ_n, Π_n) set theoretic property (as formalized in some language $\mathcal{L}$ extending $\in_{\Delta_0}$) yields as output a property of the same Lévy complexity. We use this observation with some liberality in several places in this book.

To see the expressive power of $\in_{\Delta_0}$, consider the following notions and their logical complexity for $\mathsf{ZF}^-_{\Delta_0}$:

- (x is a cardinal) is the Π_1-formula (for $\in_{\Delta_0}$)

$$(x \text{ is an ordinal}) \wedge \forall f \, [(f \text{ is a function} \wedge \mathrm{dom}(f) \in x) \to \mathrm{ran}(f) \neq x].$$

- (x is $\aleph_1$) is the boolean combination of Σ_1-formulae

$$(x \text{ is a cardinal}) \wedge (\omega \in x) \wedge$$

$$\wedge \exists F \, \Big[(F : \omega \times x \to x) \wedge \forall 0 \neq \alpha \in x \, (F \restriction \omega \times \{\alpha\} \text{ is a surjection on } \alpha) \Big].$$

- CH is the Σ_2-sentence

$$\exists f \, [(f \text{ is a function} \wedge \mathrm{dom}(f) \text{ is } \aleph_1) \wedge \forall r \subseteq \omega \, (r \in \mathrm{ran}(f))],$$

and $\neg$CH is the boolean combination of Π_2-sentences[5]

$$\exists x \, (x \text{ is } \aleph_1) \wedge \forall f \, [(\mathrm{dom}(f) \text{ is } \aleph_1 \wedge f \text{ is a function}) \to \exists r \subseteq \omega \, (r \notin \mathrm{ran}(f))].$$

- (x is $\aleph_2$) is the Σ_2-formula given by the conjunction of the Π_1-formula

$$(x \text{ is a cardinal})$$

---

[5] We let $\neg$CH include the Σ_2-sentence $\exists x \, (x \text{ is } \aleph_1)$, for otherwise its failure could be witnessed by the assertion that there is no uncountable cardinal, a statement that holds true in H_{ω_1}, regardless of whether CH or its negation is true in the corresponding universe of sets.

with the Σ_2-formula:

$$\exists F \, \exists y$$

$$[$$

$$(y \text{ is } \aleph_1) \wedge (y \in x) \wedge (F : y \times x \to x) \wedge \forall 0 \neq \alpha \in x \, (F \upharpoonright y \times \{\alpha\} \text{ is a surjection on } \alpha)$$

$$].$$

- $2^{\aleph_0} > \aleph_2$ is the boolean combination of Π_2-sentences

$$\exists x \, (x \text{ is } \aleph_2) \wedge \forall f \, [(f \text{ is a function} \wedge \operatorname{dom}(f) \text{ is } \aleph_2)$$

$$\to \exists r \, (r \subseteq \omega \wedge r \notin \operatorname{ran}(f))].$$

- $2^{\aleph_0} \leq \aleph_2$ is the Σ_2-sentence

$$\exists f \, [(f \text{ is a function}) \wedge \operatorname{dom}(f) \text{ is } \aleph_2 \wedge \forall r \, (r \subseteq \omega \to r \in \operatorname{ran}(f))].$$

A.3   Absoluteness Between Transitive Models of Set Theory

The following lemma and definition sum up the results so far obtained and show that for transitive models of suitable fragments of ZF all the $\Delta_1(\mathsf{ZF}^-_{\Delta_0})$-properties are computed the same way.

Definition A.3.1 Given an $\in$-structure (N, E) with E a binary relation on N, $M \subseteq N$ is an E-transitive substructure if for all $a \in M$ and $b \in N$, $b \, E \, a$ entails $b \in M$.
   When E is $\in \cap N^2$, we just say that M is a transitive substructure of N.[6]

Lemma A.3.2 *Assume (N, E) is an $\in$-structure and M is an E-transitive substructure of N. Then all Δ_0-formulae are absolute for (N, E) and $(M, E \cap M^2)$. Therefore if both are models of* ZF$^-$, *all $\in$-formulae that are $\Delta_1(\mathsf{ZF}^-_{\Delta_0})$ are absolute for (N, E) and $(M, E \cap M^2)$.*

Proof Left to the reader. Easy corollary of the results of the previous sections.   □

   We now give the usual set theoretic definition of absoluteness for class subsets of V which is a special instantiation of Definition A.3.1 in case $\mathfrak{N}$ is $(V, \in)$.
   We extend the scope of the inclusion symbol $\subseteq$ to denote the inclusion relation also among subclasses M, N of V.

---

[6] Note that this does not mean that either M or N is transitive: For example, if $\gamma > \omega_1$ and $(M, \in) \prec (N, \in) \prec (V_\gamma, \in)$ with M, N both countable, then neither M nor N is transitive as $\omega_1 \in M$ by elementarity but $\omega_1 \not\subseteq N$, as N is countable.

Definition A.3.3 Given $(V, \in)$ model of ZF^-:

- $R \subseteq V^n$ is a definable class if there exists a formula $\phi_R(x_1, \ldots, x_n, y_1, \ldots, y_{m_R})$ and $b_1^R, \ldots, b_{m_R}^R \in V$ (with the number m_R of parameters depending on R) such that

$$R = \left\{ (a_1, \ldots, a_n) \in V^n : (V, \in) \models \phi_R(a_1, \ldots, a_n, b_1^R, \ldots, b_{m_R}^R) \right\}.$$

- $R \subseteq A^2$ for some $A \subseteq V$ is set-like if for all[7] $a \in A$

$$\mathrm{pred}_R(a) = \{ b \in A : R(a, b) \} \in V.$$

- $R \subseteq A^2$ for some $A \subseteq V$ is well founded if for all $Z \subseteq A$ non-empty, there exists $b \in Z$ such that $\mathrm{pred}_R(b) \cap Z$ is empty.
- Given classes $M \subseteq N \subseteq V$, a definable $R \subseteq V^n$ is absolute between M and N if and only if for some formula ϕ_R defining R in V with parameters $b_1^R, \ldots, b_{m_R}^R \in M$ it holds that for all $a_1, \ldots, a_n \in M$

$$(N, \in) \models \phi_R(a_1, \ldots, a_n, b_1^R, \ldots, b_{m_R}^R)$$

  if and only if

$$(M, \in) \models \phi_R(a_1, \ldots, a_n, b_1^R, \ldots, b_{m_R}^R).$$

- Given classes $M \subseteq N \subseteq V$, a definable $G : V^n \to V$ is absolute for M, N if the graph of G is absolute for M and N and for a formula ϕ_G and $b_1^G, \ldots, b_{m_G}^G$ as in the previous item, M, N, and V model the formula

$$\forall x_1 \ldots \forall x_n \exists! y \phi_G(x_1, \ldots, x_n, y, b_1^G, \ldots, b_{m_G}^G).$$

- A relation R is absolute for M if it is absolute for M and V. Similarly we define the notion of a class function G being absolute for M.
- Given a definable class $M \subseteq V$ and a formula $\phi(x_1, \ldots, x_m)$, $\phi^M(x_1, \ldots, x_m)$ denotes the relativization of ϕ to M and is the formula obtained from $\phi(x_1, \ldots, x_m)$ replacing all subformulae of ϕ of type $\forall x (\ldots)$ (with $(\ldots)$ being the piece of the string ϕ under the scope of the quantifier $\forall x$) by the block of symbols

$$\forall x \, (\phi_M(x, b_1^M, \ldots, b_{m_M}^M) \to \ldots),$$

---

[7] The equality symbol in the equation is meant as a definition of the operator pred_R for R an arbitrary binary relation on A.

and all subformulae of ϕ of type $\exists x\,(\dots)$ (with the same convention for $(\dots)$ as above) by the block of symbols

$$\exists x\,(\phi_M(x, b_1^M, \dots, b_{m_M}^M) \wedge \dots).$$

That is, $\phi^M(x_1, \dots, x_m)$ is the formula obtained from ϕ restricting all its quantifiers to range over elements of M.

- Given a definable relation $R \subseteq V^n$ and some $M \subseteq V$ with $b_1^R, \dots, b_{m_R}^R \in M$,

$$R^M = \{(a_1, \dots, a_n) \in M : (M, \in) \models \phi_R(a_1, \dots, a_n, b_1^R, \dots, b_{m_R}^R)\} =$$

$$= \{(a_1, \dots, a_n) \in M : (V, \in) \models \phi_R^M(a_1, \dots, a_n, b_1^R, \dots, b_{m_R}^R)\}.$$

Fact A.3.4 *Given $(V, \in)$ well-founded model of ZF^-, assume $R \subseteq A^n$, M, $A \subseteq V$ are all definable classes in V, with A and R defined by formulae with parameters in M and M transitive.*
Then:

- $R^M = R \cap M^n$ *if and only if R is absolute for M.*
- *If $R \subseteq A^2$ and $b \in A$,*

$$\mathrm{pred}_R(b) = \{a \in V : V \models \phi_A(a, b_1^A, \dots, b_{m_A}^A) \wedge \phi_R(a, b, b_1^R, \dots, b_{m_R}^R)\}$$

is a definable class in the parameters $b_1^A, \dots, b_{m_A}^A, b, b_1^R, \dots, b_{m_R}^R$.

- $R \subseteq A^2$ *is well founded if and only if the following formula holds in $(V, \in)$:*

$$\forall z(\exists x\phi_A(x, b_1^A, \dots, b_{m_A}^A) \wedge x \in z) \rightarrow$$

$$\rightarrow \Big[\exists x\,(\phi_A(x, b_1^A, \dots, b_{m_A}^A) \wedge x \in z$$

$$\wedge \forall y(\phi_R(y, x, b_1^R, \dots, b_{m_R}^R) \rightarrow y \notin z))\Big].$$

- $R \subseteq A^2$ *is set-like if and only if the following formula holds in $(V, \in)$:*

$$\forall x$$

$$\Big[$$

$$\phi_A(x, b_1^A, \dots, b_{m_A}^A) \rightarrow$$

$$\exists y\,\forall z\,[z \in y \leftrightarrow (\phi_A(z, b_1^A, \dots, b_{m_A}^A) \wedge \phi_R(z, x, b_1^R, \dots, b_{m_R}^R))]$$

$$\Big].$$

- If $R \subseteq A^2$ is set-like, $\mathrm{pred}_R : A \to V$ is a definable class given by

$$\mathrm{pred}_R = \{(a, b) : V \models \forall z [z \in b \leftrightarrow (\phi_A(z, b_1^A, \ldots, b_{m_A}^A)$$
$$\wedge\, \phi_R(z, a, b_1^R, \ldots, b_{m_R}^R))]\}.$$

Proof Left to the reader.   □

Lemma A.3.5 *Assume the following:*

- $(V, \in)$ *models* ZF^-.
- $M \subseteq V$ *is transitive and is a model of* ZF^-.
- $A \subseteq V$ *and* $R \subseteq A^2$ *are definable classes defined by parameters in* M.
- M *models that* R^M *is set-like and well founded.*

Then M *models that there is a unique definable class function*

$$\mathrm{rk}_R^M : A \to \mathrm{Ord}$$

$$a \mapsto \mathrm{rk}_R^M(a) = \sup\{\mathrm{rk}_R^M(b) + 1 : b \in \mathrm{pred}_R(a)\}.$$

Proof By Lemmata A.2.10 and A.3.2, the map on which the recursion to define the rank function is based has a $\Delta_1(\mathsf{ZF}_{\Delta_0}^-)$-graph.   □

Lemma A.3.6 *Assume:*

- $(V, \in)$ *models* ZF^- *and is well founded.*
- M *is transitive and is a model of* ZF^-.
- $A \subseteq V$ *and* $R \subseteq A^2$ *are definable classes and are absolute between* M *and* V.
- R *is set-like and such that* pred_R *is absolute between* M *and* V.

Then $R \cap M^2 = R^M$ *is well founded in* V *if and only if* R^M *is well founded in* M.

Proof Assume $R^M = R \cap M^2$ is well founded in V. Pick a non-empty $Z \in M$ such that $M \models Z \subseteq A^M$. Since $Z \in M$ and M is transitive, we have that $Z \subseteq M$, and since A is absolute for M, we have that $A^M = A \cap M$. In particular $V \models \emptyset \neq Z \subseteq A^M = A \cap M$. Since R^M is well founded in V, Z has an R^M-minimal element a in V. Since $Z \subseteq M$ and $a \in M$ and since $R^M = R \cap M^2$, we have that M models that a is R^M-minimal for Z. Since this holds for all $Z \in M$, we conclude that M models that R^M is well founded.

Conversely assume that $R^M = R \cap M^2$ is well founded in M. Toward a contradiction assume $R \cap M^2$ is not well founded in V. Then there exists a sequence $(a_n : n \in \omega) \in V$ such that $a_j \in M$ and $R(a_{j+1}, a_j)$ holds in V for all $j \in \omega$.

Now $M \models R^M$ is well founded, and M is a transitive model of ZF^-. Thus in M we can define $\mathrm{rk}_R^M : A^M \to \mathrm{Ord}^M$ such that

$$\mathrm{rk}_R^M(a) = \sup\{\mathrm{rk}_R^M(b) + 1 : b \in \mathrm{pred}_R^M(a)\}.$$

Observe that $R(a, b)$ entails that $\mathrm{rk}_R^M(a) \in \mathrm{rk}_R^M(b)$. Since $a_j \in M$ for all $j \in \omega$, we get that for all $j \in \omega$

$$M \models \mathrm{rk}_R^M(a_j) > \mathrm{rk}_R^M(a_{j+1}).$$

But $\mathrm{rk}_R^M(a_j)$ is an ordinal of M and thus really an ordinal in V, since M is transitive. This gives that $(\alpha_j = \mathrm{rk}_R^M(a_j) : j < \omega) \in V$ is a strictly decreasing sequence in $\mathrm{Ord}^M = \mathrm{Ord} \cap M$, which contradicts the validity of the Foundation axiom in V. □

Using Lemmata A.2.10 and A.3.2 one can check that many notions that are defined using transfinite recursion over well-founded relation are absolute between transitive models of ZF^-.

Lemma A.3.7 *Assume:*

- $(V, \in)$ *models* ZF^- *and is well founded.*
- M *is transitive and* $(M, \in \restriction M)$ *is a model of* ZF^-.
- $A \subseteq V$ *and* $R \subseteq A^2$ *are definable classes and are absolute between* M *and* V.
- R *is set-like and such that* pred_R *is absolute between* M *and* V.
- R *is well founded in* V.
- $F : A \times V \to V$ *is a definable class function which is absolute for* M.

Then $G : A \to V$ *given by* $G(a) = F(a, G \restriction \mathrm{pred}_R(a))$ *is absolute between* M *and* V.

Proof By Lemmata A.2.10 and A.3.2. □

Examples on how to employ the above lemma are given by the following:

Lemma A.3.8 *Assume* $(V, \in)$ *models* ZF^-. *Let* $\mathrm{trcl} : V \to V$ *be the class function mapping a set a to its transitive closure* $\mathrm{trcl}(a)$, *i.e., the intersection of all transitive sets* $b \supseteq a$. *Then* trcl *is absolute for any M which is a transitive model of* ZF^-.

Proof It can be checked that $\mathrm{trcl}(a) = \bigcup \{\bigcup^n(x) : n \in \omega\}$, where $\bigcup^0(x) = x$ and $\bigcup^{n+1}(x) = \bigcup(\bigcup^n(x))$. We show that this definition of trcl can be given by applying the transfinite recursion theorem on the well-founded order $(\omega, \in)$. Such an order relation is well founded and absolute for transitive models of ZF^-. To define $\mathrm{trcl}(a)$, consider the function $F(x, y, z)$ defined as follows:

$$\begin{cases} F(x+1, y, z) = \bigcup y(x, z) \text{ if } x \in \omega \text{ and } y \text{ is a function and } \mathrm{dom}(y) = x \times \{z\}, \\ F(x, y, z) = z \text{ otherwise.} \end{cases}$$

We leave to the reader to check that $F(x, y, z) = w$ can be defined by means of a Δ_0-formula.

Now apply the transfinite recursion theorem in M and in V to $(\omega, \in)$, to get that $G(n, a) = F(n, G \restriction n \times \{a\}, a)$ is absolute between M and V and $G(n, a) = \bigcup^n a$. Finally apply replacement in M and in V to get that $\mathrm{trcl}(a) = \bigcup G[\omega \times \{a\}]$. Check that the formula $\phi(z, y, w, t)$ stating that

$$\phi(z, y, w, t) \equiv (t = \omega) \wedge (w \text{ is a function}) \wedge (\text{dom}(w) = t \times \{z\}) \wedge$$
$$\wedge (y = \bigcup \text{ran}(w)) \wedge (\forall n \in t \, (w(n, z) = F(n, w \upharpoonright n \times \{z\}, z))$$

is expressible by a Δ_0-formula and that $\text{trcl}(a) = b$ if and only if

$$V \models \forall w \forall y \, (\phi(a, y, w, \omega) \rightarrow y = b)$$

if and only if

$$V \models \exists w \phi(a, b, w, \omega).$$

Moreover this checking amounts to give a proof in ZF^- of the formula:

$$\forall t \forall u \forall z \, [(z = \omega) \rightarrow [(\forall w \forall y \phi(t, y, w, z) \rightarrow y = u) \leftrightarrow \exists w \phi(t, u, w, z)]].$$

In particular $\exists w \phi(t, u, w, \omega) \equiv \forall w \forall y \phi(t, y, w, \omega) \rightarrow y = u$ is $\Delta_1(\mathsf{ZF}^-_{\Delta_0})$. Thus

$$\exists w \phi(t, u, w, \omega) \equiv \forall w \forall y \phi(t, y, w, \omega) \rightarrow y = u$$

defines a property which is absolute for transitive models of ZF^-.   □

We now give a second example in which the recursion is done on a more complex relation which one can prove that it is well founded in M and then argue that it remains well founded in V using Lemma A.3.6.

Lemma A.3.9 *Let M be a transitive model of ZF^- and $R \in M$ be a well-founded relation in M on some set $A \in M$.*

Then R is well founded in V, and the Mostowski collapsing map $\pi_R : A \rightarrow V$ given by $\pi_R(a) = \pi_R[\text{pred}_R(a)]$ is absolute between M and V.

Proof Since R is a set, $R^M = R$ is set-like and such that the class function $\text{pred}_R = \text{pred}_R^M \in M$ is absolute between M and V. Since M models that R is well founded, V models that R^M is well founded by Lemma A.3.6. Now π_R is defined by induction on $R = R^M$ using the class function $F : A \times V \rightarrow V$ defined by $(a, g) \mapsto g[\text{pred}_R(a)]$, since

$$\pi_R(a) = \pi_R[\text{pred}_R(a)] = F(a, \pi_R).$$

It can be checked by means of the standard methods that $F(a, g) = c$ is definable by a formula $\phi_F(x, y, z)$ which is absolute for M. By Lemma A.3.7, we get that $\pi_R = \pi_R^M$.   □

Lemma A.3.10 *Let $\text{rk} : V \rightarrow \text{Ord}$ be the class function mapping a set a to its rank. Then rk is absolute for any M which is a transitive model of ZF^-.*

Proof Left to the reader (HINT: Use the definition $\mathrm{rk}(a) = \sup\{\mathrm{rk}(b)+1 : b \in a\}$).
□

Lemma A.3.11 *Given a set $X \in V$, let $\pi_X : X \to N_X$ be the Mostowski collapsing map defined by*

$$\pi(a) = \pi[a \cap X]$$

for $a \in X$.
   Then:

- N_X *is transitive.*
- $\pi_X(a) = a$ *whenever* $\mathrm{trcl}(a) \subseteq X$.

Proof Exercise for the reader.   □

A.4   Syntax and Semantics Inside V

We can code the syntax and the semantics of any first order language inside V in a $\Delta_1(\mathsf{ZF}^-_{\Delta_0})$-manner. We limit ourselves to describe how to code the syntax and the semantics of the language for **ZFC** with two binary relation predicates for $=$ and $\in$ in an absolute manner. The reader can check that what we are really proving is that our coding is expressible as a $\Delta_1(\mathsf{ZF}^-_{\Delta_0})$-property. Furthermore the reader can verify that these results can be generalized to any countable language and (with a bit more care) to any set-sized language.

A.4.1   Syntax

- The set of natural numbers ω stands for the set of free variables $\{x_n : n \in \omega\}$ of the language.
- $\langle i, j, 0\rangle$ with $i, j \in \omega$ stands for the formula $x_i \in x_j$.
- $\langle i, j, 1\rangle$ with $i, j \in \omega$ stands for the formula $x_i = x_j$.
- Given formulae ϕ, ψ:

   - $\langle \phi, \psi, 2\rangle$ stands for the formula $\phi \vee \psi$.
   - $\langle \phi, \psi, 3\rangle$ stands for the formula $\phi \wedge \psi$.
   - $\langle \phi, \emptyset, 4\rangle$ stands for the formula $\neg\phi$.
   - $\langle \phi, \emptyset, 2n + 5\rangle$ stands for the formula $\exists x_n\phi$.
   - $\langle \phi, \emptyset, 2n + 6\rangle$ stands for the formula $\forall x_n\phi$.

Formally we define the set $\mathsf{Form} \subseteq V_\omega$ by recursion on ω, letting $F : V^2 \times \omega \to V$ be defined by $F(x, y, w) = \langle x, y, w \rangle$. Now we can let

$$\mathsf{AtForm} = \{ \langle i, j, k \rangle : i \in \omega \wedge j \in \omega \wedge k \in 2 \}$$

represent in V the set of atomic formulae and

$$\mathsf{Form} = \bigcap \{ Z \subseteq V_\omega : \mathsf{AtForm} \subseteq Z \wedge \forall x \in Z \, \forall y \in Z \, \forall w \in \omega \, F(z, y, w) \in Z \}$$

represent in V the set of formulae.

Clearly AtForm is absolute for transitive models of ZF^-, and by the same methods by which one can prove that ω is absolute for transitive models of ZF^-, one can also prove that Form is absolute for transitive models of ZF^-.

Moreover the functions:

- Subform : $\mathsf{Form} \to [\mathsf{Form}]^{<\omega}$ recognizing which are the proper subformulae of a formula
- FV : $\mathsf{Form} \to [\omega]^{<\omega}$ recognizing which are the free variables of a formula

can also be shown to be absolute for transitive models of ZF^-. We leave the details to the reader.

A.4.2   Semantics

As of now we have just defined certain subsets of V_ω which are absolute for transitive models of ZF^-. In order to show that they can really represent the concept of formula as the extension of a set in V, we need to define a semantics inside V which given a Tarski structure $(M, \in, =) \in V$ shows that our definition in the metalanguage of $(M, \in, =) \models \phi(a_1, \ldots, a_n)$ given according to Tarski truth rules can be described as a definable property in V of the triple $(M, \overline{\phi}, \langle a_1, \ldots, a_n \rangle)$, where $\overline{\phi}$ is the triple $\langle z, y, j \rangle$ in Form which codes the formula ϕ as an element of Form.

Definition A.4.1 Assume $(V, \in)$ models ZF^-. The satisfaction predicate

$$\mathsf{Sat} : V \times \mathsf{Form} \times V^{<\omega} \to 3$$

(where 0 stands for false, 1 for true, and 2 for meaningless) is defined in V by the following rules:

$$
\begin{cases}
\mathsf{Sat}(Z, \overline{\phi}, \vec{a}) = 2 \text{ if } \mathrm{Freevar}(\overline{\phi}) \not\subseteq \mathrm{dom}(\vec{a}) \\[4pt]
\text{(i.e., } \vec{a} \text{ does not give an assignment to some of the free variables of } \phi\text{),} \\[4pt]
\text{or if } \vec{a} = \langle a_1, \ldots, a_n \rangle \notin Z^{<\omega}, \text{ otherwise:} \\[8pt]
\mathsf{Sat}(Z, \overline{x_i \in x_j}, \langle a_1, \ldots, a_n \rangle) = 1 \text{ if } a_i \in a_j \text{ and } 0 \text{ otherwise}, \\[4pt]
\mathsf{Sat}(Z, \overline{x_i = x_j}, \langle a_1, \ldots, a_n \rangle) = 1 \text{ if } a_i = a_j \text{ and } 0 \text{ otherwise}, \\[8pt]
\mathsf{Sat}(Z, \overline{\psi \wedge \phi}, \langle a_1, \ldots, a_n \rangle) \\[4pt]
= \mathsf{Sat}(Z, \overline{\psi}, \langle a_1, \ldots, a_n \rangle) \cdot \mathsf{Sat}(Z, \overline{\phi}, \langle a_1, \ldots, a_n \rangle), \\[4pt]
\mathsf{Sat}(Z, \overline{\psi \vee \phi}, \langle a_1, \ldots, a_n \rangle) \\[4pt]
= \max\{\mathsf{Sat}(Z, \overline{\psi}, \langle a_1, \ldots, a_n \rangle), \mathsf{Sat}(Z, \overline{\phi}, \langle a_1, \ldots, a_n \rangle)\}, \\[4pt]
\mathsf{Sat}(Z, \overline{\neg \psi}, \langle a_1, \ldots, a_n \rangle) = 1 - \mathsf{Sat}(Z, \overline{\psi}, \langle a_1, \ldots, a_n \rangle), \\[8pt]
\mathsf{Sat}(Z, \overline{\exists x_{i_j} \psi(x_{i_1}, \ldots, x_{i_k})}, \langle a_1, \ldots, a_n \rangle) = \\[4pt]
= \sup\{\mathsf{Sat}(Z, \overline{\psi(x_{i_1}, \ldots, x_{i_k})}, \langle a_1, \ldots a_{i_{j-1}}, a, a_{i_{j+1}}, \ldots, a_n \rangle) : a \in Z\}, \\[4pt]
\mathsf{Sat}(Z, \overline{\forall x_{i_j} \psi(x_{i_1}, \ldots, x_{i_k})}, \langle a_1, \ldots, a_n \rangle) = \\[4pt]
= \inf\{\mathsf{Sat}(Z, \overline{\psi(x_{i_1}, \ldots, x_{i_k})}, \langle a_1, \ldots a_{i_{j-1}}, a, a_{i_{j+1}}, \ldots, a_n \rangle) : a \in Z\}.
\end{cases}
$$

Lemma A.4.2 *Assume* $(V, \in)$ *models* ZF^-. $\mathsf{Sat} : V^3 \to 2$ *is a definable class function which is absolute for transitive models M of* ZF^- *with* $Z \in M$.

Proof We leave to the reader to check this property of Sat by means of the methods developed in the previous sections of this chapter. □

Moreover the following holds as well:

Lemma A.4.3 $(V, \in)$ *models* ZF^-. *For any formula* $\phi(x_1, \ldots, x_n)$ *and any* $Z \in V$ *and* $(a_1, \ldots, a_n) \in Z^{<\omega}$,

$$(Z, \in, =) \models \phi(a_1, \ldots, a_n)$$

if and only if

$$(V, \in, =) \models \mathsf{Sat}(Z, \overline{\phi}, (a_1, \ldots, a_n)).$$

Proof The proof is a straightforward induction on the complexity of ϕ and is left to the reader. □

Lemma A.4.4 (Downward Löwenheim-Skolem Theorem) *Assume* $(V, \in)$ *models* ZFC *and* $X \subseteq Z$ *are sets in* V. *Then there is a set* $W \in V$ *with* $X \subseteq W \subseteq Z$, *such that* $|W| = |X| + \aleph_0$ *and*

$$V \models \forall \vec{a} \in W^{<\omega} \, \forall \vec{\phi} \in \mathsf{Form} \, [\mathsf{Sat}(Z, \overline{\phi}, \vec{a}) = \mathsf{Sat}(W, \overline{\phi}, \vec{a})].$$

Proof Since V is a model of ZFC, we can run inside V the proof of the Downward Löwenheim-Skolem Theorem where we replace the notion of formula by elements of Form, and the notion of Tarski truth is interpreted by means of the class function Sat. $\square$

A.5   The H_κ-s and How to Get Countable Transitive Models of ZFC

Unless otherwise specified in this section we work in the standard universe $(V, \in)$. The results and definitions can be relativized to the other models of ZFC.

Definition A.5.1 Given a regular cardinal κ, H_κ is the family of sets whose transitive closure has size less than κ.

Lemma A.5.2 *Assume* $(V, \in)$ *models* ZFC. *Then, for any regular cardinal* κ, *the following holds:*

- $H_\kappa \subseteq V_\kappa$.
- H_κ *is a transitive set.*
- $(H_\kappa, \in)$ *is a model of* ZFC^- *if* κ *is uncountable.*
- κ *is strongly inaccessible if and only if* $(H_\kappa, \in)$ *models* ZFC, *in which case* $H_\kappa = V_\kappa$.
- $(H_{\aleph_0}, \in)$ *models* ZFC *with the exception of the Infinity Axiom.*

Proof Clearly $H_\kappa \subseteq V_\kappa$: if $x \in H_\kappa$ and $\mathrm{rk}(x) \geq \kappa$, and then the rank function on $\mathrm{trcl}(x)$ is surjective on κ; this gives that $\mathrm{trcl}(x)$ has size larger than κ. Hence H_κ is a set. It is also immediately checked to be transitive, as $x \in y$ entails $\mathrm{trcl}\, x \subseteq \mathrm{trcl}\, y$. Hence $(H_\kappa, \in)$ models Extensionality and Foundation. As for the other axioms, Pair and Union are almost immediate:

$$|\mathrm{trcl}(\{x, y\})| = |\mathrm{trcl}(x)| + 2 + |\mathrm{trcl}(y)| < \kappa,$$

while Separation follows from the fact that $x \subseteq y$ entails $\mathrm{trcl}(x) \subseteq \mathrm{trcl}(y)$.

The axiom of choice holds since any $X \in H_\kappa$ is contained in the bijective image of a set of size less than κ. This gives a well-order on X almost right away.

Now for Collection assume $R \subseteq X \times H_\kappa$ is such that $\mathrm{dom}(R) = X$. By choice applied in V, we can find $F : X \to H_\kappa$ such that $a R F(a)$ for all $a \in X$. Since X has size less than κ, so does F. Now $F[X]$ is a set with size less than κ consisting of sets of hereditary size less than κ; hence (by the regularity of κ) $\mathrm{trcl}(F[X])$ has size less than κ. We can therefore infer that also $\mathrm{trcl}(F)$ has size less than κ. Hence F belongs to H_κ.

Regarding the Powerset Axiom note that for all $\lambda < \kappa$ $\mathcal{P}(\lambda) \subseteq H_\kappa$; hence $(H_\kappa, \in)$ models the Powerset Axiom if and only if $\mathcal{P}(\lambda)^{H_\kappa} = \mathcal{P}(\lambda)$ for all $\lambda < \kappa$ if and only if κ is strong limit.

Now if κ is strongly inaccessible, any set in V_κ has transitive closure of size less than κ and hence belongs to H_κ. The other inclusion holds for any κ.

Finally if $\kappa = \omega$, κ is strong limit and regular but $\omega \notin H_\kappa$. $\qquad\square$

The models of type $(H_\kappa, \in)$ provide an alternative stratification of V as the increasing union of them:

Fact A.5.3 *Assume V models* ZFC. *Then $V = \bigcup_{\lambda \in \mathsf{Card}} H_\lambda$ and $H_\kappa = \bigcup_{\lambda \in \mathsf{Card} \cap \kappa} H_\lambda$ for all limit cardinals κ.*

Proof Since any set has a transitive closure and a size, any set X belongs to $H_{|\mathrm{trcl}(X)|^+}$. Clearly $H_\lambda \subseteq H_\kappa$ for $\lambda < \kappa$, and any set X in H_κ for κ a limit cardinal belongs to any H_λ with $|\mathrm{trcl}(X)| < \lambda < \kappa$. $\qquad\square$

A.5.1   Getting Countable Transitive Models of ZFC

We use the previous results to argue the following:

Lemma A.5.4 *Assume $(V, \in)$ models* ZFC, *and there is a strongly inaccessible cardinal $\kappa \in V$. Then there is a countable transitive $M \in V$ which is a model of* ZFC.

Proof By Lemma A.5.2, $V_\kappa \models$ ZFC. Apply Lemma A.4.4 to $X = \emptyset$ to get some countable W such that

$$V \models \forall \vec{a} \in W^{<\omega} \, \forall \overline{\phi} \in \mathsf{Form}\,[\mathsf{Sat}(V_\kappa, \overline{\phi}, \vec{a}) = \mathsf{Sat}(W, \overline{\phi}, \vec{a})].$$

Then $\in \cap W^2$ is extensional and well founded, since $(W, \in)$ models the Axiom of Extensionality, being a subset of V.

This gives that the Mostowski collapsing map $\pi_W : W \to V$ of the well-founded relation $\in \cap W^2$ on W^2 is an isomorphism with its image $N_W = \pi_W[W]$ and that $N_W \in V$ is transitive and countable, being the image of the set $W \in V$. Since $\pi_W \in V$, we get that

$$V \models \pi_W : W \to N_W \text{ is an isomorphism of } (W, \in) \text{ with } (N_W, \in).$$

This gives that for all $\overline{\phi} \in \mathsf{Form}$ and $\langle a_1, \ldots, a_n \rangle \in W^{<\omega}$

$$V \models \mathsf{Sat}(W, \overline{\phi}, \langle a_1, \ldots, a_n \rangle) = \mathsf{Sat}(N_W, \overline{\phi}, \langle \pi_W(a_1), \ldots, \pi_W(a_n) \rangle).$$

Therefore $(N_W, \in)$ is a model of ZFC, since for all axioms ϕ of ZFC

$$(V, \in) \models \mathsf{Sat}(N_W, \overline{\phi}, \emptyset),$$

and by Lemma A.4.3 this occurs only if

$$(N_W, \in) \models \phi.$$

$\square$

Exercise A.5.5 Sticking to the notation of the proof of the previous lemma shows that the first ordinal moved by π_W is ω_1. (HINT: Note that $\omega_1 \cap W$ is countable since W is countable; therefore $\pi_W(\omega_1) = \pi_W[\omega_1 \cap W]$ is countable. On the other hand if $\alpha \in \omega_1 \cap W$, $\alpha \subseteq W$ holds true. HINT: By elementarity there is $f : \omega \to \alpha$ surjective and in W).

Prove also that $\pi_W(r) = r$ for all $r \subseteq \omega$.

More generally show that if $\lambda < \kappa$ is a cardinal and $X \prec V_\kappa$ is elementary of size λ with $\lambda \subseteq X$, then $\pi_X(\alpha) = \alpha$ for all $\alpha \in X \cap \lambda^+$, while $\pi_X(\lambda^+) < \lambda^+$.

Appendix B
Lévy and Shoenfield's Absoluteness, the Reflection Theorem, and Basics on Constructibility

In this Appendix we prove a few classical results on the absoluteness and reflection properties of models of set theory. In particular we establish Lévy's absoluteness, a strong form of the reflection theorem for the H_κ-s, and a variation of Shoenfield's absoluteness, and we give a very sketchy presentation of the constructible universe L. None of the results presented here is needed elsewhere; however any scholar interested in set theory should know them, the earlier the better.

B.1 Lévy Absoluteness

We state and prove Lévy's absoluteness theorem under the assumption that the model of ZFC we work in is transitive, but this assumption is unnecessary.

Theorem B.1.1 *Let $(V, \in_{\Delta_0})$ be a model of ZFC_{Δ_0} and $\lambda > \kappa$ be infinite cardinals for V with λ regular.*
Assume

$$\phi_1(\vec{x}_1), \ldots, \phi_k(\vec{x}_k), \psi_1(\vec{x}_1, y), \ldots, \psi_n(\vec{x}_n, y)$$

are $\in$-formulae which are $\Delta_1(\mathsf{ZFC}^-_{\Delta_0})$ and

$$\mathsf{ZFC}^-_{\Delta_0} \models \forall \vec{x} \exists! y \, \psi_i(\vec{x}_i, y)$$

for $i = 1, \ldots, n$. Then the structure

$$(H_\lambda, \in^{H_\lambda}_{\Delta_0}, R^{H_\lambda}_{\phi_j} : j = 1, \ldots, k, \, f^{H_\lambda}_{\phi_l} : l = 1, \ldots, n, \, A : A \subseteq \mathcal{P}(\kappa)^k, \, k \in \mathbb{N})$$

M. Viale, *The Forcing Method in Set Theory*, La Matematica per il 3+2 168, https://doi.org/10.1007/978-3-031-71660-7

is Σ_1-elementary in

$$(V, \in_{\Delta_0}^V, R_{\phi_j}^V : j = 1, \ldots, k, f_{\psi_l}^V : l = 1, \ldots, n, A : A \subseteq \mathcal{P}(\kappa)^k, k \in \mathbb{N}),$$

where R_{ϕ_j} and f_{ψ_l} are interpreted by means of axioms $\mathsf{Ax}_{\phi_j}^0$ and $\mathsf{Ax}_{\psi_l}^1$ for $j = 1, \ldots, k$ and $l = 1, \ldots, n$ in both structures.

Its proof is a variant of the classical result of Lévy (which is the above theorem stated just for the language $\in_{\Delta_0}$):

Proof Let $\mathcal{L}$ be the language

$$\in_{\Delta_0} \cup \{R_{\phi_j} : j = 1, \ldots, k\} \cup \{f_{\psi_l} : l = 1, \ldots, n\},$$

$\phi(\vec{x}, y)$ be a quantifier free formula for the language under consideration, where only predicates $A_1, \ldots, A_k$ appear,[1] and $\vec{a} \in H_\lambda$ be such that

$$(V, \mathcal{L}^V, A_1, \ldots, A_k) \models \exists y \phi(\vec{a}, y).$$

Let $\alpha > \kappa$ be a regular cardinal such that for some $b \in H_\alpha$

$$(V, \mathcal{L}^V, A_1, \ldots, A_k) \models \phi(\vec{a}, b).$$

Then H_α is a model of $\mathsf{ZFC}_{\Delta_0}^-$ by Lemma A.5.2.

We claim also that

$$(H_\alpha, \mathcal{L}^{H_\alpha}, A_1, \ldots, A_k) \sqsubseteq (V, \mathcal{L}^V, A_1, \ldots, A_k). \tag{B.1}$$

Proof This is the case by Lemmas A.3.2 and A.2.10:

- V and H_α are both models of $\mathsf{ZFC}_{\Delta_0}^-$.
- $A_1, \ldots, A_k \in H_\alpha$ and hence are contained in H_α being the latter transitive.
- All other predicates and function symbols of $\mathcal{L} \setminus \in_{\Delta_0}$ are $\Delta_1(\mathsf{ZFC}_{\Delta_0}^-)$ with the ψ_i used for the function symbols provably defining graphs of functions in $\mathsf{ZFC}_{\Delta_0}^-$.

□

Consequently,

$$(H_\alpha, \mathcal{L}^{V_\alpha}, A_1, \ldots, A_k) \models \phi(\vec{a}, b).$$

---

[1] Note that $\exists x \in y A(y)$ is not a quantifier free formula and is actually equivalent to the Σ_1-formula $\exists x(x \in y \wedge A(y))$.

By the downward Löwenheim-Skolem theorem applied in V to H_α, we can find $X \subseteq H_\alpha$ which is the domain of an $\mathcal{L} \cup \{A_1, \ldots, A_k\}$-elementary substructure of

$$(H_\alpha, \mathcal{L}^{H_\alpha}, A_1, \ldots, A_k)$$

such that X is a set of size κ containing κ and such that $A_1, \ldots, A_k, \kappa, b, \vec{a} \in X$. Since $|X| = \kappa \subseteq X$, we claim the following:

Claim B.1.1.1 *The following hold:*

- $H_\lambda \cap X$ *is a transitive set.*
- κ^+ *is the least ordinal in* X *which is not contained in* X.
- $X \cap \kappa^+$ *is an ordinal.*

Proof Any $a \in H_{\kappa^+} \cap X$ is the surjective image of κ by some $g : \kappa \to \mathrm{trcl}\, a$ with $g \in H_{\kappa^+}$. By elementarity of $X \prec H_\alpha$, we can assume that g is in X. Therefore

$$a \subseteq \mathrm{trcl}(a) \subseteq g[\kappa] \subseteq M \cap H_{\kappa^+}.$$

$\kappa^+ \nsubseteq X$ since it has size bigger than X, while any $\alpha \in X \cap \kappa^+$ is a subset of X by the above argument; in particular $X \cap \kappa^+$ is a transitive set linearly ordered by $\in$, hence an ordinal.   □

Let M be the transitive collapse of X via the Mostowski collapsing map π_X.

Claim B.1.1.2 $\pi_X(a) = a$ *for all* $a \in H_{\kappa^+} \cap X$; κ^+ *is the first ordinal moved by* π_X.

Proof By induction on the rank of $a \in H_{\kappa^+} \cap X$, we show that $a \subseteq M$:

$$\pi_X(a) = \pi_X[a \cap X] = \pi_X[a] = a,$$

where the last equality holds by inductive assumption on each $b \in a$ and the second equality by the previous claim.

By the previous claim for any $\alpha \in H_{\kappa^+} \cap X$, $\pi_X(\alpha) = \alpha$. On the other hand,

$$\pi_X(\kappa^+) = \pi_X[X \cap \kappa^+] < \kappa^+.$$

□

Moreover, for $A \subseteq \mathcal{P}(\kappa)^n$ in X,

$$\pi_X(A) = A \cap M. \tag{B.2}$$

We prove equation (B.2):

Proof Since $X \cap V_{\kappa+1} \subseteq X \cap H_{\kappa^+}$, π_X is the identity on $X \cap H_{\kappa^+}$, and $A \subseteq \mathcal{P}(\kappa) \subseteq V_{\kappa+1}$, we get that

$$\pi_X(A) = \pi_X[A \cap X] = \pi_X[A \cap X \cap V_{\kappa+1}] = A \cap M \cap V_{\kappa+1} = A \cap M.$$

$\square$

It suffices now to show that

$$(M, \mathcal{L}^M, \pi_X(A_1), \ldots, \pi_X(A_k)) \sqsubseteq (H_\lambda, \mathcal{L}^{H_\lambda}, A_1, \ldots, A_k). \tag{B.3}$$

Assume B.3 holds; since π_X is an isomorphism and $\pi_X(A_j) = \pi_X[A_j \cap X]$, we get that

$$(M, \mathcal{L}^M, \pi_X(A_1), \ldots, \pi_X(A_k)) \models \phi(\pi_X(b), \vec{a})$$

since

$$(X, \mathcal{L}^V, A_1 \cap X, \ldots, A_k \cap X) \models \phi(b, \vec{a}).$$

By (B.3) we get that

$$(H_\lambda, \mathcal{L}^{H_\lambda}, A_1, \ldots, A_k) \models \phi(\pi_X(b), \vec{a}),$$

and we are done.

We prove (B.3):

Proof Since $(M, \in)$ is a transitive model of ZFC^- with $M \subseteq H_\lambda$, any atomic $\mathcal{L}$-formula holds true in $(M, \mathcal{L}^M)$ if and only if it holds in $(H_\lambda, \mathcal{L}^{H_\lambda})$ (again by Lemmas A.3.2 and A.2.10 as H_λ is a model of $\mathsf{ZFC}^-_{\Delta_0}$).

It remains to argue that the same occurs for the formulae of type $A_j(x)$, i.e., that $A_j \cap M = \pi_X(A_j)$ for all $j = 1, \ldots, n$, which is the case by (B.2). $\square$

$\square$

B.2   The Reflection Theorem

We show that for any fixed n for most cardinals κ it holds that $(H_\kappa, \in_{\Delta_0})$ is Σ_n-elementary in $(V, \in_{\Delta_0})$.

Notation B.2.1 Let $(V, \in_{\Delta_0})$ be a model of $\mathsf{ZF}^-_{\Delta_0}$.

- A definable class C consisting of ordinals in V is:

  - *Unbounded* if $C \setminus \alpha$ is non-empty for all $\alpha \in \mathrm{Ord}$.
  - *Closed* if for all sets $X \subseteq C$ $\sup(X) = \bigcup X \in C$.
  - *Club* if it is closed and unbounded.

- Card denotes the class of infinite cardinals.

Exercise B.2.2 Card is club and is defined by a Π_1-formula $\psi_{\mathsf{Card}}(x)$ without parameters for signature $\in_{\Delta_0}$.

Theorem B.2.3 *For every natural number n, there is a club C_n such that:*

1. C_n *is contained in* Card.
2. C_n *is defined by a Π_n-formula $\phi_n(x)$ for $\in_{\Delta_0}$ in no parameters and displayed free variable.*
3. $C_1 = \mathsf{Card} \setminus \{\aleph_0\}$.
4. *For all $\kappa < \lambda$ both in C_n $(H_\kappa, \in_{\Delta_0})$ are Σ_n-elementary in $(H_\lambda, \in_{\Delta_0})$.*
5. *For all Σ_n-formula $\phi(\vec{x})$ for $\in_{\Delta_0}$, all $\lambda \in C_n$, and all $\vec{a} \in H_\lambda^{<\omega}$*

$$(H_\lambda, \in_{\Delta_0}) \models \phi(\vec{a})$$

if and only if

$$(V, \in_{\Delta_0}) \models \phi(\vec{a}).$$

In particular every cardinal $\lambda \in C_n$ is such that $(H_\lambda, \in_{\Delta_0})$ is Σ_n-elementary in $(V, \in_{\Delta_0})$.

The proof requires to analyze carefully the syntactic complexity of the formalization in language $\in_{\Delta_0}$ of several set theoretic concepts.

Remark B.2.4 The class

$$\big\{(\kappa, H_\kappa) : \kappa \in \mathsf{Card}\big\}$$

is definable by a Π_1-formula in no parameters.

Proof $z \in H_\kappa$ if and only if there is some $u \in H_\kappa$ such that

$$\phi(z, u, \kappa) := (u \text{ is a function}) \wedge \mathrm{dom}(u) \in \kappa \wedge \mathrm{ran}(u) \supseteq \mathrm{trcl}(z).$$

The above statement is provably $\Delta_1(\mathsf{ZF}^-_{\Delta_0})$ in parameter κ and free variables z, u (exercise for the reader).

Therefore $x = H_y$ is formalized by

$$(y \text{ is a cardinal}) \wedge \forall z \, (\exists u \, \phi(z, u, y) \rightarrow z \in x) \wedge \forall z \in x \, \exists u \in x \, \phi(z, u, y).$$

The first two conjuncts are Π_1-formulae for $\in_{\Delta_0}$, the third is provably $\Delta_1(\mathsf{ZF}^-_{\Delta_0})$, and therefore $x = H_y$ is the extension of a Π_1-formula in any model of $\mathsf{ZF}^-_{\Delta_0}$. □

We now prove the theorem:

Proof The proof is split in several claims.

Claim B.2.4.1 *Assume X is a set of ordinals satisfying items 1, 4, and 5 of the theorem for some natural number n.*
Then $X \cup \{\sup(X)\}$ also does.

Proof Let $\kappa = \sup(X)$.
First of all we need the following:

Exercise B.2.5 Let X be a set of cardinals and $\kappa = \sup(X) = \bigcup X$. Then κ is a cardinal and

$$H_\kappa = \bigcup_{\lambda \in X} H_\lambda.$$

1 By Exercise B.2.5.
4 By 4 for X, $\{H_\lambda : \lambda \in X\}$ is a chain of $\in_{\Delta_0}$-structures under inclusion such that for $\lambda_1 < \lambda_2$ in X $(H_{\lambda_1}, \in_{\Delta_0})$ is Σ_n-elementary in $(H_{\lambda_2}, \in_{\Delta_0})$.
    It is then a standard model theoretic fact that $(H_\lambda, \in_{\Delta_0})$ is Σ_n-elementary in $(H_\kappa, \in_{\Delta_0})$ for all $\lambda \in X$. The details are again a useful exercise for the reader (on basic model theory).
5   Note that 5 is a convoluted way to assert that $(H_\lambda, \in_{\Delta_0})$ is Σ_n-elementary in $(V, \in_{\Delta_0})$ for all $\lambda \in X$.
    It is then a standard model theoretic fact that $(H_\kappa, \in_{\Delta_0})$ is Σ_n-elementary in $(V, \in_{\Delta_0})$. The details are again a useful exercise for the reader (on basic model theory).

□

Now for $n = 1$ any set of successor cardinals satisfies 4 and 5 (by Lévy's absoluteness Theorem B.1.1 and Exercise A.1.6). Therefore all the uncountable limit cardinals also satisfy 4 and 5 (by the above claim).

Set $C_1 = \mathsf{Card} \setminus \{\aleph_0\}$. Then C_1 is club and is defined by a Π_1-formula in no parameters. The reader can easily check that items 1–5 of Theorem B.2.3 hold for C_1.

Now assume items 1, 2, 4, and 5 hold for C_n which is defined by a certain Π_n-formula $\phi_n(x)$ for $\in_{\Delta_0}$.

We want to find C_{n+1} club satisfying 1, 2, 4, and 5 as witnessed by a certain Π_{n+1}-formula $\phi_{n+1}(x)$ for $\in_{\Delta_0}$.

We leave to the reader to check that the set $X_m \subseteq \mathsf{Form}_{\in_{\Delta_0}}$ of codes for Σ_m-formulae for $\in_{\Delta_0}$ is recursive and that the family

$$\{X_m : m \in \omega\}$$

is the extension of a $\Delta_1(\mathsf{ZF}^-_{\Delta_0})$-property without parameters.

Given $u \in X_m$, we let $\psi_u(\vec{x})$ be the Σ_m-formula coded by u.

Let

$$F_n : C_n \to C_n$$
$$\kappa \mapsto F_n(\kappa)$$

be defined by the following two requests:

- For each $\vec{a} \in H_\kappa^{<\omega}$ and each $u \in X_{n+1}$, if there is $\lambda \in C_n$ such that

$$(H_\lambda, \in_{\Delta_0}) \models \psi_u(\vec{a}),$$

then

$$(H_{F_n(\kappa)}, \in_{\Delta_0}) \models \psi_u(\vec{a}).$$

- For all cardinals $\eta \in F_n(\kappa) \cap C_n$ with $\eta > \kappa$, there are some $\vec{a} \in H_\kappa^{<\omega}$ and some $u \in X_{n+1}$ such that

$$(H_\eta, \in_{\Delta_0}) \not\models \psi_u(\vec{a}).$$

That is, $F_n(\kappa)$ is the least λ for which the first request can be satisfied for all $u \in X_{n+1}$ and $\vec{a} \in H_\kappa^{<\omega}$.

We show that F_n is definable by a Π_{n+1}-formula.

First of all note that:

- $\phi_n(x)$ is a Π_n-formula expressing $(x \in C_n)$.
- $y = H_z$ is a Π_1-formula by Remark B.2.4.
- $x \in \mathsf{Card}$ is Π_1.

Let $\psi^*(u, \vec{w}, v)$ be the following formula in displayed free variables:[2]

$$\psi^*(u, \vec{w}, v) :=$$

$$u \in X_{n+1}$$

$$\wedge$$

---

[2] $\vec{w} \in w$ should rather be expressed as $\bigwedge_{i=0}^{k} w_i \in w$, where $\vec{w} = \langle w_0, \ldots, w_k \rangle$.

$$\phi_n(v)$$

$$\wedge$$

$$\forall w \, (w = H_v \to \vec{w} \in w).$$

We can observe that:

- $\psi^*(u, \vec{w}, v)$ formalizes the concept:

  v is a cardinal such that H_v is Σ_n-elementary in V and $\vec{w} \in H_v$, and u is a Σ_{n+1}-formula.

- $\psi^*(u, \vec{w}, v)$ is a Π_m-formula for $\in_{\Delta_0}$ for $m = \max\{2, n\}$ in displayed free variables: $u \in X_{n+1}$ is Δ_0 in free variable u, $\phi_n(v)$ is Π_n in free variable v, while $\forall w \, (w = H_v \to \vec{w} \in w)$ is Π_2 in free variables $v, \vec{w}$.

Now let $\psi_0(u, \vec{w}, v)$ be the formula in displayed free variables:

$$\psi^*(u, \vec{w}, v)$$

$$\wedge$$

$$\forall y \forall z$$

$$[$$

$$(\phi_n(z) \wedge y = H_z \wedge (v \in z \vee v = z))$$

$$\to$$

$$\neg\mathsf{Sat}(y, u, \vec{w})$$

$$].$$

We can observe that:

- For[3] $u \in \mathsf{Form}_{\in_{\Delta_0}}$, $\lambda \in \mathsf{Card}$, and $\vec{a} \in V$ $\psi_0(u, \vec{a}, \lambda)$ formalizes the concept:

  u is a code for the Σ_{n+1}-formula $\psi_u(\vec{x})$, $\vec{a} \in H_\lambda$, and no $\eta \in C_n \setminus \lambda$ is such that

  $$(H_\eta, \in_{\Delta_0}) \models \psi_u(\vec{a}).$$

  That is, when $\psi_u(\vec{x})$ is Σ_{n+1} and $\vec{a} \in H_\lambda$, $\psi_u(\vec{a})$ fails in all structures $(H_\eta, \in_{\Delta_0})$ with $\eta \in C_n \setminus \lambda$.

- $\psi_0(u, \vec{w}, v)$ is Π_{n+1} in free variables $u, \vec{w}, v$: $\psi^*(u, \vec{w}, v)$ is a Π_m-formula for $m = \max\{2, n\}$, while the second conjunct is the universal quantification of an implication whose premise is a Π_n-formula and whose thesis is a $\Delta_1(\mathsf{ZF}^-_{\Delta_0})$-formula, i.e., it is a Π_{n+1}-formula.

---

[3] Here u denotes a set in V not a free variable; we run short of letters.

Let also

$$\psi_1(u, \vec{w}, v) :=$$

$$\psi^*(u, \vec{w}, v)$$

$$\wedge$$

$$\forall y$$

$$[$$

$$(y = H_v)$$

$$\rightarrow$$

$$\text{Sat}(y, u, \vec{w})$$

$$].$$

We can observe that:

- For $u \in \mathsf{Form}_{\in_{\Delta_0}}$, $\lambda \in \mathsf{Card}$, and $\vec{a} \in V$, $\psi_1(u, \vec{a}, \lambda)$ formalizes the concept:

  u is a code for the Σ_{n+1}-formula $\psi_u(\vec{x})$, $\lambda \in C_n$, $\vec{a} \in H_\lambda^{<\omega}$, and

  $$(H_\lambda, \in_{\Delta_0}) \models \psi_u(\vec{a}).$$

- $\psi_1(u, \vec{w}, v)$ is a Π_m-formula in displayed free variables for $m = \max\{2, n\}$: $\psi^*(\vec{w}, v)$ is Π_m in displayed free variables, and the second conjunct is Π_2 in free variables $u, \vec{w}, v$.

  Given some $\kappa \in C_n$, let $R_\kappa(\langle u, \vec{a} \rangle, \vec{b})$ hold if:

- Either $\psi_0(u, \vec{a}, \kappa)$ holds (i.e., for all $\lambda \in C_n$ with $\vec{a} \in H_\lambda$ $(H_\lambda, \in_{\Delta_0}) \not\models \psi_u(\vec{a})$) and $\vec{b} = \emptyset$
- Or for some $\lambda \geq \kappa$ $\psi_1(u, \vec{a}, \lambda)$ holds (i.e., $\lambda \in C_n$, $\vec{a} \in H_\lambda$ and $(H_\lambda, \in_{\Delta_0}) \models \psi_u(\vec{a})$), and $\vec{b}$ is a witness in $H_\lambda^{<\omega}$ of the truth in the structure $(H_\lambda, \in_{\Delta_0})$ of $\psi_u(\vec{a})$.

For each cardinal $\kappa \in C_n$, $R_\kappa \subseteq (X_{n+1} \times H_\kappa^{<\omega}) \times V$ is clearly definable in V and is total, i.e., $\mathrm{dom}(R) = X_{n+1} \times H_\kappa^{<\omega}$.

By the Collection Principle applied in V to R_κ, one has that there is some $\lambda \in C_n$ such that for all $\vec{a} \in H_\kappa^{<\omega}$ and u code for a Σ_{n+1}-formula, if $R(\langle u, \vec{a} \rangle, \vec{b})$ holds, it does so as witnessed by $\vec{b} \in H_\lambda^{<\omega}$. Thus, for each $\kappa \in C_n$, we can set $F_n(\kappa)$ to be the least such λ.

$F_n(x) = z$ is formalized by the Π_{n+1}-formula in free variables x, z:

$$\phi_n(x)$$

$$\wedge$$

$$\phi_n(z)$$

$$\wedge$$

$$\forall y \forall w$$

$$[$$

$$(y = H_x \wedge w = H_z)$$

$$\rightarrow$$

$$\forall \vec{a} \in y^{<\omega} \, \forall u \in X_{n+1} \, (\psi_0(u, \vec{a}, y) \vee \psi_1(u, \vec{a}, w))$$

$$]$$

$$\wedge$$

$$\forall w^* \forall w \forall z^* \forall y$$

$$(z^* \in z \wedge x \in z^* \wedge \phi_n(z^*) \wedge w^* = H_{z^*} \wedge w = H_z \wedge y = H_x)$$

$$\rightarrow$$

$$\exists \vec{a} \in y^{<\omega} \, \exists u \in X_{n+1} \, (\neg\psi_1(u, \vec{a}, w^*) \wedge \psi_1(u, \vec{a}, w))$$

$$].$$

Now we let

$$C_{n+1} = \left\{ \theta \in \mathsf{Card} : \ F_n(\theta) = \theta \right\}.$$

We leave to the reader to fill in the missing details in checking that:

- C_{n+1} is the extension of a Π_{n+1}-formula in one free variable (HINT: Check that $\exists \vec{a} \in y^{<\omega} \, \exists u \in X_{n+1} \, (\psi_1(u, \vec{a}, w) \wedge \neg\psi_1(u, \vec{a}, w^*))$ is the conjunction of a Σ_n-formula with a Π_n-formula; apply the Collection Principle to absorb the bounded existential quantifiers $\exists \vec{a} \in y^{<\omega} \, \exists u \in X_{n+1}$ in the rightmost unbounded universal quantifier of $\psi_1(u, \vec{a}, w) \wedge \neg\psi_1(u, \vec{a}, w^*)$).
- C_{n+1} is unbounded: Given $\alpha \in \mathrm{Ord}$ pick $\alpha_0 \in C_n$ with $\alpha_0 > \alpha$; set $\alpha_{m+1} = F_n(\alpha_m)$ for $m \in \omega$, and $\alpha_\omega = \sup_{m \in \omega} \alpha_m$; then $\alpha_\omega \in C_{n+1}$ (HINT: By inductive assumption 4 applied to the club C_n, $\alpha_\omega \in C_n$ and $H_{\alpha_m} \prec_n H_{\alpha_\omega}$ for all n; also given $u \in X_{n+1}$ any $\vec{a} \in H_{\alpha_\omega}$ is in H_{α_m} for some $m < \omega$, so if a witness for $\psi_u(\vec{a})$ exists, there is one in $H_{\alpha_{m+1}} \subseteq H_{\alpha_\omega}$).
- C_{n+1} is closed.
- C_{n+1} satisfies items 4 and 5 of the theorem.

Therefore C_{n+1} is exactly the class we were looking for.

This completes the proof of the theorem.   □

B.3 Shoenfield's Absoluteness

We present in this section a (an apparently weaker) form of Shoenfield's absoluteness which shows that the Σ_1-theory of $(H_{\aleph_1}, \in_{\Delta_0})$ is forcing invariant.

Lemma B.3.1 (Cohen's Absoluteness) *Assume $\phi(x, a)$ is a quantifier free formula for $\in_{\Delta_0}$ in the parameter $\vec{a} \in H^n_{\omega_1}$. Then the following are equivalent:*

1. *$H_{\omega_1} \models \exists x \phi(x, \vec{a})$.*
2. *For all complete boolean algebra, B $[\![\exists x \phi(x, \check{a}_1, \ldots, \check{a}_n)]\!] = 1_{\mathsf{B}}$.*
3. *There is a complete boolean algebra B such that $[\![\exists x \phi(x, \check{a}_1, \ldots, \check{a}_n)]\!] > 0_{\mathsf{B}}$.*

We mention also (without proof) the following reinforcement of Cohen's absoluteness (assuming large cardinal axioms):

Theorem B.3.2 *(Woodin) Assume there is a proper class of Woodin cardinals.*
Let $\phi(x_1, \ldots, x_n)$ be any formula for $\in_{\Delta_0}$ in displayed free variables, and $(a_1, \ldots, a_n) \in H^n_{\omega_1}$. Then the following are equivalent:

1. *$H_{\omega_1} \models \phi(a_1, \ldots, a_n)$.*
2. *For all complete boolean algebra, B $[\![\phi(\check{a}_1, \ldots, \check{a}_n)]\!] = 1_{\mathsf{B}}$.*
3. *There is a complete boolean algebra B such that $[\![\phi(\check{a}_1, \ldots, \check{a}_n)]\!] > 0_{\mathsf{B}}$.*

Woodin cardinals are a particular type of large cardinals (see the last Appendix E for some information on this topic). The assumptions require that above any cardinal α one can find a Woodin cardinal κ. See [21, Cor. 3.1.16] for a proof.

We now prove Lemma B.3.1:

Proof We shall actually prove the following slightly stronger formulation of the nontrivial direction in the three equivalences above:

$H_{\omega_1} \models \exists x \phi(x, \vec{a})$ if

$$[\![\exists x \phi(x, \check{a}_1, \ldots, \check{a}_n)]\!] > 0_{\mathsf{B}}$$

for some complete boolean algebra $\mathsf{B} \in V$.

To simplify the exposition, we prove this statement under the further assumption that there exists an inaccessible cardinal $\kappa > \mathsf{B}$.[4]

Assume $\phi(x, \vec{y})$ is a Σ_0-formula and $[\![\exists x \phi(x, \check{a}_1, \ldots, \check{a}_n)]\!] > 0_{\mathsf{B}}$ for some complete boolean algebra $\mathsf{B} \in V$ with parameters $\vec{a} \in (H_{\omega_1})^n$. Pick a model $M \in V$ such that $M \prec (H_\kappa)^V$, M is countable in V, and $\mathsf{B}, \vec{r} \in M$. Note that $\omega \subseteq M$ as ω and all its elements are definable in $M \prec H_\kappa$ without parameters.

Let $\pi_M : M \to N$ be its transitive collapse (i.e., $\pi_M(a) = \pi_M[a \cap M]$ for all $a \in M$) and $\mathsf{Q} = \pi_M(\mathsf{B})$. By the same arguments of Lévy's absoluteness Theorem B.1.1

---

[4] The large cardinal assumption can be removed using the reflection theorem. The argument can be retrieved from the content of the last section of Appendix C.

$\pi_M(a) = a$ for all $a \in H_{\omega_1}$ since $M \cap H_{\omega_1}$ is transitive, being ω a subset of M in M.

Since π_M is an isomorphism of M with N,

$$N \models \mathsf{ZFC} \wedge (b = [\![\exists x \phi(x, \check{a}_0, \ldots, \check{a}_n)]\!] > 0_{\mathsf{Q}}).$$

Now let $G \in V$ be N-generic for Q with $b \in G$ (G exists since N is countable); then by Cohen's theorem of forcing applied in V to N, we have that $N[G] \models \exists x \phi(x, \vec{a})$. So we can pick $u \in N[G]$ such that $N[G] \models \phi(u, \vec{a})$. Since $N, G \in (H_{\omega_1})^V$, we have that V models that $N[G] \in H_{\omega_1}^V$, and thus V models that u as well belongs to $H_{\omega_1}^V$. Since $\phi(x, \vec{y})$ is a quantifier free $\in_{\Delta_0}$-formula, V models that $\phi(u, \vec{a})$ is absolute between the transitive sets $N[G] \subset H_{\omega_1}$ to which $u, \vec{a}$ belong. In particular u witnesses in V that $H_{\omega_1}^V \models \exists x \phi(x, \vec{a})$.   □

B.3.1   Projective Sets and Definable Subsets of H_{ω_1}

We relate the first order theory of the structure $(H_{\omega_1}, \in_{\Delta_0})$ to second order arithmetic. We assume the reader of this section is familiar with the required facts on Polish spaces. The reference text for unexplained notions is [18].

Definition B.3.3 A topological space (X, τ) is Polish if it admits a separable and completely metrizable topology.

$\Delta_1^1(X^n)$ (the Borel sets on X^n) is the smallest σ-algebra of sets[5] contained in $\mathcal{P}(X^n)$ and containing the product topology on X^n induced by τ.

$\mathbb{R}^n$ with euclidean topology, 2^ω and ω^ω with product topology, and the separable Banach spaces are all examples of Polish spaces.

Recall that all uncountable Polish spaces are Borel isomorphic, i.e., if X and Y are uncountable and Polish, there is a Borel map[6] $f : X \to Y$ which is a bijection (see for example [18, Thm. 15.6]).

Definition B.3.4 Let (X, τ) be a Polish space:

- $\Sigma_1^1(X^n)$ is given by those $Y \subseteq X^n$ such that $Y = f[Z]$ for some $Z \in \Delta_1^1(X^m)$ and some continuous $f : X^m \to X^n$.
- Given the family of $\Sigma_n^1(X^m)$-sets, the family of $\Pi_n^1(X^m)$-sets is given by those Z such that $Z = X^m \setminus Y$ for some $Y \in \Sigma_n^1(X^m)$.
- Given the family of $\Pi_n^1(X^k)$-sets, the $\Sigma_{n+1}^1(X^m)$-sets are those Y such that $Y = f[Z^k]$ for some continuous $f : X^k \to X^m$ and some $Z \in \Pi_n^1(X^k)$.
- The *projective* sets are those in $\Sigma_n^1(X^m)$ for some natural numbers n, m.
- The *analytic* sets are those in $\Sigma_1^1(X^m)$ for some natural number m.

---

[5] A σ-algebra is a boolean algebra of sets which admits countable suprema and countable infima.
[6] A Borel map is an f such that $f^{-1}[Z]$ is in $\Delta_1^1(X)$ for all $Z \in \Delta_1^1(Y)$.

- The *coanalytic* sets are those in $\Pi_1^1(X^m)$ for some natural number m.

It can be shown that the projective sets are (up to a Borel isomorphism) those definable with parameters in the structure

$$(X, B : B \in \Delta_1^1(X))$$

for any uncountable Polish space X. When $X = 2^\omega \cong \mathcal{P}(\mathbb{N})$, the above is a standard model for second order number theory.

In the sequel we analyze the projective sets using the space ω^ω endowed with product topology. Here we prove the following:

Lemma B.3.5 *Assume* $A \in \Sigma_k^1((\omega^\omega)^n)$; *then* A *is definable in the structure* $(H_{\omega_1}, \in_{\Delta_0})$ *by a* Σ_k*-formula* $\phi_A(\vec{x}, \vec{r})$ *with* $\vec{r} \in (2^{\omega^{<\omega}})^{n+k}$ *and* $\vec{x} = \langle x_1, \ldots, x_n \rangle$ *an n-tuple of free variables.*

The optimal result connecting projective sets to definable classes in H_{ω_1} is given below. The reader will find its proof in [16].

Theorem B.3.6 (Lemma 25.25 [16]) *A set* $A \in \Sigma_{k+1}^1((\omega^\omega)^n)$ *if and only if it is definable in the structure* $(H_{\omega_1}, \in_{\Delta_0})$ *by a* Σ_k*-formula* $\phi_A(\vec{x}, \vec{r})$ *with* $\vec{r} \in (\omega^\omega)^{<\omega}$ *and* $\vec{x} = \langle x_0, \ldots, x_{n-1} \rangle$ *an n-tuple of free variables.*

We now prove B.3.5:

Proof Consider the set U given by the tuples

$$(x_1, \ldots, x_n, y_1, \ldots, y_n) \in (\omega^\omega)^n \times (2^{\omega^{<\omega}})^n$$

such that for some $s_1, \ldots, s_n \in \omega^{<\omega}$ it holds that

$$\bigwedge_{j=1\ldots n} y_j(s_j) = 1 \wedge s_j \subseteq x_j.$$

In [18] it is shown that U is universal for the open sets of $(\omega^\omega)^n$, in the sense that any open set of $(\omega^\omega)^n$ is a section of U along some $(y_1, \ldots, y_n) \in (2^{\omega^{<\omega}})^n$. Let $x \in \omega^\omega$ stand for the Δ_0-formula

$$(x \text{ is a function}) \wedge \text{dom}(x) = \omega \wedge \text{ran}(x) \subseteq \omega.$$

Notice that U is the extension in H_{ω_1} of the Δ_0-formula

$$\phi_n(x_1, \ldots, x_n, y_1, \ldots, y_n, \omega, \omega^{<\omega})$$

given by

$$\bigwedge_{j=1,\ldots,n} (x_j \in \omega^\omega)$$

$$\wedge$$

$$\bigwedge_{j=1,\ldots,n} [(y_j \text{ is a function}) \wedge \mathrm{dom}(y_j) = \omega^{<\omega} \wedge \mathrm{ran}(y_j) = 2]$$

$$\wedge$$

$$\bigwedge_{j=1,\ldots,n} (\exists s \in \omega^{<\omega} (s \subseteq x_j \wedge y_j(s) = 1)).$$

Consider the set $A_{n-1} \subseteq (\omega^\omega)^{n-1} \times (2^{\omega^{<\omega}})^n$ defined by the Σ_1-formula

$$\psi_{n-1}^1(x_2, \ldots, x_n, y_1, \ldots, y_n, \omega, \omega^{<\omega})$$

given by

$$\exists x_1[(x_1 \in \omega^\omega) \wedge \neg \phi_n(x_1, x_2, \ldots, x_n, y_1, \ldots, y_n, \omega, \omega^{<\omega})].$$

By the results of [18], A_{n-1} is a universal set for $\Sigma_1^1((\omega^\omega)^{n-1})$ and each A_{n-1} is defined by a Σ_1-formula, i.e., every $B \in \Sigma_1^1((\omega^\omega)^{n-1})$ is a section of A_{n-1}: By choosing the parameters $(a_1, \ldots, a_n) \in (2^{\omega^{<\omega}})^n$, one defines all possible sections of A_{n-1} by the Σ_1-formulae $\psi_{n-1}^1(x_2, \ldots, x_n, a_1, \ldots, a_n, \omega, \omega^{<\omega})$; these sections give all the possible analytic subsets of $(\omega^\omega)^{n-1}$.

Hence we have shown that all analytic subsets of $(\omega^\omega)^n$ are Σ_1-definable in $(H_{\omega_1}, \in_{\Delta_0})$ by the formula $\psi_n^1(x_2, \ldots, x_{n+1}, y_1, \ldots, y_{n+1}, \omega, \omega^{<\omega})$ with parameters in H_{ω_1}.

Now assuming the above holds for all n, an induction on k (the details are left to the reader) shows that:

1. $\Pi_{k+1}^1((\omega^\omega)^n)$-sets are defined by the Π_{k+1}-formulae

$$\phi_n^{k+1}(\vec{x}, \vec{r}, \omega, \omega^{<\omega}) \equiv \neg \psi_n^{k+1}(\vec{x}, \vec{r}, \omega, \omega^{<\omega})$$

as $\vec{r}$ ranges in $(2^{\omega^{<\omega}})^{n+k+1}$.

2. $\Sigma_{k+1}^1((\omega^\omega)^n)$-sets are defined by the Σ_{k+1}-formula

$$\exists x_{k+1} (x_{k+1} \in \omega^\omega) \wedge \phi_n^k(x_{k+1}, x_{k+2}, \ldots, x_{n+k+1}, \vec{r}),$$

as $\vec{r}$ ranges in $(2^{\omega^{<\omega}})^{n+k+1}$.

The proof is completed. □

B.4 The Constructible Universe L

We give here a compact presentation of L proving its basic properties, i.e.:

- L is the smallest class sized transitive model of ZFC.
- L satisfies GCH.

Definition B.4.1 Given a set X,

$$\mathsf{Def}(X) = \left\{ Y \subseteq X : \exists \phi \in \mathsf{Form}_\in, \vec{b} \in X^{<\omega} \left(Y = \left\{ z \in X : \mathsf{Sat}(X, \phi, \vec{b}^\frown z) \right\} \right) \right\}.$$

That is, $\mathsf{Def}(X)$ gives the one-dimensional definable subsets of the structure $(X, \in)$.

Proposition B.4.2 *The map $X \mapsto \mathsf{Def}(X)$ is $\Delta_1(\mathsf{ZF}^-_{\Delta_0})$ and*

$$\mathsf{ZF}^-_{\Delta_0} \models \forall x \exists! y\, (y = \mathsf{Def}(x)).$$

Proof Left to the reader. □

Definition B.4.3 Let $(V, \in)$ be a model of ZF:

- $L^V_0 = \emptyset$.
- $L^V_{\alpha+1} = \mathsf{Def}(L^V_\alpha)^V$.
- $L^V_\beta = \bigcup_{\alpha < \beta} L^V_\alpha$ for β limit.
- $L^V = \bigcup_{\alpha \in \mathrm{Ord}} L^V_\alpha$.

We write L rather than L^V when the context is clear.

Proposition B.4.4 *The map $\alpha \mapsto L_\alpha$ is defined in any model $(V, \in)$ of ZF^- by a $\Delta_1(\mathsf{ZF}^-_{\Delta_0})$-property and*

$$\mathsf{ZF}^-_{\Delta_0} \models \forall \alpha \in \mathrm{Ord} \exists! y\, (y = L_\alpha).$$

Consequently $x \in L$ is expressible by the Σ_1-formula

$$\exists y \exists \alpha\, (\alpha \in \mathrm{Ord} \wedge y = L_\alpha \wedge x \in y).$$

Proof Left to the reader. Note that the map $\alpha \mapsto L_\alpha$ is defined by transfinite recursion on a Δ_1-property. □

Proposition B.4.5 *Let $(V, \in)$ be a model of ZF.*
   Then for all $\alpha < \beta \in \mathrm{Ord}$:

- L_α *is transitive.*
- L_α *contains α.*
- $L_\alpha \in L_\beta$.

Therefore L is transitive and $L_\alpha, \alpha \in L$ for all $\alpha \in \mathrm{Ord}$.

Proof Left to the reader. Note that $L_\alpha \subseteq L_{\alpha+1}$ is trivially checked (use the formula $x \in b$ in parameter $b \in L_\alpha$), and also $L_\alpha \in L_{\alpha+1}$ (use the formula $x = x$). Now prove all the above statements by induction. $\qquad\square$

Lemma B.4.6 *Assume $(V, \in)$ is a model of* ZF *and $N \subseteq V$ is transitive and is such that $(N, \in_{\Delta_0}) \models$ ZF$^-_{\Delta_0}$.*
   Then $L^N = L^V \cap N$.

Proof Since Def is $\Delta_1($ZF$^-_{\Delta_0})$, it is computed the same way in N and V. Therefore so is $\alpha \mapsto L_\alpha$, being it also $\Delta_1($ZF$^-)$ (as it is defined using Def by transfinite recursion). $\qquad\square$

Notation B.4.7 $V = L$ is the Π_2-sentence for $\in_{\Delta_0}$

$$\forall x\, \exists \alpha\, \exists y\, (\alpha \in \text{Ord} \wedge x \in y \wedge y = L_\alpha).$$

Theorem B.4.8 *Assume $(V, \in)$ is a model of* ZF*. Then L^V is a model of* ZF $+ V = L$.

Proof L satisfies Extensionality and Foundation, since it is contained in V.

L satisfies Infinity as it contains all ordinals.

Note that $\cup a, \{a, b\}$ are in $L_{\alpha+1}$ if $a, b \in L_\alpha$. Therefore L satisfies the Union and Pairing axioms.

Now we prove that the Collection Principle holds in L: Assume $R \subseteq L^2$ is a definable class in L, then R is also definable in V, since L is definable in V. Given $X \in L$ with $\text{dom}(R) \supseteq X$, by the Collection Principle applied in V, we get some $Y \in V$ such that for all $x \in X$ there is $y \in Y$ such that $\langle x, y \rangle \in R$. Since $Y \subseteq L$, for some $\alpha, Y \subseteq L_\alpha$. Since $L_\alpha \in L$, L_α witnesses the Collection Principle in L for R, X.

As for the Separation axiom, given $\phi(x, y, \vec{z})$ Σ_n-formula, $b \in L$ and parameters $\vec{a}$ in $L^{<\omega}$, find λ such that H_λ is Σ_n-elementary in V and $L^{H_\lambda} = L^V \cap H_\lambda$ with $\vec{a}$ in H_λ. This is possible by the reflection theorem. Note that $L^{H_\lambda} = L_\lambda$.

Therefore for all $c \in b$,

$$(L_\lambda, \in_{\Delta_0}) \models \phi(c, b, \vec{a})$$

if and only if

$$(H_\lambda, \in_{\Delta_0}) \models \phi^L(c, b, \vec{a})$$

if and only if

$$(V, \in_{\Delta_0}) \models \phi^L(c, b, \vec{a})$$

if and only if

$$(L, \in_{\Delta_0}) \models \phi(c, b, \vec{a}).$$

This gives that

$$\{c \in b : (L, \in_{\Delta_0}) \models \phi(c, b, \vec{a})\} = \{c \in b : (L_\lambda, \in_{\Delta_0}) \models \phi(c, b, \vec{a})\} \in L_{\lambda+1} \subseteq L.$$

Therefore L satisfies the Separation axiom.

Regarding the Powerset Axiom, one has that for all $X \in L$

$$\mathcal{P}(X)^L = \mathcal{P}(X) \cap L \subseteq L_\gamma$$

for some γ, being $\mathcal{P}(X)^L$ a set in V which is contained in L.

We may assume that γ is large enough so that $X \in L_\gamma$, therefore obtaining $\mathcal{P}(X)^L \in L_{\gamma+1} \subseteq L$.

Clearly L satisfies $V = L$.   □

Theorem B.4.9 *There is a definable class well-ordering $<_L$ which is*

$$\Delta_1(\mathsf{ZF}^-_{\Delta_0} + V = L).$$

Therefore $(L, \in)$ models the Axiom of Choice.

Proof We need the following exercise on well-orders:

Exercise B.4.10 Assume $(X, <)$ is a well-order, then $<^*$ is a well-order on $X^{<\omega}$, where[7]

$s <^* t$ if $|s| < |t|$ or $|s| = |t|$ and $s <_{\text{lex}} t$, where $|s|$ denotes the length -or size- of s).

Furthermore the map $R \mapsto R^*$ defined on total orderings is $\Delta_1(\mathsf{ZF}^-_{\Delta_0})$ (it is actually definable by a Δ_0-formula).

Now we define by transfinite recursion the well-ordering $<_L$ by the following inductive procedure:

- $<_0 = \emptyset$.
- $<_\beta = \bigcup_{\alpha < \beta} <_\alpha$ for β limit.
- Given $<_\alpha$ well-ordering of L_α, we extend it to $<_{\alpha+1}$ well-ordering of $L_{\alpha+1}$ by setting (according to the unique relevant case which can verify some of the clauses below) for $a, b \in L_{\alpha+1}$:

  - $a <_{\alpha+1} b$ if $a, b \in L_\alpha$ and $a <_\alpha b$.
  - $a <_{\alpha+1} b$ if $a \in L_\alpha$ and $b \in L_{\alpha+1} \setminus L_\alpha$.

---

[7] $<_{\text{lex}}$ denotes the lexicographic order on $X^{|s|}$ induced by $<$.

- Assume $a, b \in L_{\alpha+1} \setminus L_\alpha$. Then for each $u \in L_{\alpha+1} \setminus L_\alpha$, we let $\vec{c}_u$ be the least parameter according to $<_\alpha^*$ such that for some $\in$-formula ϕ, $u \in L_{\alpha+1}$ as witnessed by ϕ, i.e., $u \in L_{\alpha+1}$ because

$$u = \{d \in L_\alpha : (L_\alpha, \in) \models \phi(d, \vec{c}_u)\}.$$

We also let ϕ_u be the least $\in$-formula ϕ (according to a recursive well-ordering $<_\in$ of $\mathsf{Form}_\in$ fixed in advance) such that

$$u = \{d \in L_\alpha : (L_\alpha, \in) \models \phi(d, \vec{c}_u)\}.$$

  Then we let $a <_{\alpha+1} b$ if:

  * Either $\vec{c}_a <_\alpha^* \vec{c}_b$
  * Or $\vec{c}_a = \vec{c}_b$ and $\phi_a <_\in \phi_b$.

- $<_L = \bigcup_{\alpha \in \mathrm{Ord}} <_\alpha$.

We leave to the reader to check that this procedure can be formalized in $\mathsf{ZF}^-_{\Delta_0} + V = L$ by a transfinite recursion on a $\Delta_1(\mathsf{ZF}^-_{\Delta_0} + V = L)$-property and thus results in a definition of $<_L$ which is $\Delta_1(\mathsf{ZF}^-_{\Delta_0} + V = L)$.

We also leave to the reader to check that $<_L$ is a well-order as:

- Each $<_\alpha$ is a well-order of L_α.
- For $\alpha < \beta$,

$$<_\beta \cap L_\alpha^2 = <_\alpha .$$

□

Theorem B.4.11 *Assume* $(V, \in)$ *models* ZF. *Then* L *models* GCH, *i.e.,*

$$L \models \forall \lambda \, (\lambda \text{ is an infinite cardinal} \rightarrow 2^\lambda = \lambda^+).$$

The key to the proof of this theorem is the following:

Lemma B.4.12 (Condensation Lemma) *Let* ZF^-_n *denote the theory* ZF^- *where the Collection Principle and Separation hold only for* Σ_n-formulae for $\in_{\Delta_0}$.

Let n *be large enough such that whenever* $\kappa \in C_n^L$ *is given by the Reflection Theorem B.2.3 applied in* L, H_κ^L *reflects* $V = L$.

Given $\kappa \in C_n^L$, *let* $X \prec L_\kappa$ *be an elementary substructure* (X *in* V *or in* L) *of* $(L_\kappa, \in_{\Delta_0})$ *and* $\pi_X : X \rightarrow N_X$ *be its transitive collapsing map.*

Then $N_X = L_{\mathrm{otp}(X \cap \mathrm{Ord})}$.

Proof Note that $(L_\kappa, \in_{\Delta_0})$ satisfies $V = L + \mathsf{ZF}^-_{\Delta_0}$, by choice of κ (which is possible by the reflection theorem applied in L, being L a model of ZFC).

Therefore so does N_X, since $X \prec L_\kappa$. Therefore $N_X = L^{N_X}$. Since N_X is transitive and models $\mathsf{ZF}^-_{\Delta_0}$, $L^{N_X} = L_{N_X \cap \mathrm{Ord}}$ by Lemma B.4.6.

□

We can now prove Theorem B.4.11.

Proof Working in L, given $a \subseteq \lambda$, find κ as in the proof of the Condensation Lemma B.4.12 such that $a \in L_\kappa$.

Pick $X \prec (L_\kappa, \in)$ elementary substructure such that:

- $X \in L$.
- $a \in X$.
- $\lambda \subseteq X$.
- $|X|^L = \lambda$ holds in L.

This is possible by an application of the downward Löwenheim-Skolem theorem in L to the structure $(L_\kappa, \in)$, since L models ZFC and $(L_\kappa, \in)$ is in L.

Use the same methods of Exercise A.5.5 to show that $\pi_X(\lambda) = \lambda$ and $\pi_X(a) = a$. Note also that

$$\gamma = \mathrm{Ord} \cap N_X = \pi_X[X \cap \kappa]$$

is an ordinal of size $\lambda = |X|^L$ in L (i.e., there is $f : \gamma \to \lambda$ bijection with f in L).

Since $N_X \models \mathsf{ZF}^-_{\Delta_0} + V = L$,

$$N_X = L^{N_X} = L_\gamma.$$

Therefore we get that $a \in L_\gamma$ with $\gamma < (\lambda^+)^L$.

Exercise B.4.13 For all $\alpha \geq \omega$ L models that L_α is a surjective image of α.

(HINT: The proof goes by induction. The limit case is simpler. For the successor case note that $L_{\alpha+1} = \mathsf{Def}(L_\alpha)$ has its cardinality controlled by $\omega \times L_\alpha^{<\omega}$ provably in L. Now apply the inductive assumptions, and use that L models ZFC. The combination of the two gives that $|\alpha|^L = |\omega \times L_\alpha^{<\omega}|^L$ holds in L).

Since L models ZFC, L models that $|L_\alpha|^L = |\alpha| + \aleph_0$.

Therefore L models that $|L_{\lambda^+}|^L = (\lambda^+)^L$.

We have already shown that $L_{(\lambda^+)^L}$ contains $\mathcal{P}(\lambda)^L$.

The theorem is proved.   □

Remark B.4.14 We filled our proofs with notation that makes explicit the relativization to L of many definable concepts. This is advisable since these concepts may be computed differently in L and in V.

For example:

- $\mathcal{P}(\omega)^L \neq \mathcal{P}(\omega)^V$ if CH fails in V.
- ω_1^L might be a countable ordinal in V (see the following exercise).

Exercise B.4.15 Assume M is a countable transitive model of ZFC. Let G be M-generic for $\mathsf{RO}(\mathsf{Fn}(\omega, \omega_1^M))^M$. Then $(\omega_1^L)^{M[G]}$ is countable in $M[G]$.

(HINT: G adds a surjection $g : \omega \to \omega_1^M$. Note that $L^M = L^{M[G]}$ and $(\omega_1^L)^M \leq \omega_1^M$.)

Appendix C
More on Forcing

In this Appendix we compare the approach to forcing of the present book with that of Kunen's [19]. This will allow the reader to follow without problems the approach to forcing pursued in that book and in most of the literature on the topic.

We also analyze in more detail the syntactic complexity of the forcing relation.

C.1 Forcing with Posets Versus Forcing with Boolean Valued Models

We compare in this section the approach to forcing of the present book with that of [19].

Definition C.1.1 Let $(V, \in^V)$ be a model of ZFC and $(P, \leq) \in V$ be a preorder with a maximum 1_P. Define (by transfinite recursion in $(V, \in^V)$):

- $V_0^P := \emptyset$.
- $V_{\alpha+1}^P := \{ \dot{q} : \dot{q} \text{ is a subset of } V_\alpha^P \times P \}$.
- $V_\lambda^P := \bigcup \{ V_\alpha^P : \alpha \in \mathrm{Ord}(V) \}$ for λ limit.
- $V^P = \bigcup_{\alpha \in \mathrm{Ord}^V} V_\alpha^P$.
- $\check{x} := \{ \langle \check{y}, 1_P \rangle : (V, \in^V) \models y \in x \}$ for any $x \in V$.

We leave to the reader to check that V^P is the extension of a formula in the parameter P.

Given a V-generic filter G for P, we define the interpretations of P-names recursively:

Definition C.1.2 Let $(V, \in)$ be a model of ZFC, M in V a *transitive* substructure of V, $(P, \leq)$ in M a preorder with a maximum 1_P, and $G \subseteq P$ also in V an M-generic

M. Viale, *The Forcing Method in Set Theory*, La Matematica per il 3+2 168, https://doi.org/10.1007/978-3-031-71660-7

filter for P. By transfinite recursion we define in V for all $\dot{q} \in M^P$

$$\dot{q}_G = \{\dot{r}_G : \exists p \in G(\langle \dot{r}, p \rangle \in \dot{q})\}.$$

Let us now introduce the external forcing relation for forcing with a partial order.

Definition C.1.3 (External Forcing Relation) Let $(V, \in^V)$ be a model of ZFC, M in V a *transitive* substructure of V, and $(P, \leq)$ in M a preorder with a maximum 1_P.

Let $p \in P$, $\varphi(x_1, \ldots, x_n)$ be a formula with displayed free variables and $\dot{q}_1, \ldots, \dot{q}_m$ be P-names in M^P.

$$p \Vdash_P \varphi(\dot{q}_1, \ldots, \dot{q}_m),$$

if and only if for any $G \subseteq P$ which is an M-generic filter for P with $p \in G$, we have

$$(M[G], \in^V \restriction M[G]) \models \varphi((\dot{q}_1)_G, \ldots, (\dot{q}_n)_G).$$

There is also an internal (to M) definition of the forcing relation for partial orders, but for the atomic formulae $x \in y, x \subseteq y, x = y$ it is rather convoluted (see [19, Def. VI.3.3]). We decide to sidestep this issue in building the connection between Cohen's approach to forcing and the approach given in the present book, by defining an RO(P)-valued structure on V^P for the language $\{\in, \subseteq, =\}$ and establishing a boolean isomorphism between V^P as in Definition C.1.1 and $V^{\mathsf{RO}(P)}$ as in Definition 7.1.25.

The key properties established by Cohen for the external definition of forcing are the truth and definability lemmata:

Lemma C.1.4 (Definability Lemma) *Let $(V, \in^V)$ be a model of* ZFC, *M in V a transitive substructure of V, and $(P, \leq)$ in M a preorder with a maximum 1_P.*

Then, for any $\in$-formula $\varphi(x_1, \ldots, x_m)$ with displayed free variables, the relation

$$p \Vdash_P \varphi(\dot{q}_1, \ldots, \dot{q}_m)$$

is definable in $(M, \in^V \restriction M)$ with parameters $p, \dot{q}_1, \ldots, \dot{q}_m, P$.

Lemma C.1.5 (Truth Lemma) *Let $(V, \in^V)$ be a model of* ZFC, *M in V be a transitive substructure of V, and $(P, \leq)$ in M be a preorder with a maximum 1_P.*

Assume G is M-generic for P. Then

$$M[G] \models \varphi((\dot{q}_1)_G, \ldots, (\dot{q}_n)_G)$$

if and only if there is $p \in G$ such that

$$p \Vdash_P \varphi(\dot{q}_1, \ldots, \dot{q}_m) \text{ holds in } (M, \in).$$

We will give a proof of the above lemmata relating V^P to $V^{\mathsf{RO}(P)}$ in such a way that the external definition of forcing for P as given in Definition C.1.3 is a predicate with exactly the same extension of the predicate given by the internal definition of forcing for $V^{\mathsf{RO}(P)}$ as given in Definition 7.1.25 (and in Definition C.1.6 to follow).

We now connect the presentation of forcing for set sized posets in models of ZFC (as done, e.g., in [19]) with the boolean valued approach to forcing outlined in the present book.

Definition C.1.6 Let V be a model of ZFC, P be a partial order in V, and $i : P \to \mathsf{RO}(P)$ be a dense embedding of P into its boolean completion,

$$V^P = \left\{ \dot{a} \in V : (V, \in^V) \models \dot{a} \subseteq V^P \times P \right\}.$$

We set:

- $\in^i_P (\dot{b}_0, \dot{b}_1) = [\![\dot{b}_0 \in \dot{b}_1]\!]^i_P = \bigvee \left\{ [\![\dot{a} = \dot{b}_0]\!]^i_P \wedge i_P(b) : \langle \dot{a}, b \rangle \in \dot{b}_1 \right\}.$
- $\subseteq^i_P (\dot{b}_0, \dot{b}_1) = [\![\dot{b}_0 \subseteq \dot{b}_1]\!]^i_P = \bigwedge \left\{ \neg i_P(b) \vee [\![\dot{a} \in \dot{b}_1]\!]^i_P : \langle \dot{a}, b \rangle \in \dot{b}_0 \right\}.$
- $=^i_P (\dot{b}_0, \dot{b}_1) = [\![\dot{b}_0 = \dot{b}_1]\!]^i_P = [\![\dot{b}_0 \subseteq \dot{b}_1]\!]^i_{\mathsf{B}} \wedge [\![\dot{b}_1 \subseteq \dot{b}_0]\!]^i_P.$

Remark C.1.7 Once again the definition of the classes V^P, $=^i_P$, $\subseteq^i_P$, $\in^i_P$ is a shorthand for a recursive definition by rank and (apparently) depends on the choice of $i : P \to \mathsf{RO}(P)$. Furthermore a direct proof that the above definition is well posed and gives a (definable class) $\mathsf{RO}(P)$-valued model of ZFC requires an argument of the same complexity and length as that of Theorem 7.1.27. We proceed with a different approach by establishing directly a boolean isomorphism of $(V^P, \in^i_P, \subseteq^i_P, =^i_P)$ with $(V^{\mathsf{RO}(P)}, \in^V_{\mathsf{RO}(P)}, \subseteq^V_{\mathsf{RO}(P)}, =^V_{\mathsf{RO}(P)})$. This also outlines that different choices of $i : P \to \mathsf{RO}(P)$ produce isomorphic boolean valued models.

The following holds:

Theorem C.1.8 *Assume $(V, \in)$ is a transitive model of ZFC and $P \in V$ is a partial order. Let $\mathsf{B} = \mathsf{RO}(P)^V \in V$ and $i : P \to \mathsf{B}$ be a dense embedding. Then V models that*

$$\left\langle V^{\mathsf{B}}, =_{\mathsf{B}}, \in_{\mathsf{B}}, \subseteq_{\mathsf{B}} \right\rangle$$

and

$$\left\langle V^P, =^i_P, \in^i_P, \subseteq^*_P \right\rangle$$

are isomorphic B-valued models for $\mathcal{L} = \{\in, \subseteq, =\}$.

More precisely the map

$$\hat{i} : V^P \to V^{\mathsf{B}}$$

$$a \mapsto f_{\dot{a}} = \left\{ \left\langle f_{\dot{b}}, \bigvee \{ i(p) : \langle \dot{b}, p \rangle \in \dot{a} \} \right\rangle : \langle \dot{b}, p \rangle \in \dot{a} \right\}$$

is the desired Id-*isomorphism (according to Def. 6.1.6).*

Proof We proceed by induction on the rank in the square order of the pair $(\dot{a}, \dot{b})$ to establish that

$$[\![\dot{a} \ R \ \dot{b}]\!]^i_P = [\![f_{\dot{a}} \ R \ f_{\dot{b}}]\!]^{V^{\mathsf{B}}}_{\mathsf{B}}$$

for any $R \in \{=, \subseteq, \in\}$. This will show that $\hat{i}$ is an Id-embedding of B-valued models for $\{\in, \subseteq, =\}$.

We have that:

$$[\![\dot{b}_0 \in \dot{b}_1]\!]^i_P = \bigvee \left\{ [\![\dot{a} = \dot{b}_0]\!]^i_P \wedge i(p) : \langle \dot{a}, p \rangle \in \dot{b}_1 \right\}$$

$$= \bigvee_{\dot{a} \in \mathrm{dom}(\dot{b}_1)} \left(\bigvee \left\{ [\![f_{\dot{a}} = f_{\dot{b}_0}]\!]^{V^{\mathrm{RO}(P)}}_{\mathrm{RO}(P)} \wedge i(p) : \langle \dot{a}, p \rangle \in \dot{b}_1 \right\} \right)$$

$$= \bigvee_{\dot{a} \in \mathrm{dom}(\dot{b}_1)} \left(\bigvee \left\{ [\![f_{\dot{a}} = f_{\dot{b}_0}]\!]^{V^{\mathrm{RO}(P)}}_{\mathrm{RO}(P)} \wedge i(p) : i(p) \leq f_{\dot{b}_1}(f_{\dot{a}}) \right\} \right)$$

$$= \bigvee_{\dot{a} \in \mathrm{dom}(\dot{b}_1)} \left([\![f_{\dot{a}} = f_{\dot{b}_0}]\!]^{V^{\mathrm{RO}(P)}}_{\mathrm{RO}(P)} \wedge f_{\dot{b}_1}(f_{\dot{a}}) \right)$$

$$= [\![f_{\dot{b}_0} \in f_{\dot{b}_1}]\!]^{V^{\mathrm{RO}(P)}}_{\mathrm{RO}(P)} ;$$

while:

$$[\![\dot{b}_0 \subseteq \dot{b}_1]\!]^i_P = \bigwedge \{ [\![\dot{a} \in \dot{b}_1]\!]^i_P \vee \neg i(p) : \langle \dot{a}, p \rangle \in \dot{b}_0 \}$$

$$= \bigwedge_{\dot{a} \in \mathrm{dom}(\dot{b}_0)} \left(\bigwedge \left\{ [\![f_{\dot{a}} \in f_{\dot{b}_1}]\!]^{V^{\mathrm{RO}(P)}}_{\mathrm{RO}(P)} \vee \neg i(p) : \langle \dot{a}, p \rangle \in \dot{b}_0 \right\} \right)$$

$$= \bigwedge_{\dot{a} \in \mathrm{dom}(\dot{b}_0)} \left(\bigwedge \left\{ [\![f_{\dot{a}} \in f_{\dot{b}_1}]\!]^{V^{\mathrm{RO}(P)}}_{\mathrm{RO}(P)} \vee \neg i(p) : i(p) \leq f_{\dot{b}_0}(f_{\dot{a}}) \right\} \right)$$

$$= \bigwedge_{\dot{a} \in \mathrm{dom}(\dot{b}_0)} \left([\![f_{\dot{a}} \in f_{\dot{b}_1}]\!]^{V^{\mathrm{RO}(P)}}_{\mathrm{RO}(P)} \vee \neg f_{\dot{b}_0}(f_{\dot{a}}) \right)$$

$$= \left[\!\left[f_{\dot{b}_0} \subseteq f_{\dot{b}_1} \right]\!\right]_{\mathrm{RO}(P)}^{V^{\mathrm{RO}(P)}}.$$

In both cases the second equality uses the inductive assumptions and the before-last one the definition of $\hat{i}$.

This proves that $\hat{i}$ is a boolean Id_B-embedding. It remains to argue that it is boolean surjective to conclude that it is a boolean isomorphism.

We prove the following:

Claim For all $\tau \in V^{\mathsf{B}}$ there exists $\dot{a}_\tau \in V^P$ such that

$$\left[\!\left[\tau = f_{\dot{a}_\tau} \right]\!\right]_{\mathsf{B}} = 1_{\mathsf{B}}.$$

Proof First of all we leave to the reader to check the following:

Exercise C.1.9 for all $\tau \in V^{\mathsf{B}}$, letting

$$\tau^* = \left\{ \left\langle \sigma, \left[\!\left[\sigma \in \tau \right]\!\right]_{\mathsf{B}}^{V^{\mathsf{B}}} \right\rangle : \sigma \in \mathrm{dom}(\tau) \right\},$$

we have that

$$\left[\!\left[\tau^* = \tau \right]\!\right]_{\mathsf{B}}^{V^{\mathsf{B}}} = 1_{\mathsf{B}}, \tag{C.1}$$

$$\left[\!\left[\sigma \in \tau^* \right]\!\right]_{\mathsf{B}}^{V^{\mathsf{B}}} = \tau^*(\sigma)\, \text{for all } \sigma \in \mathrm{dom}(\tau). \tag{C.2}$$

Now we proceed by induction on the rank of τ.

For $\tau \in V^{\mathsf{B}}$, let $\dot{a}_\tau$ be given by

$$\left\{ \langle \dot{a}_\sigma, p \rangle : \sigma \in \mathrm{dom}(\tau) : i(p) \leq \left[\!\left[\sigma \in \tau \right]\!\right]_{\mathsf{B}} \right\}.$$

Note that for $\sigma \in \mathrm{dom}(\tau)$

$$f_{\dot{a}_\tau}(f_{\dot{a}_\sigma}) = \bigvee \{ i(p) : \langle \dot{a}_\sigma, p \rangle \in \dot{a}_\tau \} = \left[\!\left[\sigma \in \tau \right]\!\right].$$

We prove by induction on ranks that:

$$\left[\!\left[f_{\dot{a}_\tau} = \tau \right]\!\right]_{\mathsf{B}}^{V^{\mathsf{B}}} = 1_{\mathsf{B}}.$$

Assume it holds for all σ of rank smaller than τ, then:

$$\left[\!\left[f_{\dot{a}_\tau} \subseteq \tau \right]\!\right]_{\mathsf{B}}^{V^{\mathsf{B}}} = \bigwedge_{\sigma \in \mathrm{dom}(\tau)} \left(f_{\dot{a}_\tau}(f_{\dot{a}_\sigma}) \to \left[\!\left[f_{\dot{a}_\sigma} \in \tau \right]\!\right]_{\mathsf{B}}^{V^{\mathsf{B}}} \right)$$

$$= \bigwedge_{\sigma \in \mathrm{dom}(\tau)} \left(\left[\!\left[\sigma \in \tau \right]\!\right] \to \left[\!\left[f_{\dot{a}_\sigma} \in \tau \right]\!\right]_{\mathsf{B}}^{V^{\mathsf{B}}} \right)$$

$$\geq \bigwedge_{\sigma \in \mathrm{dom}(\tau)} [\![\sigma \in \tau]\!] \rightarrow ([\![f_{\dot{a}_\sigma} = \sigma]\!]_{\mathsf{B}}^{V^{\mathsf{B}}} \wedge [\![\sigma \in \tau]\!]_{\mathsf{B}}^{V^{\mathsf{B}}})]$$

$$= \bigwedge_{\sigma \in \mathrm{dom}(\tau)} ([\![\sigma \in \tau]\!] \rightarrow [\![\sigma \in \tau]\!]_{\mathsf{B}}^{V^{\mathsf{B}}})$$

$$= 1_{\mathsf{B}}.$$

Conversely:

$$[\![\tau \subseteq f_{\dot{a}_\tau}]\!]_{\mathsf{B}}^{V^{\mathsf{B}}} = [\![\tau^* \subseteq f_{\dot{a}_\tau}]\!]_{\mathsf{B}}^{V^{\mathsf{B}}}$$

$$= \bigwedge_{\sigma \in \mathrm{dom}(\tau)} (\tau^*(\sigma) \rightarrow [\![\sigma \in f_{\dot{a}_\tau}]\!]_{\mathsf{B}}^{V^{\mathsf{B}}})$$

$$= \bigwedge_{\sigma \in \mathrm{dom}(\tau)} ([\![\sigma \in \tau]\!] \rightarrow [\![\sigma \in f_{\dot{a}_\tau}]\!]_{\mathsf{B}}^{V^{\mathsf{B}}})$$

$$\geq \bigwedge_{\sigma \in \mathrm{dom}(\tau)} [f_{\dot{a}_\tau}(f_{\dot{a}_\sigma}) \rightarrow ([\![f_{\dot{a}_\sigma} = \sigma]\!]_{\mathsf{B}}^{V^{\mathsf{B}}} \wedge [\![f_{\dot{a}_\sigma} \in f_{\dot{a}_\tau}]\!]_{\mathsf{B}}^{V^{\mathsf{B}}})]$$

$$= \bigwedge_{\sigma \in \mathrm{dom}(\tau)} (f_{\dot{a}_\tau}(f_{\dot{a}_\sigma}) \rightarrow [\![f_{\dot{a}_\sigma} \in f_{\dot{a}_\tau}]\!]_{\mathsf{B}}^{V^{\mathsf{B}}})$$

$$= 1_{\mathsf{B}}.$$

Remark C.1.10 The above result can be relativized to any (transitive) model M which is a model of **ZFC**. We leave the details to the reader. Clearly combining Theorem C.1.8 with Theorem 7.2.5, one gets a proof of the definability and truth lemmata.

C.2   Syntactic Analysis of Forcing

We briefly sketch that there is a correspondence in Lévy complexity between the amount of **ZFC** satisfied by the ground model and that satisfied in the forcing extension. This allows removing the assumption of the existence of an inaccessible cardinal we freely used in our presentation of the forcing method.

Lemma C.2.1 Let B be a cba and $\tau_0 \ldots \tau_n \in V^{\mathsf{B}}$. Then for all formulae ϕ

$$[\![\exists x \in \tau_0 \, \phi(x, \tau_0, \ldots, \tau_n)]\!]_{\mathsf{B}} =$$

$$= \bigvee_{\sigma \in \mathrm{dom}(\tau_0)} (\tau_0(\sigma) \wedge [\![\phi(\sigma, \tau_0, \ldots, \tau_n)]\!]_{\mathsf{B}})$$

$$= \bigvee_{\sigma \in \mathrm{dom}(\tau_0)} [\![\sigma \in \tau_0 \wedge \phi(\sigma, \tau_0, \ldots, \tau_n)]\!]_B \,.$$

Proof By definition

$$[\![\exists x \in \tau_0 \, \phi(x, \tau_0, \ldots, \tau_n)]\!]_B = \bigvee_{\tau \in V^B} [\![\tau \in \tau_0 \wedge \phi(\tau, \tau_0, \ldots, \tau_n)]\!]_B \,.$$

Note that for all $\tau \in V^B$

$$[\![\tau \in \tau_0 \wedge \phi(\tau, \tau_0, \ldots, \tau_n)]\!]_B =$$

$$= \bigvee_{\sigma \in \mathrm{dom}(\tau_0)} ([\![\tau = \sigma]\!]$$

$$\wedge \tau_0(\sigma) \wedge [\![\phi(\tau, \tau_0, \ldots, \tau_n)]\!])$$

$$\leq \bigvee_{\sigma \in \mathrm{dom}(\tau_0)} (\tau_0(\sigma) \wedge [\![\phi(\sigma, \tau_0, \ldots, \tau_n)]\!]).$$

From the above, the nontrivial inequality

$$[\![\exists x \in \tau_0 \, \phi(x, \tau_0, \ldots, \tau_n)]\!]_B \leq \bigvee_{\sigma \in \mathrm{dom}(\tau_0)} (\tau_0(\sigma) \wedge [\![\wedge \phi(\sigma, \tau_0, \ldots, \tau_n)]\!])$$

follows.

The other inequalities are left to the reader.   □

Lemma C.2.2 *Let* B *be a cba and* $\tau_0 \ldots \tau_n \in V^B$.
Then for each Σ_n-*formula* $\phi(x_0, \ldots, x_n)$ *for* $\in_{\Delta_0}$ *in displayed free variables,*

$$[\![\phi(\tau_0, \ldots, \tau_n)]\!]_B$$

is Σ_n *for* $\in_{\Delta_0}$ *in parameters* $\tau_0 \ldots \tau_n,$ B.

Proof Left to the reader. Use the previous lemma combined with Definition 7.1.29
for $[\![\phi(\tau_0, \ldots, \tau_n)]\!]_B$ and Remark 7.1.30.   □

Theorem C.2.3 *Assume* M *is a countable transitive set.* B $\in M$ *is a cba in* M, *and*
M *satisfies all axioms of* ZFC *with the exception of the Collection Principle and the*
Separation axiom.
   Let G *be* M-*generic for* B.
   Then for all $n \geq 1$, *if* M *satisfies the Collection Principle and the Separation*
axiom for Σ_n-*formulae, so does* $M[G]$.

Proof Left to the reader. Note that in the proof of the forcing theorem to establish
an axiom of ZFC in $M[G]$, we used some axioms of ZFC in M. Check that to assert

Separation or Collection for a Σ_n-formula in $M[G]$, it suffices to use Σ_n-instances of the same axiom in M for the corresponding formulae needed in that proof.   □

Now we can remove the inaccessibility assumption in our presentation of the forcing method: Suppose we want to prove CH is independent from ZFC. If it were not, by compactness, there would be $\phi_1, \ldots, \phi_m$ axioms of ZFC which prove CH (or its negation).

Now for some n these axioms are part of the theory ZFC_n which asserts the Collection Principle and the Separation axiom just for the $\in_{\Delta_0}$-formulae of complexity Σ_n. By the reflection theorem, there is some H_λ such that

$$(H_\lambda, \in_{\Delta_0}) \prec_n (V, \in_{\Delta_0})$$

and the B* needed to establish ¬CH in a forcing extension is in H_λ.

Now we can follow the same pattern we already pursued. Take N countable elementary submodel of H_λ with B* $\in N$. Let M be its transitive collapse and B be the collapse of B*. Then M models ZFC_n and also that $[\![\neg\mathsf{CH}]\!]_B = 1_B$.

Let G be M-generic for B. Then $M[G]$ models ZFC_n and ¬CH. Therefore ZFC_n cannot prove CH.

The same argument can be used for any other statement formalizable in the $\in$-signature which we want to prove consistent with the axioms of ZFC.

Appendix D
More on Topology

We develop some central topics in topology which are not strictly needed for the comprehension of the other parts of the book. We present a few classical results in general topology whose proofs are rooted in set theoretic methods, notably Tychonoff's theorem on the preservation of compactness under arbitrary products and the Stone-Čech compactification for Tychonoff spaces. We also give a natural characterization of extremely disconnected compact Hausdorff spaces which is hardly traceable in the literature. We believe these results complement very well the material of Chaps. 2, 3, 4, and 5.

D.1 Nets and Convergence via Nets

Definition D.1.1 Let $(P, \leq_P)$ be a downward directed preorder (i.e., any finite subset of P has a lower bound) and (X, τ) a topological space.

- A *P-net on* (X, τ) is a map $\Lambda : P \to X$.
- A *P*-net Λ is *frequently* in $U \subseteq X$ if all $p \in P$ there is $r \leq p$ such that $\Lambda(r) \in U$.
- A *P*-net Λ is *eventually* in $U \subseteq X$ if there is $r \in P$ such that $\Lambda(p) \in U$ for all $p \leq r$.
- A *P*-net Λ on (X, τ) *converges to* $x \in X$ (equivalently *x is a limit point of* Λ) if Λ is eventually in U for all $U \in F_{x,\tau} = \{V \in \tau : x \in V\}$.
- $x \in X$ is an *accumulation point for the P-net* Λ *on* (X, τ) if Λ is frequently in U for all $U \in F_{x,\tau}$.
- Given a downward directed preorder $(Q, \leq_Q)$, a *Q*-net $\Gamma : Q \to X$ is a *subnet* of $\Lambda : P \to X$ if there is $j : Q \to P$ order preserving and downward cofinal[1] such that $\Gamma = \Lambda \circ j$.

---

[1] $j : Q \to P$ is downward cofinal if for all $p \in P$ there is $q \in Q$ such that $j(q) \leq_P p$.

© The Author(s), under exclusive license to Springer Nature Switzerland AG 2024
M. Viale, *The Forcing Method in Set Theory*, La Matematica per il 3+2 168,
https://doi.org/10.1007/978-3-031-71660-7

Lemma D.1.2 *Let (X, τ), (Y, σ) be topological spaces and $f : X \to Y$ be a function. The following are equivalent for all $x \in X$:*

1. *f is continuous at x.*
2. *For every downward directed preorder $(P, \leq)$ and every P-net Γ converging to x, $f \circ \Gamma$ converges to $f(x)$.*

Proof

1 implies 2   Assume f is continuous at x, and let $\Gamma : P \to X$ be a net converging to x. Given U open neighborhood of $f(x)$, find V open neighborhood of x such that $f[V] \subseteq U$. Then for some $p \in P$ and all $q \leq_P p$ $\Gamma(q) \in V$, hence $f \circ \Gamma(q) \in U$ for all $q \leq_P p$.

2 implies 1   Assume f is not continuous at x and find U open neighborhood of $f(x)$ such that $f[V] \not\subseteq U$ for any $V \in F_{x,\tau}$. Then let $\Gamma : F_{x,\tau} \to X$ be such that $\Gamma(V)$ is some $x_V \in V$ with $f(x_V) \notin U$. Then Γ converges to x, but $f \circ \Gamma$ has range disjoint from U and hence cannot converge to $f(x)$.

$\square$

Lemma D.1.3 *Let (X, τ) be a topological space. The following are equivalent:*

1. *(X, τ) is Hausdorff.*
2. *Every convergent P-net on X has at most one limit point.*

Proof

1 implies 2   Assume $x \neq y$ are limit points for the net $\Gamma : P \to X$. Find U, V disjoint open neighborhoods of x, y. Then there are $p, q \in P$ such that for all $r \leq p$ $\Gamma(r) \in U$ and for all $r \leq q$ $\Gamma(r) \in V$. This contradicts the downward directedness of P.

2 implies 1   Assume $x \neq y$ are such that $U \cap V \neq \emptyset$ for all U, V open neighborhoods of x, y. Consider the net

$$\Gamma : P = \{U \cap V : U \in F_{x,\tau}, V \in F_{y,\tau}\} \to X$$

(with P ordered by inclusion) such that $\Gamma(U \cap V) \in U \cap V$ for all $U \in F_{x,\tau}, V \in F_{y,\tau}$. Then x, y are easily seen to be both limit points of Γ.

$\square$

Lemma D.1.4 *Let (X, τ) be a topological space. The following are equivalent:*

1. *(X, τ) is compact.*
2. *Every P-net on X has a convergent subnet.*

Proof

1 implies 2   Given $\Gamma : P \to X$, let $C_p = \mathrm{Cl}(\{\Gamma(q) : q \leq_P p\})$ for $p \in P$. Since P is downward directed, C_p is a family with the finite intersection property. By compactness there is a point $x \in \bigcap_{p \in P} C_p$. Then for all $p \in P$ and $U \in$

$F_{x,\tau}$, $U \cap \{\Gamma(q) : q \leq_P p\}$ is non-empty: First of all $x \in U \cap C_p$ witnesses that $U \cap C_p$ is an open non-empty subset of C_p in the subspace topology. Now $\{\Gamma(q) : q \leq_P p\}$ is dense in C_p, and hence $U \cap \{\Gamma(q) : q \leq_P p\}$ must also be non-empty, as $U \cap C_p$ is open in C_p.

Consider $Q = \{\langle p, U \rangle : U \in F_{x,\tau},\ \Gamma(p) \in U\}$ with its natural product order, and let $h : Q \to P$ map $\langle p, U \rangle$ to p. Clearly h is downward cofinal and order preserving: For any $p \in P$ $h(p, X) = p \leq p$. Furthermore Q is downward directed: If $\langle r_1, U_1 \rangle, \ldots, \langle r_n, U_n \rangle \in Q$, let $p \leq r_1, \ldots, r_n$ and $U = \bigcap_{i=1}^{n} U_i$; since $\{\Gamma(q) : q \leq_P p\} \cap U$ is non-empty, we can find $r \leq p$ with $\Gamma(r) \in U$. Then $\langle r, U \rangle$ refines $\langle r_1, U_1 \rangle, \ldots, \langle r_n, U_n \rangle \in Q$.

We claim that x is a limit point for $\Gamma \circ h$: Assume not and let $U \in F_{x,\tau}$ be such that for all $t = \langle p, V \rangle \in Q$, there is $s_t = \langle r_t, W_t \rangle \leq \langle p, V \rangle$ such that $\Gamma \circ h(s_t) = \Gamma(r_t) \notin U$. Then, given any $p \in P$ with $\Gamma(p) \in U$, for $t = \langle p, U \rangle$ we have that $\Gamma(r_t) \in W_t \subseteq U$, while $\Gamma(r_t) = \Gamma \circ h(s_t) \notin U$, a contradiction.

2 implies 1   Assume $\mathcal{H} = \{F_i : i \in I\}$ is a family with finite intersection property consisting of closed sets for τ. We must show that $\bigcap \mathcal{H}$ is non-empty.

Let $\mathcal{F}$ be a maximal filter of closed sets containing $\mathcal{H}$ (ordered by inclusion). Let $\Gamma : \mathcal{F} \to X$ map each $F \in \mathcal{F}$ to some $x_F \in F$. Let $h : P \to \mathcal{F}$ be cofinal and order preserving such that $\Gamma \circ h$ is convergent to some $x \in X$. Consider $\mathrm{Cl}(\{x\})$, and assume that $C \cap \mathrm{Cl}(\{x\})$ is empty for some $C \in \mathcal{F}$. Then $U = X \setminus C \in F_{x,\tau}$ is an open neighborhood of x such that $\Gamma \circ h(q) \notin U$ for any q such that $h(q) \subseteq C$, which contradicts the convergence of $\Gamma \circ h$ to x.

We claim that $\mathcal{F}$ is the filter generated by the closed supersets of $\mathrm{Cl}(x)$. If not for some $D \in \mathcal{F}$, $\emptyset \neq E = D \cap \mathrm{Cl}(x)$ is a proper closed subset of $\mathrm{Cl}(x)$ to which x cannot belong (otherwise $E \supseteq \mathrm{Cl}(x)$). Then $U = X \setminus E \in F_{x,\tau}$ is such that $\Gamma \circ h(q) \notin U$ for any $q \in Q$ with $h(q) \subseteq E$, contradicting the convergence to x of $\Gamma \circ h$.

Therefore $x \in \mathrm{Cl}(\{x\}) = \bigcap \mathcal{F} \subseteq \bigcap \mathcal{H}$ as was to be shown.

$\square$

D.2   Tychonoff's Theorem for Products of Compact Spaces

Tychonoff's theorem on the preservation of compactness by infinitary product plays a fundamental role across various mathematical domains. We give below a proof which uses essentially the notion of net.

Theorem D.2.1 (Tychonoff) *Let (X_i, τ_i) be compact spaces for $i \in I$ and (X, τ) be the product space with product topology. The following are equivalent:*

1. *(X, τ) is compact.*
2. *For every $i \in I$, (X_i, τ_i) is compact.*

Proof The implication from 1 to 2 is left to the reader.

We prove the converse implication. Fix $\Gamma : P \to X$ a P-net for some downward directed P. We must find $\Lambda : Q \to X$ convergent subnet of Γ.

Let F be a maximal filter on $\mathcal{P}(X)$ among the family of $U \subseteq X$ such that Γ is frequently in U (i.e., for every $p \in P$ there is some $q \leq_P p$ such that $\Gamma(q) \in U$).

F exists by a standard application of Zorn's Lemma to the family $\mathcal{F}$ of filters H whose elements U are all such that Γ is frequently in U ordered by inclusion.

F is an ultrafilter on $\mathcal{P}(X)$: It is clearly a filter, and for any partition of X in two sets, at least one of them is such that Γ is frequently in it; hence at least one of them belongs to F.

Let

$$Q = \{ \langle p, U \rangle : p \in P, U \in F, \Gamma(p) \in U \}$$

be ordered by $\langle p, U \rangle \leq_Q \langle q, V \rangle$ if $p \leq_P q$ and $U \subseteq V$.

Then $(Q, \leq_Q)$ is easily seen to be a downward directed partial order: If

$$(r_1, X_1), \ldots, (r_n, X_n)$$

are in Q,

$$Y = X_1 \cap \cdots \cap X_n \in F;$$

therefore we can find $p \leq r_1, \ldots, r_n$ and $q \leq p$ such that $\Gamma(q) \in Y$; this gives that $\langle q, Y \rangle \in Q$ refines all $(r_1, X_1), \ldots (r_n, X_n)$.

Let $h : Q \to P$ map each $\langle p, U \rangle \in Q$ to p, and $\Lambda = \Gamma \circ h$. Then Λ is a subnet of Γ as $h : Q \to P$ is downward cofinal and order preserving (for any $p \in P$ $h(p, X) = p$).

To conclude the proof it is enough to show that Λ is convergent to some $x \in X$. Let $h_i : Q_i \to Q$ be such that $\Lambda_i = \pi_i \circ \Lambda \circ h_i$ is a subnet of $\pi_i \circ \Lambda$ converging to some $x_i \in X_i$. h_i and x_i exist by the compactness of X_i.

Claim D.2.1.1 $\pi_i \circ \Lambda$ converges to x_i for all $i \in I$.

Proof Let U be an open neighborhood of x_i and $U^i = \{ f \in X : f(i) \in U \}$.

We claim that $U^i \in F$: If not $Y = X \setminus U^i$ is in F, with $\pi_i[Y] = X_i \setminus U$, and with some $p \in P$ such that $\Gamma(p) \in Y$; however it holds that for some $q \in Q_i$ and all $r \leq_{Q_i} q$, $\Lambda_i(r) \in U$ (by the convergence of Λ_i to x_i); since h_i is order preserving and cofinal in Q, we have that for some $r \leq_{Q_i} q$ $h_i(r) = \langle s, Z \rangle \leq_Q \langle p, Y \rangle, h_i(q)$; and then $\Gamma(s) \in Z \subseteq Y$, which gives that

$$\Lambda_i(r) = \pi_i \circ \Lambda \circ h_i(r) = \pi_i \circ \Gamma(s) \notin U,$$

which is impossible, since $r \leq_{Q_i} q$.

Now observe that if $h_i(q), \langle r, U^i \rangle \geq \langle p, V \rangle \in Q$ for some $r \in P$ with $\Gamma(r) \in U^i$, we have that $\Gamma(p) \in V$ and $\pi_i[V] \subseteq U$. This gives that $\pi_i \circ \Lambda(p, V) = \pi_i \circ \Gamma(p) \in U$ for all $\langle p, V \rangle \in Q$ refining $h_i(q), \langle r, U^i \rangle$.

Since the above argument can be repeated for all U open neighborhoods of x_i in X_i, $\pi_i \circ \Lambda$ also converges to x_i. $\qquad\square$

Now let $x \in X$ be defined by $x(i) = x_i$. We claim that Λ converges to x: Let U be an open neighborhood of X; w.l.o.g. we can assume that for some $i_1, \ldots, i_n \in I$ and $U_j \in \tau_{i_j}$ for $j = 1, \ldots, n$,

$$U = \left\{ f \in X : f(i_j) \in U_j \text{ for } j = 1, \ldots, n \right\}.$$

Find $(r_j, X_j) \in Q$ for $j = 1, \ldots, n$ such that for all $(p, Y) \le (r_j, X_j)$, $\pi_{i_j} \circ \Lambda(p, Y)$ is in $\pi_{i_j}[U] = U_j$. Let (q, Z) refine in Q each (r_j, X_j). Then for all $(p, Y) \in Q$ refining (q, Z), $\Lambda(p, Y) = \Gamma(p)$ is in U, as for all j $\pi_{i_j} \circ \Lambda(p, Y)$ is in $\pi_{i_j}[U] = U_j$. $\qquad\square$

D.3   Stone-Čech Compactifications

Recall that the compactification of a space (X, τ) is a compact space (K, σ) together with a topological embedding $i : X \to K$ (i.e., a continuous injective map such that (X, τ) is homeomorphic to $(i[X], \sigma_{\upharpoonright i[X]})$).

The aim of this section is to characterize the Hausdorff spaces which admit at least a Hausdorff compactification. These are the Tychonoff spaces. We will show that for these spaces it is always possible to build their "largest possible" compactification: Exploiting the category theoretic terminology, this compactification will be characterized in terms of a universal property giving a precise mathematical content to the—for now—vague notion of "largeness."

Definition D.3.1 Let (X, τ) be a topological space.
$C \subseteq X$ is a 0-set if there exists $f : X \to [0; 1]$ continuous such that $f^{-1}[\{0\}] = C$.

Notation D.3.2 Let (X, τ) be a topological space. τ^0 denotes the family of 0-sets of (X, τ).

Remark that clopen sets are 0-sets (as witnessed by the characteristic function of their complement) and 0-sets are closed. The basic geometric picture captured by these definitions is that 0-sets are those closed sets which can be approximated from above continuously by open supersets. On the other hand in general closed sets may not be 0-sets.

Definition D.3.3 A space (X, τ) is Tychonoff if singletons of points are closed sets and for all $x \in X$ and C closed with $x \notin C$, we can find $f : X \to [0; 1]$ continuous and such that $f(x) = 0$, $f \upharpoonright C = 1$.

The following is fundamental in the arguments to follow:

Proposition D.3.4 *Let (X, τ) be a topological space and C_0, C_1 be closed subsets of X.*

1. *Assume C_0, C_1 are 0-sets. Then $C_0 \cap C_1$ and $C_0 \cup C_1$ are also 0-set.*
2. *Assume C_0, C_1 are disjoint 0-sets. Then there are open sets $V_i \supseteq C_i$ for $i = 0, 1$ with disjoint closures and which are the complement of 0-sets.*

Proof Let f_i witness that C_i is a 0-set for $i = 0, 1$. Then:

1. $h = \frac{f_1 + f_0}{2}$ and $k = f_1 \cdot f_0$ are continuous and witness that $C_0 \cap C_1$ and $C_1 \cup C_2$ are 0-sets.
2. $g = \frac{f_0}{f_0 + f_1}$ is continuous and such that $g^{-1}[\{i\}] = C_i$. $V_0 = g^{-1}[[0; 1/3)]$ and $V_1 = g^{-1}[(2/3; 1]]$ are the complements of 0-sets (as witnessed by $g_0(x) = 1 - \min\{1, 3g(x)\}$ for V_0 and $g_1(x) = \max\{0, 3g(x) - 2\}$ for V_1 which are both continuous) such that $C_i \subseteq V_i$ and $\mathsf{Cl}(V_0) \cap \mathsf{Cl}(V_1)$ is empty. $\qquad\square$

The notion of 0-set has been introduced to get the separation property given by the second item above: In general for a Hausdorff topological space, it is not true that disjoint closed sets can be separated by disjoint open sets, and on the other hand for disjoint 0-sets, this is always possible. This separation property of 0-sets will be used to define βX and to prove that it is Hausdorff.

Recall that a topological space (X, τ) is *normal* if and only if any two closed disjoint sets can be separated by disjoint open sets.

Lemma D.3.5 (Urysohn Lemma) *(X, τ) is normal if and only if for every pair of closed disjoint sets C_0, C_1, there is $f : X \to [0, 1]$ continuous such that $f^{-1}[\{i\}] \supseteq C_i$ for $i = 0, 1$.*

Proof See [24, Thm. 1.5.6]. $\qquad\square$

The outcome is that for spaces that are not normal, the 0-sets define a large collection of closed sets Γ which is closed under finite unions and intersections, contains the clopen sets, and satisfies the property that any two disjoint sets in Γ can be separated by disjoint open sets whose complement is in Γ. As we will see below, these are the key properties one needs to prove that the space of maximal filters on Γ is Hausdorff and compact.

Fact D.3.6 *Locally compact Hausdorff spaces are Tychonoff.*

Proof Left to the reader. $\qquad\square$

Definition D.3.7 Given a topological space (X, τ), βX is the family of maximal filters of 0-sets in the partial order $(\tau^0 \setminus \{\emptyset\}, \supseteq)$.

The following is a fundamental easy outcome of Proposition D.3.4.

Proposition D.3.8 *Let (X, τ) be a topological space.*

- *$C, D \in \tau^0$ are compatible in the partial order $(\tau^0 \setminus \{\emptyset\}, \supseteq)$ if and only if $C \cap D$ is non-empty.*

- *If $\mathcal{F}$ is a filter in the partial order $(\tau^0 \setminus \{\emptyset\}, \supseteq)$ and $C_1, \ldots, C_n \in \mathcal{F}$, then $C = \bigcap_{i=1}^n C_i \in \mathcal{F}$; hence C is non-empty.*
- *If $\mathcal{F}$ is a maximal filter in the partial order $(\tau^0 \setminus \{\emptyset\}, \supseteq)$, $C \in \mathcal{F}$, and $C = C_1 \cup \cdots \cup C_n$ with each C_i a 0-set, then at least one C_i is in $\mathcal{F}$.*
- *Assume X is normal. Then βX coincides with the family of maximal filters of closed sets.*
- *Assume X is Tychonoff. Then, for all $x \in X$, the set*

$$\mathcal{F}_x = \{F : F \subseteq X \text{ is a 0-set with } x \in F\}$$

is a non-empty maximal filter in $(\tau^N \setminus \{\emptyset\}, \supseteq)$.

Proof We prove just the last assertion: Assume $C \cap D \neq \emptyset$ for all $D \in \mathcal{F}_x$ but $x \notin C$. Then there exists $f : X \to [0; 1]$ continuous such that $f(x) = 0$ and $f \upharpoonright C = 1$. Hence $C \cap f^{-1}[\{0\}] = \emptyset$, but $f^{-1}[\{0\}] \in \mathcal{F}_x$ since $x \in f^{-1}[\{0\}]$, a contradiction. $\square$

Definition D.3.9 Given a topological space (X, τ) and $Y \subseteq X$,

$$\beta^c Y = \{\mathcal{F} \in \beta X : \exists C \in \mathcal{F}\,(C \subseteq Y)\}.$$

$$\beta^o Y = \{\mathcal{F} \in \beta X : \forall C \in \mathcal{F}\,(C \cap Y \neq \emptyset)\}.$$

Proposition D.3.10 *Assume (X, τ) is a topological space. Then, for all $Y_1, \ldots, Y_n \subseteq X$, the following hold:*

1. *For any 0-set E $\beta^c E = \emptyset$ if and only if $E = \emptyset$.*
2. *$\beta^c Y_1 \subseteq \beta^o Y_1$.*
3. *$\beta^o Y_1 = \beta X \setminus \beta^c (X \setminus Y_1)$.*
4. *$\beta^c Y_1 = \beta X \setminus \beta^o (X \setminus Y_1)$.*
5. *$\beta^c Y_1 \cap \cdots \cap \beta^c Y_n = \beta^c (Y_1 \cap \cdots \cap Y_n)$.*
6. *$\beta^o Y_1 \cup \cdots \cup \beta^o Y_n = \beta^o (Y_1 \cup \cdots \cup Y_n)$.*
7. *If E is a 0-set, $\beta^o E = \beta^c E$. Hence $\mathcal{F} \notin \beta^c E$ if and only if some $D \in \mathcal{F}$ is disjoint from E, and $\beta^o (X \setminus E) = \beta^c (X \setminus E)$.*

Proof

1. Given $E \in \tau^0$ and non-empty, extend $\{E\}$ to a maximal filter.
2. Trivial by definition.
3. Unraveling the definitions

$$\beta X \setminus \beta^c (X \setminus Y_1) = \beta X \setminus \{\mathcal{F} : \exists C \in \mathcal{F}\,(C \subseteq (X \setminus Y_1))\} =$$

$$= \{\mathcal{F} : \forall C \in \mathcal{F}\,(C \nsubseteq (X \setminus Y_1))\} = \{\mathcal{F} : \forall C \in \mathcal{F}\,(C \cap Y_1 \neq \emptyset)\} = \beta^o (Y_1).$$

4. Again unraveling the definitions

$$\beta X \setminus \beta^o (X \setminus Y_1) = \beta X \setminus \{\mathcal{F} : \forall C \in \mathcal{F} (C \cap (X \setminus Y_1) \neq \emptyset)\}$$
$$= \{\mathcal{F} : \exists C \in \mathcal{F} (C \cap (X \setminus Y_1) = \emptyset)\} = \{\mathcal{F} : \exists C \in \mathcal{F} (C \subseteq Y_1)\} = \beta^c (Y_1).$$

5. $\mathcal{F} \in \beta^c Y_1 \cap \cdots \cap \beta^c Y_n$ if and only if there are $C_i \in \mathcal{F}$ such that $C_i \subseteq Y_i$ for all $i = 1, \ldots, n$, which gives that $C = \bigcap_{i=1}^n C_i \subseteq Y_1 \cap \cdots \cap Y_n$ is a 0-set in $\mathcal{F}$. We conclude that $\beta^c Y_1 \cap \cdots \cap \beta^c Y_n \subseteq \beta^c (Y_1 \cap \cdots \cap Y_n)$. The converse inclusion is trivial.
6. By the previous items,

$$\beta X \setminus (\beta^o Y_1 \cup \cdots \cup \beta^o Y_n) = (\beta X \setminus \beta^o Y_1) \cap \cdots \cap (\beta X \setminus \beta^o Y_n)$$
$$= \beta^c (X \setminus Y_1) \cap \cdots \cap \beta^c (X \setminus Y_n)$$
$$= \beta^c ((X \setminus Y_1) \cap \cdots \cap (X \setminus Y_n))$$
$$= \beta^c (X \setminus (Y_1 \cup \cdots \cup Y_n))$$
$$= \beta X \setminus \beta^o (Y_1 \cup \cdots \cup Y_n),$$

hence the thesis.
7. Assume $\mathcal{F} \in \beta^o C$. Observe that

$$\mathcal{G} = \left\{ E \in \tau^N : \exists D \in \mathcal{F} (E \supseteq D \cap C) \right\}$$

is a filter on $(\tau^0 \setminus \{\emptyset\}, \supseteq)$ containing $\mathcal{F} \cup \{C\}$. By maximality of $\mathcal{F}$, $C \in \mathcal{F}$, hence $\mathcal{F} \in \beta^c C$.

$\square$

Definition D.3.11 Given a topological space (X, τ), we let β_τ be the topology on βX generated by the family

$$\left\{ \beta^o U : X \setminus U \in \tau^0 \right\}$$

(i.e., β_τ is the weakest topology containing all $\beta^o U$ with $X \setminus U \in \tau^0$).

By the previous propositions $\emptyset = \beta^o \emptyset$, $\beta X = \beta^o X$, $\beta^o U = \beta^c U$ for all $X \setminus U \in \tau^0$ and

$$\beta^o (U_1 \cap \cdots \cap U_n) = \beta^c (U_1 \cap \cdots \cap U_n) = \beta^c U_1 \cap \cdots \cap \beta^c U_n = \beta^o U_1 \cap \cdots \cap \beta^o U_n.$$

Therefore the following holds:

Fact D.3.12 *Let (X, τ) be a topological space. Then $\left\{ \beta^o U : X \setminus U \in \tau^0 \right\}$ is a base for β_τ, and any closed set for β_τ is the intersection of a family of basic closed sets of the form $\beta^c E$ with $E \in \tau^0$.*

We are ready to prove the main properties of the Stone-Čech compactification of a topological space.

Theorem D.3.13 *Assume (X, τ) is a topological space. Then $(\beta X, \beta_\tau)$ is a compact Hausdorff space. Moreover assume (X, τ) is a Tychonoff space; then:*

- *The map*

$$i_X : X \mapsto \mathcal{F}_x = \left\{ C \in \tau^0 : x \in C \right\}$$

  is a topological embedding.
- *Any continuous $f : X \to K$ with K compact Hausdorff admits a unique continuous extension to a $\beta f : \beta X \to K$ such that $f = \beta f \circ i_X$.*
- *$(\beta X, \beta_\tau)$ is unique up to homeomorphisms with these properties. In particular any compactification of (X, τ) is the continuous image of $(\beta X, \beta_\tau)$.*
- *(X, τ) is locally compact and normal if and only if $i_X[X]$ is a dense open subset of βX.*

Proof We divide the proof of the theorem in several distinct steps:

βX **is Hausdorff.**   Pick $\mathcal{F}_1 \neq \mathcal{F}_0 \in \beta X$, and let $C \in \mathcal{F}_1 \setminus \mathcal{F}_0$. By Proposition D.3.10 applied to $\mathcal{F}_0$ and C, there is $D \in \mathcal{F}_0$ such that $C \cap D$ is empty. By Proposition D.3.4, find U and V complements of 0-sets and disjoint such that $C \subseteq U$ and $D \subseteq V$. Then $\mathcal{F}_1 \in \beta^o U$, $\mathcal{F}_0 \in \beta^o V$, and $\beta^o U \cap \beta^o V = \beta^c U \cap \beta^c V = \beta^c(U \cap V) = \emptyset$.

βX **is compact.**   Fix a family $\mathcal{H}$ of closed sets of βX with the finite intersection property. We can assume $\mathcal{H}$ consists of basic closed sets of the form $\beta^c E$ with E a 0-set (by the same argument we used in the proof of the compactness of $\mathrm{St}(\mathsf{B})$ in 3.8.2, since any closed set is the intersection of a family of sets of type $\beta^c E$ with $E \in \tau^0$).

Consider the family $\mathcal{H}_0$ given by the 0-sets E such that $\beta^c E \in \mathcal{H}$. Then $\mathcal{H}_0$ is non-empty (since $X \in \mathcal{H}_0$) and has the finite intersection property: Fix $C_1, \ldots, C_n \in \mathcal{H}_0$. Then

$$\beta^c(C_1 \cap \cdots \cap C_n) = \beta^c(C_1) \cap \cdots \cap \beta^c(C_n) \neq \emptyset;$$

hence $C_1 \cap \cdots \cap C_n \neq \emptyset$.

Find $\mathcal{F}$ maximal filter of 0-sets extending $\mathcal{H}_0$. Then $\mathcal{F} \in \bigcap \mathcal{H}$: Pick $\beta^c E \in \mathcal{H}$, then $E \in \mathcal{F}$, and hence $\mathcal{F} \in \beta^c E$.

i_X **is a topological embedding if (X, τ) is Tychonoff.** i_X is well defined since $\mathcal{F}_x$ is a maximal filter of 0-sets for all $x \in X$ by Proposition D.3.8.

i_X is continuous and open on its target $i_X[X]$ (seen as a subspace of βX with the inherited topology), with a dense image in βX: For all U complement of a 0-set, $\mathcal{F}_x \in \beta^c U$ if and only if $x \in U$; hence:

- $i_X[X]$ is a dense subset of βX, since it has non-empty intersection with all basic open sets.

- i_X is open on its range $i_X[X]$ and continuous as a function with target βX, since $i_X[X] \cap \beta^c U = i_X[U]$ for all basic open sets $\beta^c U$.

i_X is injective: If $x \neq y$, find $f : X \rightarrow [0; 1]$ continuous with $f(x) = 0$, and $f(y) = 1$, then we can separate $\mathcal{F}_x, \mathcal{F}_y$ with the basic open sets $\beta^o(f^{-1}[[0; 1/3)]), \beta^o(f^{-1}[(2/3; 1]])$.

Unique extension property.   We show that any continuous $f : X \rightarrow K$ with K compact Hausdorff extends uniquely to a continuous $\beta f : \beta X \rightarrow K$ such that $\beta f \circ i_X = f$. Let $\mathcal{F} \in \beta X$. Choose a net $(x_C)_{C \in \mathcal{F}}$ with $x_C \in C$ for all $C \in \mathcal{F}$. Since $(i_X(x_C))_{C \in \mathcal{F}}$ is eventually in any open neighborhood of $\mathcal{F}$ of the form $\beta^c U$ with U complement of a 0-set, we have that $(i_X(x_C))_{C \in \mathcal{F}}$ converges to $\mathcal{F}$. Since K is compact Hausdorff, we have that the image net $(f(x_C))_{C \in \mathcal{F}}$ has a convergent subnet Γ converging to some unique point $\beta f(\mathcal{F})$.

Now a key point for the arguments to follow is that the complements of 0-sets form a base for the topology on K. This holds true: K is compact Hausdorff, hence normal, hence Tychonoff—by Urysohn's Lemma.

Claim D.3.13.1 *The net $(f(x_C))_{C \in \mathcal{F}}$ with domain $\mathcal{F}$ also converges to $\beta f(\mathcal{F})$.*

Proof Otherwise we can find U open neighborhood of $\beta f(\mathcal{F})$ in K such that for frequently many $C \in \mathcal{F}$ $f(x_C) \notin U$ and U is the complement of a 0-set for the topology on K as witnessed by $g : K \rightarrow [0; 1]$. This gives that $g \circ f : X \rightarrow [0; 1]$ is continuous and witnesses that $V = f^{-1}[U]$ is the complement of a 0-set for (X, τ); hence $E = X \setminus V$ is a 0-set for (X, τ).

Since for frequently many $C \in \mathcal{F}$ $f(x_C) \notin U$, we get that $E \cap F$ is non-empty for all $F \in \mathcal{F}$. This can occur only if $E \in \mathcal{F}$. But then $(f(x_C))_{C \in \mathcal{F}}$ is eventually in $K \setminus U$ contradicting the fact that the subnet Γ converges to a point in U.   □

This shows that βf is well defined, as for each $\mathcal{F} \in \beta X$ there is a unique limit point $\beta f(\mathcal{F})$ in K for the net $(f(x_C))_{C \in \mathcal{F}}$.

To prove the continuity of βf it is enough to show that if U is the complement of a 0-set for the topology on K,

$$\beta f^{-1}[U] = \beta^c(f^{-1}[U]) :\tag{D.1}$$

$\subseteq$ for D.1: $\mathcal{G} \in \beta f^{-1}[U]$ if and only if $\beta f(\mathcal{G}) \in U$ if and only if for eventually all $C \in \mathcal{G}$ $f(x_C) \in U$ (since $(f(x_C))_{C \in \mathcal{G}}$ converges to $\beta f(\mathcal{G})$). The latter gives that $C \cap f^{-1}[U]$ is non-emtpy for all $C \in \mathcal{G}$, which amounts to say that $\mathcal{G} \in \beta^c(f^{-1}[U])$.

$\supseteq$ for D.1:   Assume $\mathcal{G} \in \beta^c(f^{-1}[U])$. Since U is the complement of a 0-set in K, $f^{-1}[U]$ is the complement of a 0-set in X. Then $\mathcal{G} \in \beta^c(f^{-1}[U])$ if and only if $D \subseteq f^{-1}[U]$ for some $D \in \mathcal{G}$. This gives that for eventually all $C \in \mathcal{G}$, $f(x_D) \in U$ and we can follow backwards the chain of equivalence of the previous item to conclude that $\mathcal{G} \in \beta f^{-1}[U]$.

This gives that the preimages by βf of the complements of 0-sets for K are open in βX. Since the complement of 0-sets form a base for the topology on K, βf is continuous.

The unique extension property of βf follows from the fact that continuous functions on Hausdorff spaces are determined by their restriction to a dense subset; $i_X[X]$ is a dense subset of βX on which βf is continuous.

Uniqueness up to homeomorphism of βX.   We now show that any compact space (Y, σ) satisfying the above extension property for (X, τ) is homeomorphic to $(\beta X, \beta_\tau)$. So assume that (Y, σ) is a Hausdorff compactification of X via a topological embedding $j : X \to Y$ such that any continuous map $f : X \to K$ with K compact Hausdorff admits a unique continuous extension $f^* : Y \to K$ such that $f^* \circ j = f$.

Now consider $\beta j : \beta X \to Y$. This map is surjective since any point y in Y is the limit of a net $(j(x_\lambda))_{\lambda \in \Lambda}$; hence $y = \beta j(\mathcal{F})$, where $\mathcal{F}$ is the limit in βX of the net $(i_X(x_\lambda))_{\lambda \in \Lambda}$.

By the universal property of Y, find $i_X^* : Y \to \beta X$ extending i_X. Then $i_X^* \circ \beta j \restriction i_X[X]$ is the identity map on $i_X[X]$ (since $i_X^* \circ \beta j(i_X(x)) = i_X^*(j(x)) = i_X(x)$). The identity map on βX is a continuous extension of $i_X^* \circ \beta j \restriction i_X[X]$; hence by the uniqueness property of βX, we get that $i_X^* \circ \beta j$ is the identity map on βX. By a symmetric argument we get that $\beta j \circ i_X^*$ is the identity map on Y. Hence βj and i_X^* are homeomorphisms which invert one another.

$i_X[X]$ **is open in βX if and only if X is locally compact and Hausdorff:**   If some $C \in \mathcal{F}$ is a compact subset of X, $\mathcal{F}$ has a non-empty intersection in X. Since X is Hausdorff and $\mathcal{F}$ maximal, this intersection must be a singleton $\{x\}$. Hence $\mathcal{F} = \mathcal{F}_x$ for some $x \in X$ if and only if some $C \in \mathcal{F}$ is compact in X. We get that $i_X[X]$ is the union of $\beta^o U$ such that $\mathsf{Cl}(U)$ is compact in X; hence $i_X[X]$ is open in βX. Conversely any open subset of βX is locally compact.

$\square$

Remark D.3.14 Notice that normality is not preserved for subspaces. This is one of the reasons why one introduces the weaker Tychonoff property which uses (a weakening of) the characterization of normality given by Urysohn Lemma. Remark that the separation property for disjoint closed sets C_0, C_1 given by the existence of a continuous $f : X \to [0; 1]$ such that $f[C_i] = \{i\}$, when predicated for disjoint closed sets of which one is the singleton of a point, is strictly stronger than the assertion that closed sets can be separated from points by disjoint open sets.

Remark D.3.15 One resorts to the introduction of the notion of 0-sets to grant that βX is Hausdorff. If one defines $\beta^* X$ as the set of maximal filters of closed sets with the topology given by the corresponding definition of $\beta^o U$ for U open in X, we would run into trouble in proving the Hausdorff property for βX. It actually fails if X is not Tychonoff. The compactness part of the proof survives with these new definitions. The problem in the proof of the Hausdorff property is the separation of arbitrary disjoint closed sets C, D by means of disjoint open sets (to establish the Hausdorff property of βX we used that two disjoint 0-sets can be separated by open disjoint sets which are the complements of 0-sets). If C and D are closed but not 0-sets, this cannot always be done.

It is somewhat peculiar that one has to introduce the space $[0; 1]$ to describe a family of closed sets which are then used with almost no reference to the properties of real numbers. [10] gives a characterization of Tychonoff spaces which is purely topological and makes no reference to 0-sets.

We conclude with the following characterization of extremally disconnected compact Hausdorff spaces:

Theorem D.3.16 *Let (X, τ) be a compact Hausdorff space. Then X is extremally disconnected if and only if it is the Stone-Čech compactification of any of its dense subsets.*

The theorem follows by a combination of [37, Prop. Pag. 284 Section 10.47, Thm. Pag. 25 Section 1.46]. We give a self-contained proof below.

Proof Assume (X, τ) is compact Hausdorff but not extremally disconnected. Then there is $U \in \tau$ which is regular open but not closed. This gives that (letting $\neg U = X \setminus \mathsf{Cl}(U)$) $U \cup \neg U$ is open dense in X. Fix $x_0 \in U$ and $x_1 \in \neg U$. Then $f(x) = x_0$ if $x \in U$ and $f(x) = x_1$ if $x \in \neg U$ is continuous on $U \cup \neg U$ but cannot be extended to a continuous map on the whole of X: Note that $\mathsf{Cl}(U) \cap \mathsf{Cl}(\neg U)$ is non-empty (else U is closed); if y belongs to this set and g extends continuously f to X, $g(y)$ is either x_0 or x_1, but it can be neither of them since it is an accumulation point both for $U = f^{-1}[\{x_0\}]$ and for $\neg U = f^{-1}[\{x_1\}]$.

The converse direction is given by the next lemma.   □

Lemma D.3.17 *Let X be a compact Hausdorff extremally disconnected space. Then for every dense subset $W \subseteq X$, every compact Hausdorff space K, and every continuous function $f : W \to K$, there exists a unique continuous map $\beta(f) : X \to K$ such that $\beta(f) \restriction W = f$.*

Proof We can assume that $X = \mathsf{St}(\mathsf{B})$ where $\mathsf{RO}(\mathsf{St}(\mathsf{B})) = \mathsf{B}$ up to isomorphism. In particular we identify any G in X with the ultrafilter on $\mathsf{RO}(X)$ given by its regular open (clopen) neighborhoods. Let $G \in X$. Since W is dense, $W \cap V \neq \emptyset$ for every open neighborhood V of G. In particular, the family $\{V \cap W : V \in G\}$ has the finite intersection property. Consequently the same holds for $\{f[V \cap W] : V \in G\}$ and for $\{\mathsf{Cl}(f[V \cap W]) : V \in G\}$. Being K compact, we have

$$\bigcap_{V \in G} \mathsf{Cl}(f[V \cap W]) \neq \emptyset.$$

Claim D.3.17.1 *Let U be a non-empty open subset of K and $G \in X$. The following are equivalent:*

1. $U \cap f[V \cap W] \neq \emptyset$ for all $V \in G$,
2. $\mathsf{Reg}(f^{-1}[U]) \in G$.

Proof

1⇒2   In X it holds that $f^{-1}[U] \cap V \neq \emptyset$ for every $V \in G$. This entails that (in X) $\mathsf{Cl}(f^{-1}[U]) \cap V \neq \emptyset$ for every $V \in G$. Now (in X) $\mathsf{Reg}(f^{-1}[U])$ is a dense subset of $\mathsf{Cl}(f^{-1}[U])$, and (in $\mathsf{Cl}(f^{-1}[U])$) $V \cap \mathsf{Cl}(f^{-1}[U])$ is a non-empty open subset of $\mathsf{Cl}(f^{-1}[U])$ for any $V \in G$. Therefore $\mathsf{Reg}(f^{-1}[U]) \cap V \neq \emptyset$ for all $V \in G$, which occurs if and only if $\mathsf{Reg}(f^{-1}[U]) \in G$.

2⇒1   Conversely assume that $\mathsf{Reg}(f^{-1}[U]) \in G$. Then $\mathsf{Reg}(f^{-1}[U]) \cap Z \neq \emptyset$ for all $Z \in G$. Since $\mathsf{Reg}(f^{-1}[U])$ is dense in $\mathsf{Cl}(f^{-1}[U])$, we get that $\mathsf{Cl}(f^{-1}[U]) \cap Z \neq \emptyset$ for all $Z \in G$. Since $f^{-1}[U]$ is dense in $\mathsf{Cl}(f^{-1}[U])$ and $\mathsf{Cl}(f^{-1}[U]) \cap Z \neq \emptyset$ is open in $\mathsf{Cl}(f^{-1}[U])$ for all $Z \in G$, we get that $f^{-1}[U] \cap Z \neq \emptyset$ for all $Z \in G$. This gives that $U \cap f[Z \cap W] \neq \emptyset$ for all $Z \in G$, as desired.

Claim D.3.17.2 *For any $G \in X$, there is exactly one point in $\bigcap_{V \in G} \mathsf{Cl}(f[V \cap W])$.*

Proof   We already noted that $\{f[V \cap W] : V \in G\}$ has the finite intersection property. Hence, since K is compact, $\bigcap \{\mathsf{Cl}(f[V \cap W]) : V \in G\}$ is non-empty.

By contradiction, assume that in $\bigcap_{V \in G} \mathsf{Cl}(f[V \cap W])$ there are two distinct points y_1, y_2. Being K Hausdorff, there are disjoint open sets U_1, U_2 such that $y_i \in U_i$, $i = 1, 2$. Since $y_i \in \bigcap_{V \in G} \mathsf{Cl}(f[V \cap W])$, we get that $U_i \cap f[V \cap W]$ is non-empty for $i = 0, 1$ and $V \in G$.

By Claim D.3.17.1, $\mathsf{Reg}(f^{-1}[U_1])$, $\mathsf{Reg}(f^{-1}[U_2]) \in G$ (where the regularization operation is performed in X). Hence $\mathsf{Reg}(f^{-1}[U_1]) \cap \mathsf{Reg}(f^{-1}[U_2])$ is non-empty clopen. By density of W in X, $\mathsf{Reg}(f^{-1}[U_1]) \cap \mathsf{Reg}(f^{-1}[U_2]) \cap W$ is non-empty and clopen in W. Since $f^{-1}[U_1]$ is dense open in $\mathsf{Reg}(f^{-1}[U_1]) \cap W$ we get that $f^{-1}[U_1]$ has non-empty intersection with $\mathsf{Reg}(f^{-1}[U_2]) \cap W$ and is relatively open in it. Since $f^{-1}[U_2]$ is dense in $\mathsf{Reg}(f^{-1}[U_2]) \cap W$, we get that $f^{-1}[U_1] \cap f^{-1}[U_2]$ is non-empty, which is a contradiction.

Thus we can define $\beta(f)(G)$ as the unique point y_G in $\bigcap\{\mathsf{Cl}(f[V \cap W]) : V \in G\}$. We leave to the reader to check that $\beta(f) \restriction W = f$.

Finally, we prove the continuity of $\beta(f)$.

Claim D.3.17.3 *Assume U is an open set of K. Then $\beta(f)[\mathsf{Reg}(f^{-1}[U])] \subseteq \mathsf{Cl}(U)$.*

Proof   Assume $G \in \mathsf{Reg}(f^{-1}[U])$. Note that $G \in \mathsf{Reg}(f^{-1}[V])$ for any V open neighborhood of $\beta(f)(G)$ (by Claim D.3.17.1). Hence $\mathsf{Reg}(f^{-1}[V]) \cap \mathsf{Reg}(f^{-1}[U])$ is non-empty for all such V. As in the proof of Claim D.3.17.2, we obtain that $f^{-1}[V] \cap f^{-1}[U]$ is non-empty for all such V. Therefore so is $U \cap V$ for all V open neighborhood of $\beta(f)(G)$.

We conclude that $\beta(f)(G)$ is an accumulation point of U.   □

Now, fix $G \in X$, and take an open neighborhood A of $\beta(f)(G)$. We prove that there exists an open neighborhood U of G such that $\beta(f)[U] \subseteq A$: By normality of K (being it compact Hausdorff), we can find an open neighborhood V of $\beta(f)(G)$ such that $\mathsf{Cl}(V) \subseteq A$. By Claim D.3.17.1, $\mathsf{Reg}(f^{-1}[V])$ is an open neighborhood of G. By Claim D.3.17.3 its image is contained in $\mathsf{Cl}(V) \subseteq A$.   □

Appendix E
Further Readings

Those who gained familiarity with the content of this book will be able to follow (with more or less effort) the following reading suggestions:

Survey papers on set theory and related matters: A list of survey papers covering topics in set theory which I wrote or am fond of (these are—in my opinion—mathematically light readings):

- Gödel seminal paper *What Is Cantor's Continuum Problem* [13] presents a clear view on the philosophy of mathematics and on the ontological status of the continuum problem (it is essentially the point of view adopted in this textbook). I strongly recommend it to anyone interested in the philosophical aspects of set theory.
- More recent surveys (which I like or wrote) developing on these themes are [2, 31, 32, 34, 35, 39, 40].

More on forcing and its use in set theory: The reader who wants to delve into the technicalities of forcing must get acquainted with iterated forcing. Afterward she/he will be ready to dive into the vast sea of consistency results established by means of (iterated) forcing.

Basics on forcing: To reinforce the familiarity with this method, look also at the other already mentioned books on the topic [5, 16, 19, 20, 27, 29, 38]. Personally I learned the technique as exposed in [19, Chapter VII]. [16, Chapters 14-15] is also a valuable reference.

Iterated forcing: The easier approach to the matter is in my opinion that given in [19, Chapter VIII]. Also [16, Chapter 16] gives a complete account. The next step would be the analysis of iteration theorems for (semi)proper forcing, see in this case [16, Chapter 31] or [20, Section V.7] or (the older) [3]; the reader might also try the (as yet provisional) account given in [36] (hopefully this will expand in a sequel of this book).

Consistency results obtained by forcing: The literature is too vast to give an exhaustive bibliography. The classical results (Martin's axiom and its

M. Viale, *The Forcing Method in Set Theory*, La Matematica per il 3+2 168, https://doi.org/10.1007/978-3-031-71660-7

consequences, Suslin's hypothesis, Kurepa trees,...) are well covered in [19, Chapters II, VII] or [16, Chapters 15, 16].

Forcing axioms: This is a theme coming in pairs with all the preceding (and many of the following) ones. These axioms assert that the universe of sets satisfies the strongest possible forms of Baire's category theorem (e.g., a very naive formulation of Martin's maximum can be *for all compact Hausdorff spaces* (X, τ) *and all families* $\{D_\alpha : \alpha < \omega_1\}$ *of dense open subsets of* X, $\bigcap_{\alpha < \omega_1} D_\alpha \neq \emptyset$, *unless this is provably inconsistent*). Forcing axioms settle most of the problems which are provably undecidable on the basis of **ZFC**, for example, Martin's maximum entails that $2^{\aleph_0} = \aleph_2$. See the surveys mentioned in the preceding items for more details on ways to correctly formulate these axioms.

> **The proper forcing axioms and its applications:**   A nice account of the first major applications of the proper forcing axiom is given in [4]; I also like the presentation of the matter given in [16, Chapter 31].

> **Martin's Maximum and its applications:**   The literature in this respect is less rich; the presentation of the iteration theorem for semiproper forcing in [16, Chapter 37] is very sketchy; on the other hand the account on the applications of this axiom given in the same chapter (and in some parts of a few other chapters of the book) is more accurate.

Large cardinals: In this book we only presented the simplest type of large cardinal axiom: strong inaccessibility. We used it to simplify slightly our account of the forcing method. However large cardinal axioms are nowadays the "accepted truths" of set theory which can be put on top of **ZFC** (at least for most of those who have a platonistic stance toward set theory). Disregarding their ontological status, large cardinals play a prominent role in set theory. The literature on the topic is as vast as the one on forcing and inextricably intertwined with it. I would say that the basics for getting acquainted with the topic are covered in [16, Chapters 10, 17]. If one wishes to delve deeper in the subject, the reference text is Kanamori's monograph [17]. One can also look at [16, Chapters 18, 19, 20, 21].

Descriptive set theory (DST):   As shown by Shoenfield's absoluteness theorem, the properties of the "simply definable" subsets of the real line are unchanged by forcing. Descriptive set theory performs a systematic analysis of the "simply definable" subsets of the reals. There are various directions the topic has taken in the past decades: One of these ultimately led to the flourishing of inner model theory (IMT) and is inextricably intertwined with large cardinal axioms and forcing; the other has brought to light the relevance of set theoretic methods in the study of central problems of functional analysis/ergodic theory/topological dynamics. Below are some references on both topics:

> **Classical descriptive set theory:**   The reference text is Kechris' monograph [18]. Nice surveys (or textbooks) on the theory of Borel reductions for analytic equivalence relations—one of the parts of DST which has found striking applications in other domains—are [9, 11, 23, 26].

IMT, constructability, determinacy, and fine structure: Large   cardinals
give a very nice picture of the properties of the "simply definable" subsets of
the real line: On the one hand they allow for a neat definition of being "simply
definable," a notion captured by universal Baireness (which is a family of
sets of reals including the Borel sets and closed under continuous images,
binary intersections, and complementation); on the other hand they allow to
prove that the Axiom of Determinacy holds for the universally Baire sets.
This axiom decides a huge set of interesting problems formalizable in second
order arithmetic. There is a clear picture of the universe of sets emerging from
the results linking determinacy to large cardinals. The unfolding of these
connections came in pairs with the development of a "fine structural" analysis
of the universe of sets. This started from a careful analysis of the constructible
universe L (which Gödel introduced to prove the consistency of $\mathsf{ZFC} + \mathsf{CH}$
relative to ZF) and developed in the construction of sophisticated canonical
model for large cardinal axioms. The interested reader should approach set
theory following the textbook [27], in particular Chapter 5 and then Chapters
8–13 (the cited book also contains a complete proof of Projective Determinacy
from large cardinals).

More connections of forcing with other domains: This is a topic I would very
much like to be covered more extensively. Below a list of papers were forcing
is used either to tackle problems in domains distinct from set theory, or to
outline similarities between set theory and other mathematical fields: [33] on
how to use forcing to prove a weak form of a conjecture of Schanuel in
number theory, [30] shows that spaces of the form $C(X)$ with X compact,
Hausdorff, extremely disconnected—when seen as boolean valued models—give
an alternative description of the corresponding $\mathsf{RO}(X)$-name in $V^{\mathsf{RO}(X)}$ for the
set of complex numbers; the forthcoming [25] outlines the connection between
boolean valued models with the mixing property and sheaves for the dense
Grothendieck topology. A theme I want to bring to light in more accessible forms
is the relation of forcing with Topos theory. An approach to forcing highlighting
some patterns of this relation is given by Bell's book [5].

Bibliography

1. Antos, C.: Class forcing in class theory. In: The Hyperuniverse Project and Maximality, pp. 1–16. Birkhäuser/Springer, Cham (2018). MR 3728990
2. Bagaria, J.: Natural axioms of set theory and the continuum problem. In: Proceedings of the 12th International Congress of Logic, Methodology, and Philosophy of Science, pp. 43–64. King's College London Publications, London (2005)
3. Baumgartner, J.E.: Iterated forcing. In: Surveys in Set Theory. London Mathematical Society Lecture Note Series, vol. 87, pp. 1–59. Cambridge University Press, Cambridge (1983). MR 823775
4. Baumgartner, J.E.: Applications of the proper forcing axiom. In: Handbook of Set-Theoretic Topology, pp. 913–959. North-Holland, Amsterdam (1984). MR 776640
5. Bell, J.L.: Set Theory: Boolean-Valued Models and Independence Proofs. Oxford University Press, Oxford (2005)
6. Burris, S., Sankappanavar, H.P.: A Course in Universal Algebra. Graduate Texts in Mathematics, vol. 78. Springer, Berlin (1981). MR 648287
7. Cohen, P.: The independence of the continuum hypothesis. Proc. Nat. Acad. Sci. U.S.A. **50**, 1143–1148 (1963). MR 157890
8. Enderton, H.B.: A Mathematical Introduction to Logic, 2nd edn. Harcourt/Academic, Burlington (2001). MR 1801397
9. Foreman, M.: What is a Borel reduction?. Notices Amer. Math. Soc. **65**(10), 1263–1268 (2018). MR 3837073
10. Frink, O.: Compactifications and semi-normal spaces. Amer. J. Math. **86**, 602–607 (1964). MR 166755
11. Gao, S.: Invariant Descriptive Set Theory. Pure and Applied Mathematics (Boca Raton), vol. 293. CRC Press, Boca Raton (2009). MR 2455198
12. Givant, S., Halmos, P.: Introduction to Boolean Algebras. Undergraduate Texts in Mathematics. Springer, New York (2009). MR 2466574 (2009j:06001)
13. Gödel, K.: What is Cantor's continuum problem? Amer. Math. Monthly **54**, 515–525 (1947). MR 23780
14. Hodges, W.: Model Theory. Encyclopedia of Mathematics and Its Applications. Cambridge University Press, Cambridge (1993)
15. Hrbacek, K., Jech, T.: Introduction to Set Theory. Monographs and Textbooks in Pure and Applied Mathematics, vol. 220, 3rd edn. Dekker, New York (1999). MR 1697766
16. Jech, T.: Set Theory. Spring Monographs in Mathematics, 3rd edn. Springer, Berlin (2003)

© The Author(s), under exclusive license to Springer Nature Switzerland AG 2024
M. Viale, *The Forcing Method in Set Theory*, La Matematica per il 3+2 168,
https://doi.org/10.1007/978-3-031-71660-7

17. Kanamori, A.: The Higher Infinite: Large Cardinals in Set Theory from Their Beginnings. Springer Monographs in Mathematics, 2nd edn. Springer, Berlin (2009). Paperback reprint of the 2003 edition. MR 2731169

18. Kechris, A.S.: Classical Descriptive Set Theory. Graduate Texts in Mathematics, vol. 156. Springer, New York (1995). MR 1321597

19. Kunen, K.: Set Theory: An Introduction to Independence Proofs. Studies in Logic and the Foundations of Mathematics, vol. 102. North-Holland Publishing, Amsterdam (1980). MR 597342 (82f:03001)

20. Kunen, K.: Set Theory. Studies in Logic (London), vol. 34. College Publications, London (2011). MR 2905394

21. Larson, P.B.: The Stationary Tower. University Lecture Series, vol. 32. American Mathematical Society, Providence (2004). Notes on a course by W. Hugh Woodin. MR 2069032

22. Marker, D.: Model Theory: An Introduction. Graduate Texts in Mathematics, vol. 217. Springer, New York (2002). MR 1924282

23. Motto Ros, L.: Can we classify complete metric spaces up to isometry? Boll. Unione Mat. Ital. **10**(3), 369–410 (2017). MR 3691805

24. Pedersen, G.K.: Analysis Now. Graduate Texts in Mathematics, vol. 118. Springer, New York (1989). MR 971256

25. Pierobon, M., Viale, M.: Boolean Valued Models, Presheaves, and étalé Spaces (2023)

26. Ros, L.M.: Classification Problems from the Descriptive Set Theoretical Perspective (2021)

27. Schindler, R.: Set Theory: Exploring Independence and Truth. Universitext. Springer, Cham (2014). MR 3243739

28. Shoenfield, J.R.: Mathematical Logic. Association for Symbolic Logic, Urbana, IL. A K Peters, Natick (2001). Reprint of the 1973 second printing. MR 1809685 (2001h:03003)

29. Smullyan, R.M., Fitting, M.: Set Theory and the Continuum Problem. Oxford Logic Guides, vol. 34. The Clarendon Press, Oxford University Press, New York (1996). Oxford Science Publications. MR 1433595

30. Vaccaro, A., Viale, M.: Generic absoluteness and boolean names for elements of a Polish space. Boll. Unione Mat. Ital. **10**, 293–319 (2017)

31. Venturi, G., Viale, M.: New axioms in set theory. Mat. Cult. Soc. Riv. Unione Mat. Ital. (I) **3**(3), 211–236 (2018). MR 3888477

32. Venturi, G., Viale, M.: What model companionship can say about the continuum problem. Rev. Symb. Logic **17**(2), 546–585 (2024)

33. Viale, M.: Forcing the truth of a weak form of Schanuel's conjecture. Confluentes Math. **8**(2), 59–83 (2016)

34. Viale, M.: Useful axioms. IfCoLog J. Log. Appl. **4**(10), 3431–3465 (2017). MR 4542241

35. Viale, M.: Strong Forcing Axioms and the Continuum Problem (Séminaire BOURBAKI Avril 2023, 75e année, 2022–2023, no. 1207)

36. Viale, M., Audrito, G., Steila, S.: A Boolean Algebraic Approach to Semiproper Iterations (2014)

37. Walker, R.C.: The Stone-Čech Compactification. Ergebnisse der Mathematik und ihrer Grenzgebiete, Band 83. Springer, New York (1974). MR 0380698

38. Weaver, N.: Forcing for Mathematicians. World Scientific Publishing, Hackensack (2014). MR 3184751

39. Woodin, W.H.: The continuum hypothesis. I. Notices Amer. Math. Soc. **48**(6), 567–576 (2001). MR 1834351

40. Woodin, W.H.: The continuum hypothesis. II. Notices Amer. Math. Soc. **48**(7), 681–690 (2001)